AF552718

Climate Change Effects on Agriculture and Economy

Climate Change Effects on Agriculture and Economy

Editors
Prof. D.R. Khanna
Prof. A.K. Chopra
Dr. Vikas Singh
Dr. Rakesh Bhutiani
Dr. Gagan Matta

Associate Editor
Dr. Ajendra Kumar

2013
BIOTECH BOOKS®

© 2013 EDITORS
ISBN 978-81-7622-278-5

All rights reserved. Including the right to translate or to reproduce this book or parts thereof except for brief quotations in critical reviews.

The views expressed in various chapters are those of the authors and not or publisher of the book.

Published by: **BIOTECH BOOKS®**
4762-63/23, Ansari Road, Darya Ganj,
New Delhi - 110 002
Phone: +91-011-23262132
E-mail: biotechbooks@yahoo.co.in

Printed at: **Chawla Offset Printers**
Delhi - 110 052

PRINTED IN INDIA

Preface

According to the United Nations Framework Convention on Climate Change (UNFCCC), the term Climate Change means a going on climatic change due to direct or indirect activities of humankind which is continuously altering the global atmosphere resulting in the variation in the climate observed over comparable time periods. The climate of our world is undergoing a dramatic change. Global warming is increasing rapidly and there is widespread consensus that the current trend is caused by increased emissions of various greenhouse gases such as carbon dioxide, hydrofluorocarbons, perfluorocarbons, sulphur hexafluoride, methane and nitrous oxide. Due to unobstructed growth of greenhouse gas emissions, the earth's temperature has raised resulting in melting of glaciers, more precipitation, extreme weather events and changing pattern in seasons.

The accelerating pace of climate change, combined with global population and income growth, threatens food security everywhere. Agriculture is tremendously at risk due to climate change. Higher temperatures eventually diminish yields of desirable crops while encouraging weed and pest proliferation. Changes in precipitation patterns increase the likelihood of short-run crop failures and long-run production declines. Although there will be gains in some crops in some regions of the world, the overall impacts of climate change on agriculture are expected to be negative, threatening global food security. In India, substantial work has been done in last decade aimed at understanding the nature and magnitude of change in yield of different crops due to projected climate change.

This book presents an overview of the impacts of climate change and global warming on the current agricultural production and food safety. We acknowledge special thanks to the authors for contributing their valuable work in the form of research papers in this book. This will definitely be helpful to the young researchers working in the concern field and hope best for this book in fulfilling the purpose of making awareness among social community.

This publication has been prepared as a tool for a better communication, information exchange and understanding of "WCMANU, 2011". Articles appeared in this Proceedings are basically *received based, non-reviewed* papers that have been submitted by the pre-registered persons to the "WCMANU, 2011". The reader who intends to refer the article in this Proceeding is recommended to contact directly to the author.

Editors

Contents

2013, Climate Change Effects on Agriculture and Economy *Pages* **1–15**
Editors: **D.R. Khanna, A.K. Chopra, Vikas Singh, Rakesh Bhutiani & Gagan Matta**
Published by: **BIOTECH BOOKS, NEW DELHI**

Chapter 1

Hydrochemical Study of River Ganga in a Stretch from Haridwar to Bijnor: A Case Study

Shikha Chamoli[1], M.K. Sharma[2] and A.K. Chopra[1]
[1]Department of Zoology and Environmental Science, Gurukula Kangri University, Haridwar, Uttarakhand
[2]National Institute of Hydrology, Roorkee, Uttarakhand

River Ganga has been revered by millions of people in India as a symbol of spiritual purity but due to industrialization, building boom, tourist influx, corpse disposals, dumping of chemical wastes and human wastes *etc.* has affected the environment of river water. Also high concentration of undesirable contaminants in the river water directly or indirectly put negative impacts on its flora and fauna. In the present investigation, hydrochemical studies of River Ganga has been carried out by collecting river water samples from four sites *viz.* Chandighat, Balawali, Shukartal and Barrage situated between Haridwar and Bijnor and analyzing them for physico-chemical parameters *viz;* Total Dissolved Solids (TDS), pH, Dissolved Oxygen (DO), Biological Oxygen Demand (BOD), Chemical Oxygen Demand (COD), Total Hardness, Chloride, Sulphate, Alkalinity, Phosphate, Nitrate, Sodium, Potassium, Calcium and Magnesium for a period of one year. The data was processed as per Bureau of Indian Standards and WHO standards (1984) to examine the suitability of river water for drinking purpose and other designated uses. The results revealed that all analyzed physico-chemical parameters were observed within the permissible limit prescribed by BIS (1991) for drinking purpose. The suitability of river water was assessed for irrigation purpose using conductivity values, per cent Na, SAR and RSC value and was found of excellent quality for irrigation.

Keywords: *River Ganga, Physico-chemical parameters, BIS, Irrigation.*

Introduction

Water is the most important compound on earth and is vital for all forms of life. Quality of water depends on its physical, chemical and biological properties. Increasing human population and their necessities is directly responsible for the deterioration of river water along with surface as well as groundwater. River Ganga (Ganges) of India has been held in high esteem since time immemorial and Hindus from all over the world cherish the idea of a holy dip in the river under the faith that by doing so they will get rid of their sins of life. More than 400 million people live along the Ganges River. During the past three decades, the country's explosive growth industrialization and rapid urbanization have put unyielding pressure on the sacred stream. Ganga, the most sacred of rivers for Hindus, has become polluted for some years now.

Therefore it is necessary that the quality of river water should be checked at regular interval of time in order to save its aquatic life as well as human population who directly or indirectly use or consume its water.

The present study involves the hydrochemical study of River Ganga in a stretch from Haridwar to Bijnor. River water is generally used for drinking and irrigation of agricultural land situated nearby areas. People also consumes the water of river Ganges as there is a belief that man become pure by its touch. The quality of the river water has been assessed by comparing water quality parameter as per Bureau of Indian Standards and WHO standards to examine the suitability of river water for drinking purpose and other designated uses.

Study Area

The Ganga is the largest and the most important river of India. The ice cave of Gaumukh at the snout of Gangotri glacier, some 7010 m above sea leveling Uttarkashi district of Uttarakhand, is recognized as the traditional source of Ganga. It descends down at the valley of Devprayag where Alaknanda, another hill stream rising from the Bhagirathi Kharak and the Sonpanth twin glaciers joins it. After the confluence with Alaknanda, the combined stream is called Ganga. It flows through Uttarakhand, Uttar Pradesh, Bihar, Jharkhand and West Bengal and finally falls into Bay of Bengal.

Water Quality Sampling and Experimental Methodology

The water samples were collected from River Ganga from 4 sites *viz.* Chandighat (S1), Balawali (S2), Shukratal (S3) and Barrage at Bijnor (S4), alternate monthly during the months from April, 2009 to March, 2010 (Figure 1.1). The samples were collected in polythene bottles without any air bubbles. The bottles were rinse thoroughly before sampling and tightly sealed after collection and labeled. Samples were brought to the laboratory and kept in the refrigerator maintained at 4.0 °C.

Analysis was carried out for various water quality parameters such as Electrical conductivity, Total Dissolved Solids, pH, Dissolved Oxygen, Free Carbon dioxide, Biological Oxygen Demand, Chemical Oxygen Demand, Total Hardness, Chloride, Alkalinity, Ammonia, Phosphate, Nitrate, Sodium, Potassium, Calcium and Magnesium as per standard procedures prescribed by APHA (1985).

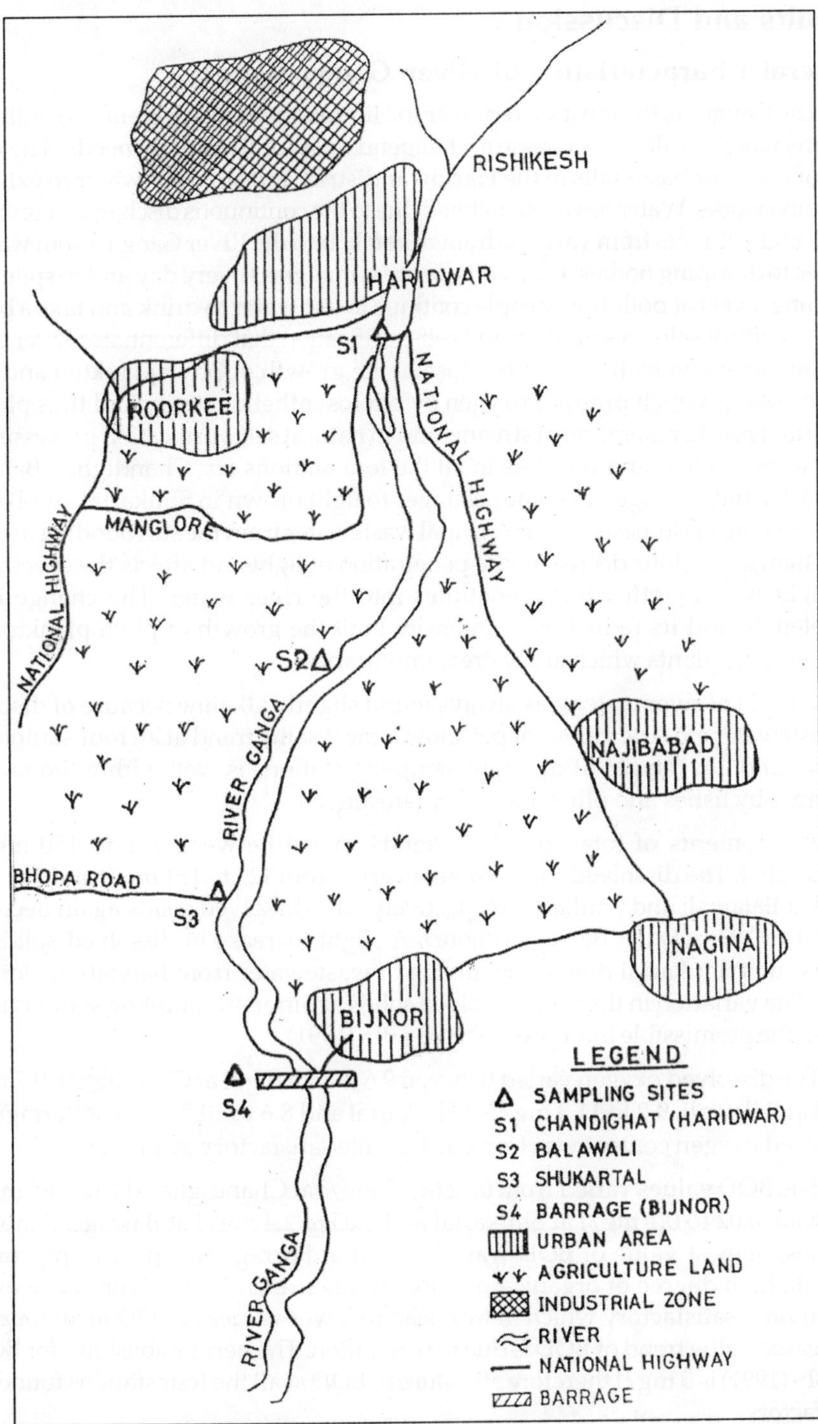

Figure 1.1: River Ganges Stretch from Haridwar to Bijnor Showing the Sampling Area. S1L: Chandighat (Haridwar); S2: Balawali; S3: Shukartal; S4: Barrage (Bijnor).

Results and Discussion

General Characteristics of River Ganga Water

The Ganges is the most sacred river to Hindus and is also a lifeline to millions of Indians who live along its course and depend on it for their daily needs. The upper part of the river basin falls in the Haridwar district and supplies water to extensive irrigation works. Water has been polluted due to the continuous discharge of untreated waste and effluents from various drains directly into the River Ganga. From washing clothes to dumping bodies, the river's purity is maligned every day and in spite of the alarming levels of pollution, people continue to use water to drink and take a bath. It is the quality of self-preservation and self-purification that differentiates River Ganga from all other rivers. It may be because of the growth of phytoplankton and other aquatic plants which produce oxygen by photosynthetic activity and thus plays an important role in reaeration of streams and in natural self purification processes. The river water is clear and odorless in all the four stations *viz.* Chandighat, Balawali, Shukartal and Barrage but colour changes to light brown in Shukartal and Barrage due to joining of domestic and municipal wastewater from the surrounding areas *etc.* The change in colour decreases the penetration of light and affects the spectrum of the light wavelength which penetrates into the river water. The change in the wavelength and its reduction in intensity limit the growth of phytoplankton and other aquatic plants which are of great importance.

The pH of river water was always found slightly alkaline because of discharge of wastewater etc. The values of pH show almost same trend at all four stations and the variation in the pH at different sampling stations is well within the range of tolerance by fishes and other aquatic organisms.

The contents of total dissolved solids varied between 121 to 150 mg/l at Chandighat. The dissolved solids content varies from 123 to 151 mg/l and 216 to 255 mg/l at Balawali and Shukartal respectively. The dissolved solids again decreased *i.e.* 144 to 193 mg/l at Barrage (Bijnor). A slight increase in dissolved solids was observed at Shukartal due to the mixing of wastewater from habitation along the river. The variation in the total dissolved solids at different sampling stations is well within the permissible limit prescribed by BIS (1991).

The dissolved oxygen varied between 9.6 to 11.5 mg/l at Chandighat, 9.7 to 10.7 mg/l at Balawali, 9.2 to 11.3 mg/l at Shukartal and 8.6 to 10.5 mg/l at Barrage. The dissolved oxygen content was found to be quite satisfactory at all sites.

The BOD values varied from 0.12 to 1.2 mg/l at Chandighat, 0.2 to 0.65 mg/l at Balawali, 0.02 to 0.1 mg/l at Shukartal and 0.05 to 2.4 mg/l at Barrage. Among all stations, highest value of BOD was observed at Barrage (Bijnor) during summer indicate high degree of organic pollution in this stretch of the river. Since oxygen condition is satisfactory which in turn lead to lower values of BOD in all four sites. Decreased value trend of BOD is due to re-aeration. The permissible limit for BOD as per BIS (1991) is 3 mg/l therefore all values of BOD at all the four stations found to be satisfactory.

The COD values ranged from 7.1 to 10.1 mg/l at Chandighat, 7.2 to 9.3 mg/l at Balawali, 7.0 to 10 mg/l at Shukartal and 5.2 to 7.5 mg/l at Barrage. The high values of COD were observed at Chandighat and Shukartal and lower at Barrage and Balawali. Because of high values of dissolved oxygen re-aeration occurs thus lowering down COD.

The amount of sulphate varied from 21 to 25 mg/l at Chandighat, 20 to 29 mg/l at Balawali, 23 to 26 mg/l at Shukartal and 14 to 17 mg/l at Barrage. Amount of sulphate is maximum during summers and minimum in rainy seasons. The maximum permissible limit of sulphate for drinking water as given by BIS (1991) is 400 mg/l. The values of all the samples were found within the maximum permissible limit. Phosphate was found to be absent or almost negligible at all four stations for whole one year. All the values were found to be less than 0.1 mg/l. Also the average values of nitrate were 0.59, 1.27, 1.44 and 1.77 mg/l in all four stations *viz.* Chandighat, Balawali, Shukartal and Barrage.

Seasonal Variation of Water Quality Parameters

Temporal variations of physical parameters of River Ganges from Haridwar to Bijnor are shown in Figure 1.2. River water showed strong seasonal pattern for maximum water quality constituents with higher concentrations during summer and lower concentration during monsoon periods. The electrical conductivity of the river water underwent seasonal fluctuations with minimum values in winter and maximum values in summer. This is because the electrical conductivity was closely related to the sum of the concentrations of four major cations (Ca^{2+}, Mg^{2+}, Na^{+} and K^{+}) in the river water as well as that of three major anions (HCO_3^-, SO_4^{2-}, Cl^-). Sodium, potassium, calcium, magnesium, chloride and sulphate ion concentrations showed clear seasonal variation, with maximum in summer months and minimum in monsoon season at almost all the sites.

Nitrate displayed no consistent seasonal behavior, which may be attributed to biological activities caused by topographic, microclimatic and different amounts of wastes discharged to the river. Further, relatively low concentrations of nitrate observed in the river throughout the year are probably due to high levels of denitrification.

Suitability for Irrigation Purpose

Water quality plays an important role in irrigated agriculture. Many problem arises during inefficient management of water for agriculture use. The concentration and composition of dissolved constituents in water determine its quality for irrigation use. Quality of water is an important consideration in any appraisal of salinity or alkali conditions in an irrigated area. Under good soil and water management practices, good quality water has the ability to cause maximum yield. The quality of irrigation water is assessed by the following characteristics:

- ☆ Salinity
- ☆ Relative Proportion of Sodium to other Cations (SAR)
- ☆ Residual Sodium Carbonate (RSC)

Salinity

Salinity is measured in terms of electrical conductivity. If the salt concentration in water increases, the soil salinity also increases, it is difficult for plants to extract water. The salts present in the water, besides affecting the growth of the plants directly, also affect the soil structure, permeability and aeration, which indirectly affect the plant growth. Soil water passes into the plant through the root zone due to osmotic pressure. As the dissolved solid content of the soil water in the root zone increases, it is difficult for the plant to overcome the osmotic pressure and the plants root membrane are able to assimilate water and nutrients. Thus, the dissolved solid content of the residual water in the root zone also has to be maintained within limits by proper leaching. The quality of water is commonly expressed by classes of relative suitability for irrigation with reference to salinity levels.

The values of total dissolved solids (TDS) in the river water varies from 121 to 150 mg/l with a mean value of 132 mg/l, 123 to 151 mg/l with a mean 135 mg/l, 216 to 255 mg/l with a mean 233 mg/l and 144 to 193 mg/l with a mean 161 mg/l at Chandighat, Balawali, Shukartal and Barrage (Bijnor), respectively. About all of the samples were found below the desirable limit. Further, the electrical conductivity values in the study area varies from 189 to 234 ìS/cm with a mean value of 207 ìS/cm, 192 to 236 ìS/cm with a mean 210 ìS/cm, 338 to 399 ìS/cm with a mean ìS/cm, and 225 to 301 ìS/cm with a mean 252 ìS/cm at Chandighat, Balawali, Shukartal and Barrage (Bijnor), respectively. Its safe use for irrigation depends upon the texture of the soil and drainage pattern.

Relative Proportion of Sodium to other Cations

The clay minerals in the soil absorb divalent cations, like calcium and magnesium ions from irrigation water. Whenever the exchange sites in clay are filled by divalent cations, the soil texture is conductive for plant growth. Sodium reacts with soil to reduce its permeability. The sodium or alkali hazard in the use of water for irrigation is determined by the absolute and relative concentration of cations and is expressed in terms of Sodium Adsorption Ratio (SAR). If the proportion of sodium is high, the alkali hazard is high; and conversely, if calcium and magnesium predominate, the hazard is less. There is a significant relationship between SAR values of irrigation water and the extent to which sodium is absorbed by the soil. If water used for irrigation is high in sodium and low in calcium, the cation-exchange complex may become saturated with sodium. This can destroy the soil structure owing to dispersion of the clay particles. A simple method of evaluating the danger of high-sodium water is the sodium-adsorption ratio, SAR (Richards, 1954):

$$SAR = \frac{Na^+}{\sqrt{(Ca^{2+} + Mg^{2+})/2}}$$

The sodium percentage is calculated as:

$$Na\% = \frac{Na^+ + K^+}{Ca^{2+} + Mg^{2+} + Na^+ + K^+} \times 100$$

Where all ionic concentrations are expressed in milliequivalent per liter. Calculation of SAR for given water provides a useful index of the sodium hazard of that water for soils and crops. A low SAR (2 to 10) indicates little danger from sodium; medium hazards are between 7 and 18, high hazards between 11 and 26, and very high hazards above that. The lower the ionic strength of the solution the greater the sodium hazards for a given SAR (Table 1.1) (Richards, 1954).

The values of sodium percentage (Na per cent), SAR and RSC in water of the river Ganga are given in Table-2. The values of SAR in the river Ganga varies from 0.14 to 0.38 with a mean value of 0.25, 0.12 to 0.15 with a mean 0.14, 0.10 to 0.51 with a mean 0.27 and 0.12 to 0.20 with a mean 0.14 at Chandighat, Balawali, Shukartal and Barrage (Bijnor), respectively. The sodium percentage in the study area was found to vary from 9 to 20 per cent with a mean value of 15 per cent, 8 to 12 per cent with a mean 10 per cent, 4 to 18 per cent with a mean 11 per cent and 7 to 10 per cent with a mean 8 per cent at Chandighat, Balawali, Shukartal and Barrage (Bijnor), respectively.

Table 1.1: Guidelines for Evaluation of Irrigation Water Quality

Water Class	*Na (per cent)*	*EC (mS/cm)*	*SAR*	*RSC (meq/l)*
Excellent	< 20	< 250	< 10	< 1.25
Good	20-40	250-750	10-18	1.25-2.0
Medium	40-60	750-2250	18-26	2.0-2.5
Bad	60-80	2250-4000	> 26	2.5-3.0
Very bad	> 80	> 4000	> 26	> 3.0

Source: CGWB and CPCB (2000).

Residual Sodium Carbonate

Water containing high concentration of carbonate and bicarbonate ions tends to precipitate calcium and magnesium as carbonate, changing the residual water to high sodium water with sodium bicarbonate in solution. As a result, the relative proportion of sodium increases and gets fixed in the soil thereby decreasing the soil permeability. This excess is denoted by Residual Sodium Carbonate (RSC) and is determined by the following formula:

$$RSC = (HCO_3^- + CO_3^{--}) - (Ca^{++} + Mg^{++})$$

Where all ionic concentrations are expressed in epm.

If the RSC exceeds 2.5 epm, the water is generally unsuitable for irrigation. Excessive RSC causes the soil structure to deteriorate, as it restricts the water and air movement through soil. If the value is between 1.25 and 2.5, the water is of marginal quality, while values less than 1.25 epm indicate that the water is safe for irrigation. The values of RSC in the river Ganga varies from -0.70 to -0.38 with a mean value of -0.54, -0.83 to -0.45 with a mea -0.65, -1.52 to -0.67 with a mean -1.09 and -0.82 to -0.61 with a mean -0.74 at Chandighat, Balawali, Shukartal and Barrage (Bijnor) respectively. RSC values in all water samples were found to be less than 1.25 epm and are found suitable for irrigation purpose.

Table 1.2: SAR, Na per cent and RSC Values in the Ganga River Water

Sl.No.	*Sampling Site*	*SAR*	*Na per cent*	*RSC*
		April, 2009		
1.	Chandighat	0.191949	11.93506	–0.70180
2.	Balawali	0.154932	10.17067	–0.77284
3.	Shukartal	0.104072	4.839250	–1.35451
4.	Barrage	0.150097	8.793893	–0.79574
		June, 2009		
5.	Chandighat	0.291542	15.58187	–0.52984
6.	Balawali	0.123806	8.429168	–0.83377
7,	Shukartal	0.118576	5.631942	–1.52115
8,	Barrage	0.154383	7.973578	–0.65057
		August, 2009		
9.	Chandighat	0.209529	15.25765	–0.41787
10.	Balawali	0.156795	12.78400	–0.45787
11.	Shukartal	0.357661	15.74404	–0.67148
12.	Barrage	0.129612	7.879528	–0.81270
		October, 2009		
13.	Chandighat	0.246549	14.60730	–0.63492
14.	Balawali	0.151851	10.68454	–0.67566
15.	Shukartal	0.229571	9.894714	–0.79754
16.	Barrage	0.122241	7.136099	–0.82361
		December, 2009		
17.	Chandighat	0.142581	9.412491	–0.63377
18.	Balawali	0.141150	10.51201	–0.65590
19.	Shukartal	0.260311	10.62182	–0.99238
20.	Barrage	0.207812	10.96481	–0.61246
		February, 2010		
21.	Chandighat	0.337346	18.24028	–0.52787
22.	Balawali	0.134322	10.55354	–0.60230
23.	Shukartal	0.315592	12.87934	–1.22344
24.	Barrage	0.142515	7.862683	–0.81185
		April, 2010		
25.	Chandighat	0.383420	20.79878	–0.38590
26.	Balawali	0.145763	10.02711	–0.58213
27.	Shukartal	0.514667	18.44531	–1.11148
28.	Barrage	0.125142	7.236062	–0.69213

Figure 1.2: Variation of Physico-chemical Parameters of River Ganges Stretch form Haridwar to Bijnor

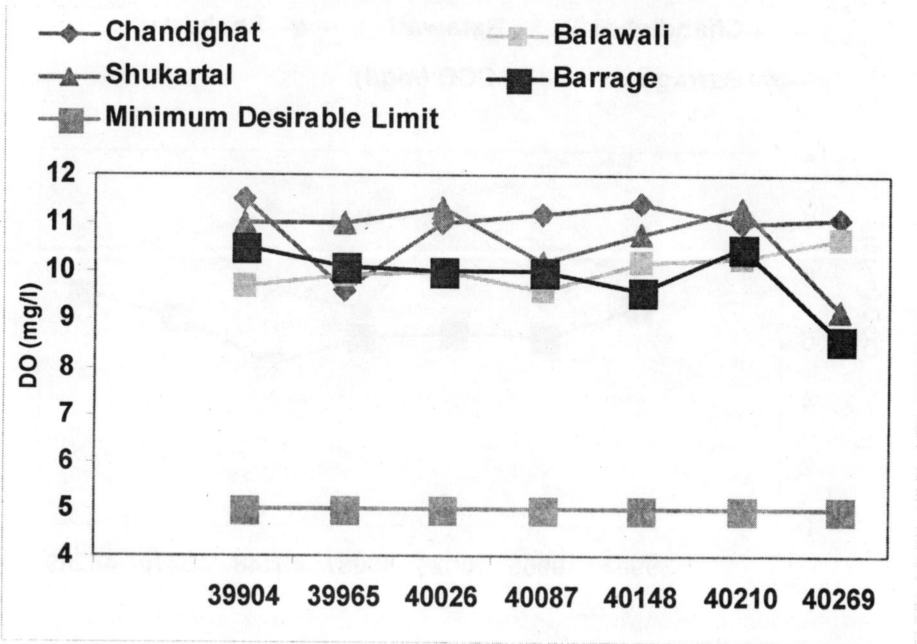

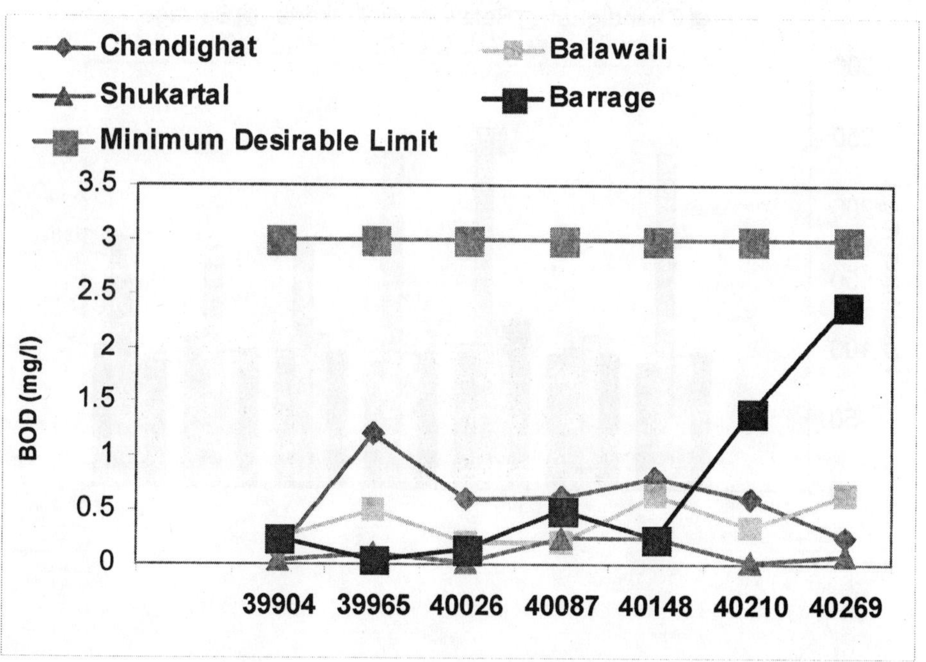

Contd...

Figure 1.2–*Contd...*

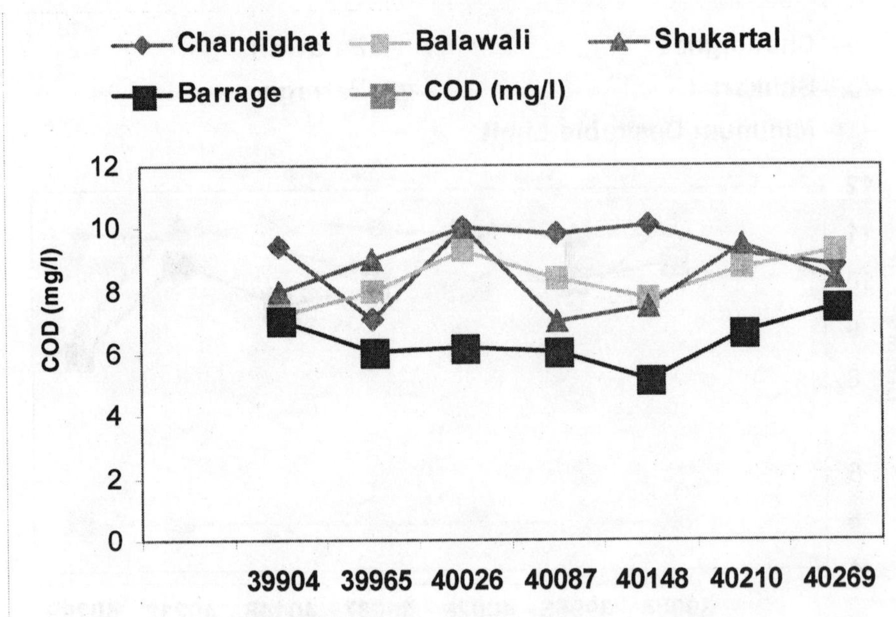

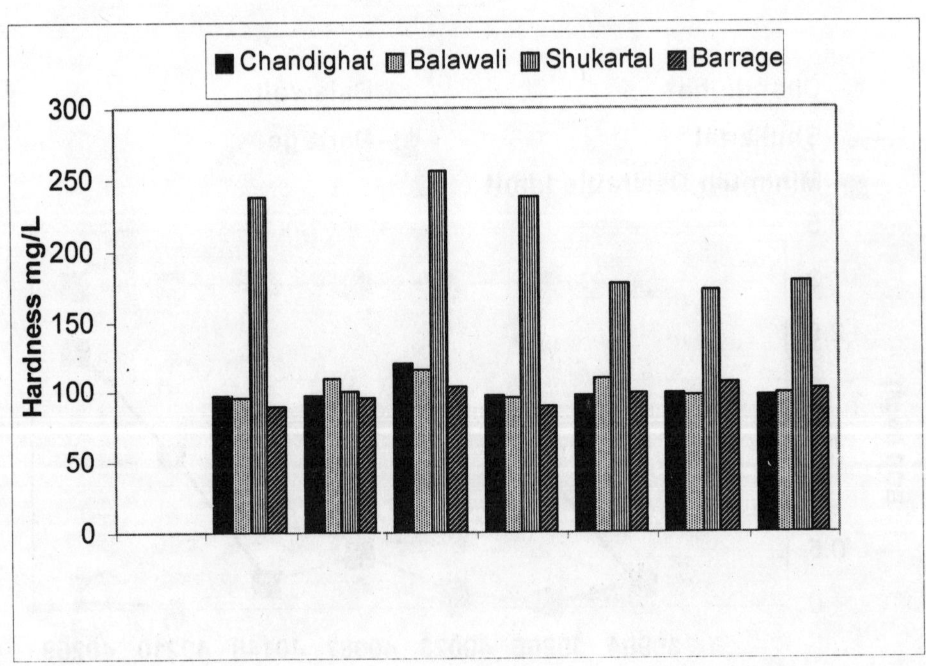

Contd...

Figure 1.2–*Contd...*

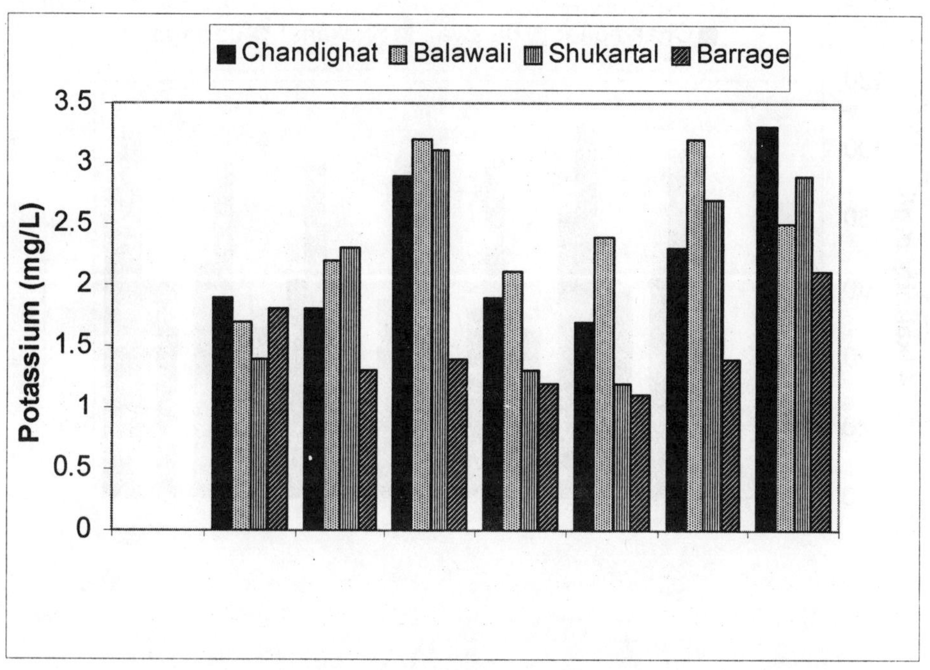

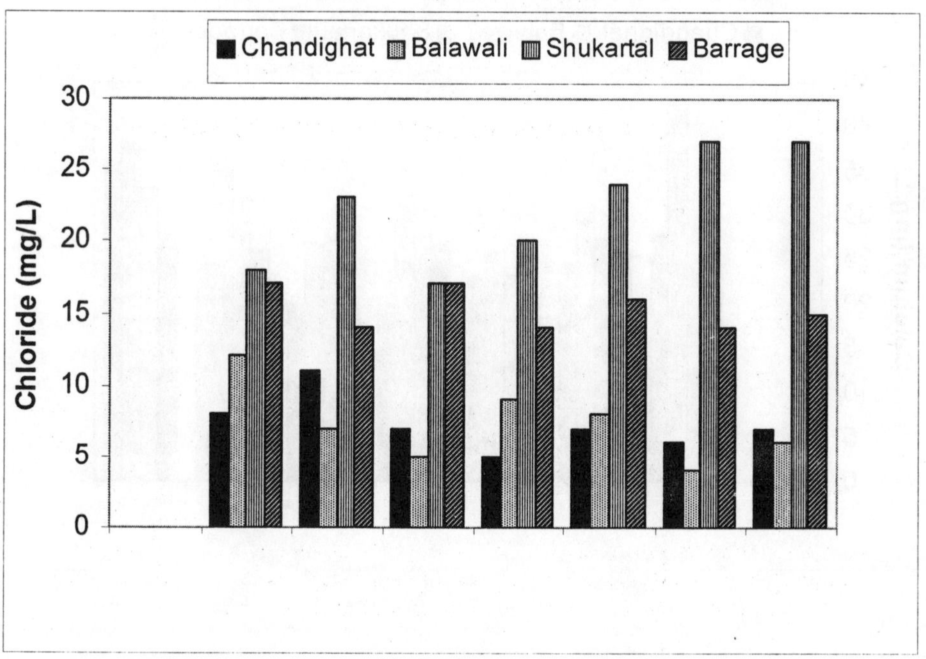

Contd...

Figure 1.2–*Contd...*

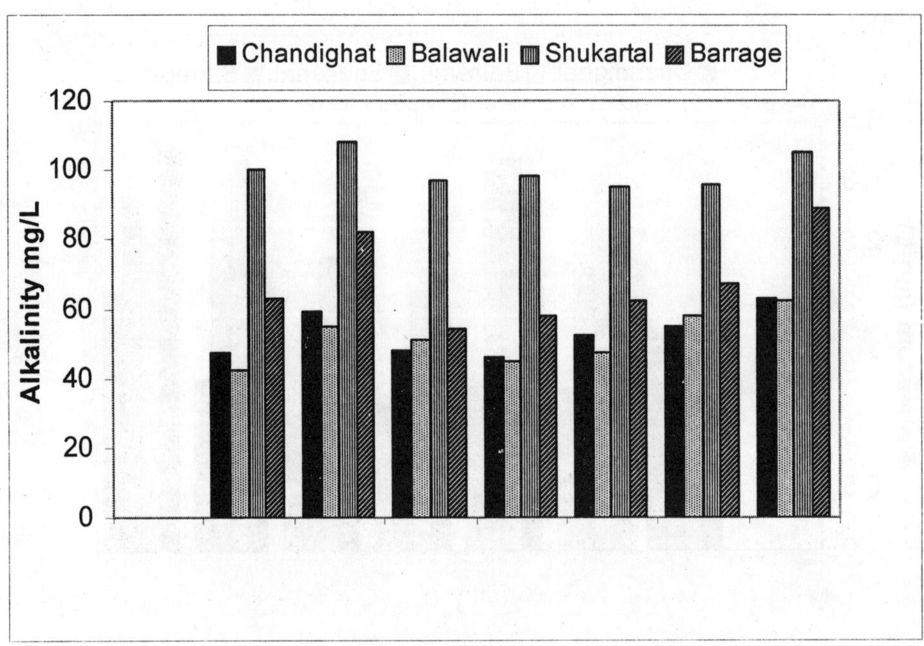

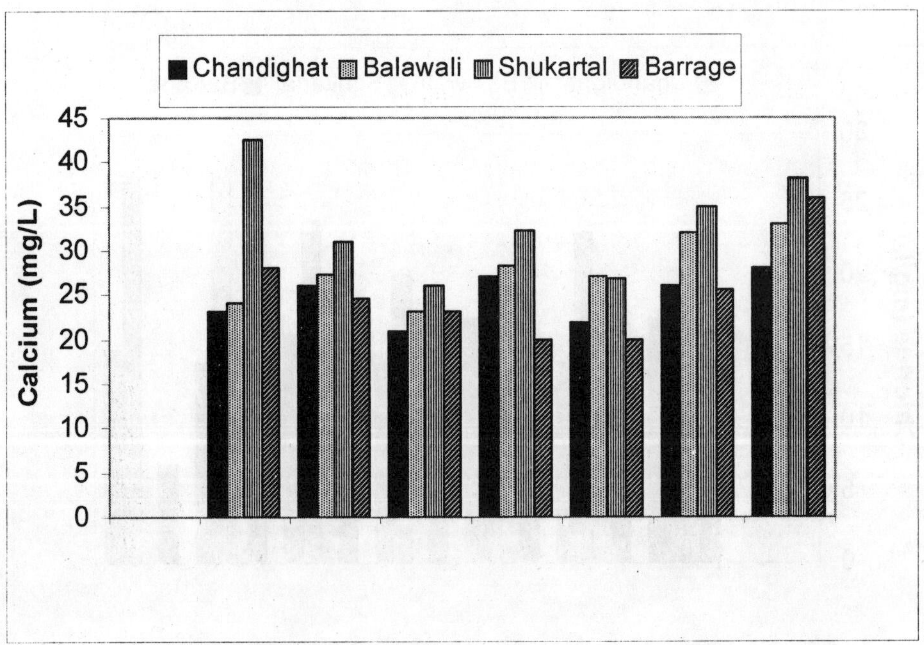

Contd...

Figure 1.2–*Contd...*

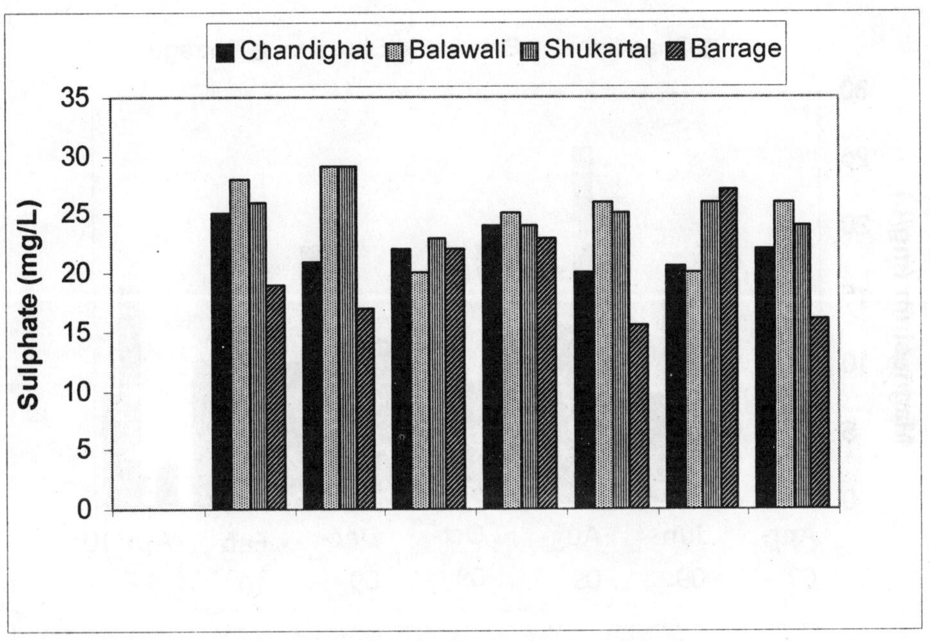

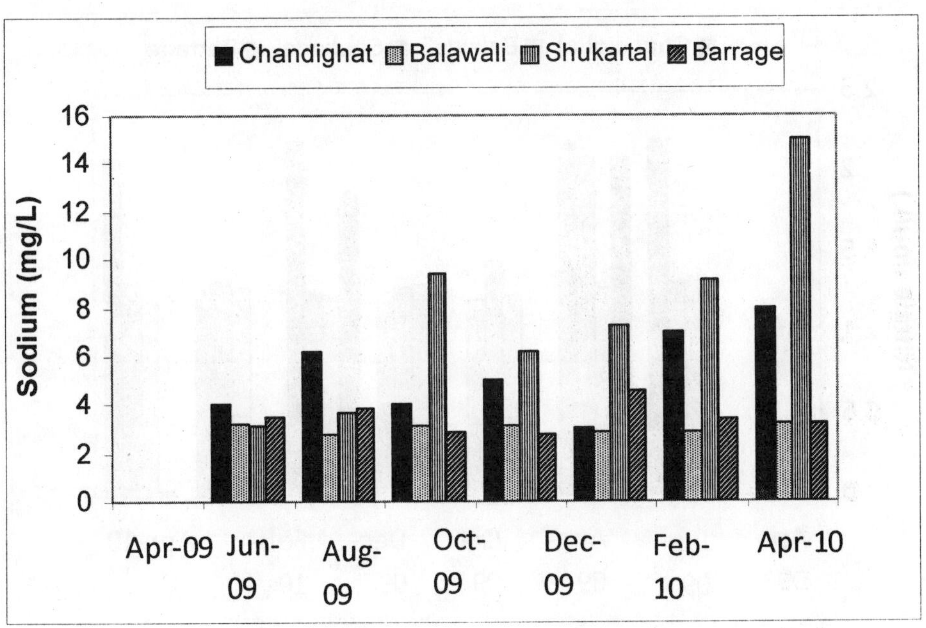

Contd...

Figure 1.2–*Contd...*

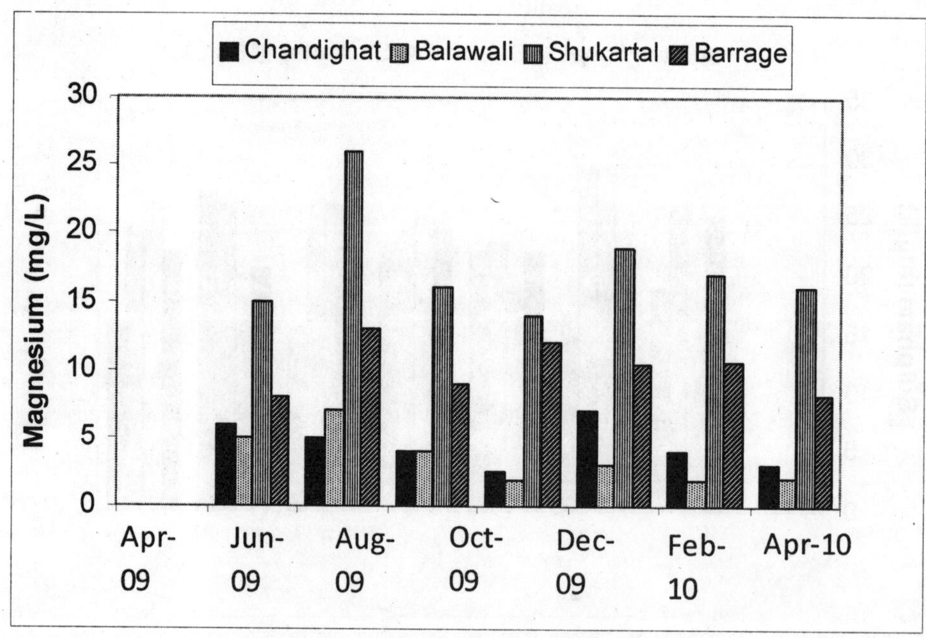

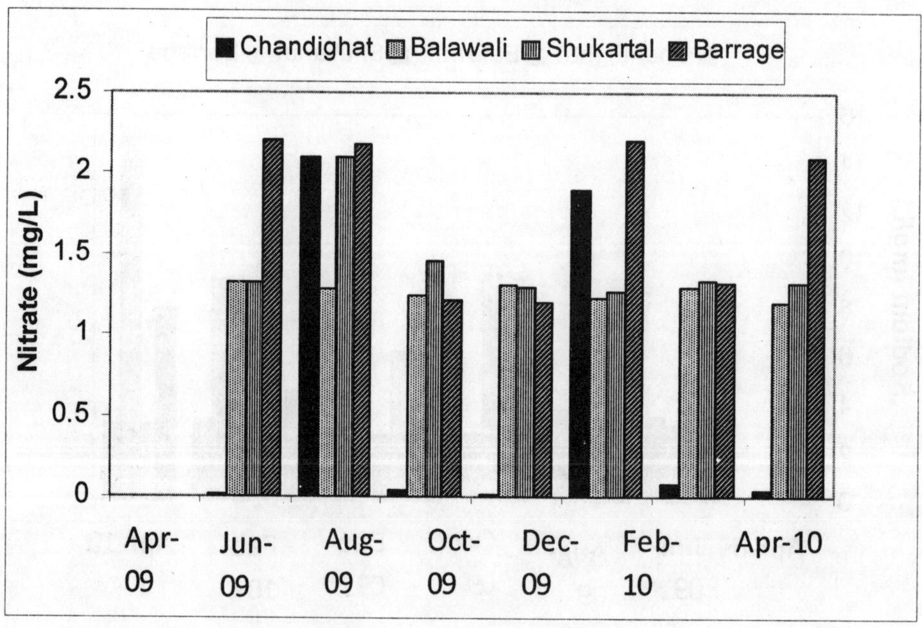

Conclusion

The analysis of the water quality parameters of river Ganges from four different stations (Chandighat, Balawali, Shukartal, Barrage) shows that all analyzed physicochemical parameters are well within the permissible limit prescribed by BIS (1991) for drinking purpose. Further the suitability of river Ganga water was assessed on the basis of conductivity values, per cent Na, SAR and RSC value and was found of excellent quality for irrigation.

References

APHA., 1985: Standard Methods for Examination of Water and Wastewater, 20th Edition, American Public Health Association, Washington D. C.

Adeleye, I.A. and Adebiyi, A.A. 2003: Physico-chemical and microbiological assessment of Oko-oba- A Nigerian Abattoir. *J. Environ. Biol.*, 24, 309-313.

BIS., 1991: Indian Standard Specification for Drinking Water. BIS, New Delhi.

CGWB and CPCB, 2000: Status of Groundwater Quality and Pollution Aspects in NCT-Delhi, January 2000.

Hujare, M. S. 2008: Seasonal variation of physico-chemical parameters in the perennial tank of

Talsande, Maharashtra. *Ecotoxicol. Environ. Monit.* 18(3): 233-242.

Jain, C.K., Singhal, D.C., and Sharma, M.K. 2003: Hydrochemical studies of the Hindon River, India: seasonal variations and quality quantity relationships. *Journal of Environmental Hydrology*, 2, 1-12.

Manjare, S.A., Vhanalakar, S.A. and Muley, D.V. 2010: Analysis of water quality using physicochemical parameters Tamdalge tank in Kolhapur district, Maharashtra. *International Journal of Advanced Biotechnology and Research* 1(2), 115-119.

Richards, L. A. (ed.). 1954: Diagnosis and improvement of saline and alkali soils, Agricultural Handbook 60, U.S. Dept. Agric., Washington, D.C., 160 p.

Shyamala, R., Shanthi, M., and Lalitha, P. 2008: Physicochemical Analysis of Borewell Water Samples of Telungupalayam Area in Coimbatore District, Tamilnadu, India. *E-Journal of Chemistry*, 5 (4), 924-929.

Tyagi, P., Buddhi, D. Chaudhary, R. and Sawhney, R.L. 2002: Degradation of groundwater quality in industrial areas of India - A review. *Ind. J.Environ. Protect.*, 20, 174- 181.

W.H.O., Guidelines for drinking water quality, Vol.1, Recommendations WHO, Geneva, 1984.

2013, Climate Change Effects on Agriculture and Economy *Pages* **17–20**
Editors: **D.R. Khanna, A.K. Chopra, Vikas Singh, Rakesh Bhutiani & Gagan Matta**
Published by: **BIOTECH BOOKS, NEW DELHI**

Chapter 2

Biodegradation of Textile Wastewater

Deepti Bhargava and Shahnaz Jahan

Department of Clothing and Textile, College of Home Science, G.B.P.U.A.&T. Pantnagar, Uttarakhand

Dyes and dyestuffs find use in a wide range of industries but are of primary importance to textile manufacturing. Enormous volumes of effluent are generated at different stages of textile manufacturing, as a result of the use of copious amounts of chemicals and dyes. Wastewaters from textile industries are a complex mixture of many polluting substances like acids, salts, organochlorine-based pesticides, heavy metals, pigments, dyes *etc.* Several tons of textiles required to meet up with societal demands are produced daily in this industry. Effluent derived from the textile and dyestuff activities can provoke serious environmental impact in the neighboring receptor water bodies because of the presence of toxic reactive dyes and dark coloration. Increasingly, the environmental and subsequent health effects of dyes released in textile industry wastewater are becoming subject to scientific scrutiny. Due to complex nature and hard-to-treat by conventional methods, textile dyeing industries are facing problems to safe discharge of wastewater. In recent years, biological decolourization using potential microorganisms capable of decolourizing and detoxifying the synthetic dyes has been considered as a promising and eco-friendly method.

Keywords: *Effluent, Biodegradation, Biological degradation, Decolourization.*

Introduction

Industrialization is considered to be the key for the development in economic terms. However, it is also recognized to be the root cause for environmental perspective. The recognition that environmental pollution is a worldwide threat to public health has given rise to new initiatives for environmental restoration for both economic and ecological reasons. The industrial effluents contain toxic and hazardous pollutants. One particular class of synthetic chemicals which is of major concern is synthetic dyes and dye intermediates.

The textile dyeing and furnishing industries use wide variety of dyestuffs due to the rapid changes in the customer's demands. More than 100,000 commercially available dyes are known and the world annual production of the dyestuffs amounts to more than 7×10^5 tonnes. Azo dyes which are used as textile colorants designed to bond covalently with cellulosic fibers (*i.e.* cotton) are extensively used because of their wide variety of color shades, high wet fastness profiles, ease of application, brilliant colors and minimum energy consumption. It has been estimated that more than 10-15 per cent of the total dyestuff used in dye manufacturing and textile industry is released in to the environment during their synthesis and dying process. Almost 2, 80,000 tonnes of textile dyes are discharged every year worldwide. Waste streams generated from textile industries are hazardous and difficult to biodegrade owing to the presence of recalcitrant dyes and pigments. As India being a major producer and exporter of textiles, the problem of environmental pollution from textile industry has a serious dimension (Sarayu and Sandhya, 2009).

Environmental problems of the textile industry are mainly caused by discharges of wastewater. The textile sector has a high water demand. Its biggest impact on the environment is related to primary water consumption (80–100 m^3/ton of finished textile) and wastewater discharge (115–175 kg of COD/ton of finished textile, a large range of organic chemicals, low biodegradability, colour and salinity). Therefore, reuse of the effluents represents an economical and ecological challenge for the overall sector. Textile processing employs a variety of chemicals, depending on the nature of the raw material and product. The effluents resulting from these processes differ greatly in composition, due to differences in processes, used fabrics and machinery. Main pollution in textile wastewater came from dyeing and finishing processes. These processes require the input of a wide range of chemicals and dyestuffs, which generally are organic compounds of complex structure. Because all of them are not contained in the final product, became waste and caused disposal problems.

Textile Wastewater Characteristics

Composite textile wastewater is characterized mainly by measurements of Biochemical Oxygen Demand (BOD), Chemical Oxygen Demand (COD), Suspended Solids (SS) and Total Dissolved Solids (TDS). Typical characteristics of textile industry wastewater are presented in Table 2.1 (Al-Kdasi *et al.*, 2005)

Dyes and dyestuffs find use in a wide range of industries but are of primary importance to textile manufacturing. Increasingly, the environmental and subsequent health effects of dyes released in textile industry wastewater are becoming subject to scientific scrutiny. Wastewater from the textile industry is a complex mixture of many polluting substances ranging from organ chlorine-based pesticides to heavy metals associated with dyes and the dyeing process. During textile processing, inefficiencies in dyeing result in large amounts of the dyestuff being directly lost to the wastewater, which ultimately finds its way into the environment.

Table 2.1: Composite Textile Industry Wastewater Characteristics

Parameters	*Values*
pH	7.0 – 9.0
Biochemical Oxygen Demand (mg/l)	80 – 6,000
Chemical Oxygen Demand (mg/l)	150 – 12,000
Total Suspended Solids (mg/l)	15 – 8,000
Total Dissolved Solids (mg/l)	2,900 – 3,100
Chloride (mg/l)	1000 – 1600
Total Kjeldahl Nitrogen (mg/l)	70 – 80

Dyes and dyestuffs find use in a wide range of industries but are of primary importance to textile manufacturing. Enormous volumes of effluent are generated at different stages of textile manufacturing, as a result of the use of copious amounts of chemicals and dyes. Wastewaters from textile industries are a complex mixture of many polluting substances like acids, salts, organochlorine-based pesticides, heavy metals, pigments, dyes *etc.* Several tons of textiles required to meet up with societal demands are produced daily in this industry. Effluent derived from the textile and dyestuff activities can provoke serious environmental impact in the neighboring receptor water bodies because of the presence of toxic reactive dyes and dark coloration. Increasingly, the environmental and subsequent health effects of dyes released in textile industry wastewater are becoming subject to scientific scrutiny. Due to complex nature and hard-to-treat by conventional methods, textile dyeing industries are facing problems to safe discharge of wastewater. In recent years, biological decolourization using potential microorganisms capable of decolourizing and detoxifying the synthetic dyes has been considered as a promising and eco-friendly method.

The textile industries produce effluents that contain several types of chemicals such as dispersants, leveling agents, acids, alkalis, carriers and various dyes. The textile factories daily discharge millions of litres of untreated effluents in the form of wastewater into public drains that eventually empty into rivers. This alters the pH, increases the biochemical oxygen demand (BOD) and chemical oxygen demand (COD), and gives the rivers intense colourations. The use of these water resources is limited and the ecosystem is affected. Several methods are used in the treatment of textile effluents to achieve decolourization. These include physiochemical methods such as filtration, specific coagulation, use of activated carbon and chemical flocculation. Some of these methods are effective but quite expensive. Biotreatment offers a cheaper and environmentally friendlier alternative for colour removal in textile effluents. The ubiquitous nature of bacteria makes them invaluable tools in effluent biotreatment.

Biodegradation of Textile Effluents

Biological treatment is the often the most economical alternatives when compared with other physical and chemical processes. Biodegradation methods such as fungal decolourization, microbial degradation, adsorption by (living or dead) microbial biomass and bioremediation systems are commonly applied to the treatment of industrial effluents because many microorganisms such as bacteria, yeast, algae and fungi are able to accumulate and degrade different pollutants. However, their application is often restricted because of technical constraint. Biological treatment requires a large land area and is constrained by sensitivity toward diurnal variation as well as toxicity of some chemicals and less flexibility in design and operation. Further, biological treatment is incapable of obtaining satisfactory colour elimination with current conventional biodegradation processes. Moreover, although many organic molecules are degraded, many others are recalcitrant due to their complex chemical structure and synthetic organic origin (Anonymous).

Bacillus sp., an effective spore forming organism revealed a result in degrading and decolourizing the textile effluent, which was one of the major environmental pollution in modern world. The continued development and application of biotechnologies for the biodegradation is limited primarily by physical factors such as pH, temperature and substrate concentration. Biotreatment offers a cheaper and environmentally friendlier alternative for colour removal in textile effluents. The ubiquitous nature of bacteria makes them invaluable tools in effluent biotreatment (Vigneeswaran *et al.*, 2010).

References

Adel, A., Azni, I., Katayon, S., Chuah Teong, G. 2004: Treatment of Textile Wastewater by Advanced Oxidation Processes – A Review. Global Nest: *The Int. J.* Vol 6(3), pp 222-230.

Anonymous: Dye: Technologies for colour removal, Retrieved on September 28, 2011 from http://publicweb.unimap.edu.my/~ppkas/home/index.php/news/articles/22-dye-technologies-for-colour-removal.

Vigneeswaran, M., Govindarajan, B. Shanmugaraja, K. and Prabakaran, V. 2010: Biotreatment for effective degradation and decoloration of textile effluent using novel spore forming *Bacillus* sp. *Journal of Ecobiotechnology*, 2 (11).

Sarayu, K. and Sandhya. S. 2009: Potential of facultative microorganisms for biotreatment of textile wastewater. *Envis Centre Newsletter* Vol.7(2).

2013, Climate Change Effects on Agriculture and Economy Pages 21–28
Editors: D.R. Khanna, A.K. Chopra, Vikas Singh, Rakesh Bhutiani & Gagan Matta
Published by: BIOTECH BOOKS, NEW DELHI

Chapter 3

Monitoring of Air Quality Parameters of Different Locations within the Holy City Kurukshetra, Haryana

Jyoti Malik, Pooja, Snehlata and Smita Chaudhry
Institute of Environmental Studies, Kurukshetra University, Kurukshetra, Haryana

The increase in mean global surface temperature by about 0.3 to 0.6 °C since the late 19th century is attributed to the result of human activities, especially the use of fossil fuels in automobiles and power plants. Air pollutants in the broadest sense are considered to be the main cause of climate change. Also, weather and climate have a major role to play in determining the air quality patterns over multiple spatial and temporal scales. The present study was done to monitor the air quality at different locations of Kurukshetra, Haryana which is a holy city and emerging centre of commercialization and tourism hence increasing air pollution in the city. Efforts have been made to find out the present status of air pollutants in terms of most commonly polluting agents *i.e.* Suspended Particulate Matter, Sulphur dioxide and Nitrogen dioxide concentrations at different locations of the city. The concentrations of the pollutants (24 hrs average) were monitored during summer season of 2009, at seven different sites, using high volume air sampler (APM 460 BL). The study revealed that TSPM as well as RSPM concentration exceeded the prescribed CPCB limit (100 µg/m^3) at all the locations with maximum value at unplanned commercial site and minimum at educational site of Kurukshetra University campus. However, the concentration of SO_2 and NO_2 was found to be well within the permissible limits (80 µg/m^3) as prescribed by CPCB. The status of air pollutants in the present study will be helpful for the regulatory agencies to identify locations where natural resources and human health could be at risk. Integrating climate change and air pollution control programmes can lead to significant cost savings and important benefits to human health and the environment.

Keywords: *Climate change, Air pollution, Suspended particulate matter, Sulphur dioxide, Nitrogen dioxide.*

Introduction

The relationship of man and environment is necessarily symbiotic and equilibrium between the two must be maintained at all costs. But rapid industrialization and development of transport network have added impetus to economic development at the cost of environment. Although such development is integral to economic growth, the problem lies in their unfettered proliferation in India, leading to severe environmental degradation, particularly since 1970. Unplanned urbanization and industrialization are causing deterioration of the environment and quality of life in developing countries (Jayanti and Krishnamurthy, 2006).

The increase in mean global surface temperature by about 0.2 to 0.6 °C since the late 19th century is attributed to the result of human activities, especially the use of fossil fuels in automobiles and power plants (IPCC, 2001; Hamburg *et al.*, 1997; Schimmel *et al.*, 1995). Climate change and air pollution are interrelated. There has been a growing interest in the effects of climate change on the future air pollution levels. Climate modeling studies predict that in future some meteorological parameters will change due to natural variability and anthropogenic changes of atmospheric composition with a general increase in temperature (Houghton, 2001). Different temperature gradients and atmospheric circulation changed because of global warming as it affects air pollutant transport and dispersion and thus affects air quality (Borrego *et al.*, 2000). The combined effect of high emissions and unfavorable weather results in air pollution (Jacob and Winner, 2009). Also, air pollutants mainly in the broadest sense are considered to be the main cause of climate change.

In India, pollution has become a great topic of debate at all levels and especially the air pollution because of the enhanced anthropogenic activities such as burning fossil fuels, *i.e.* natural gas, coal and oil-to power industrial processes and motor vehicles. Among the harmful chemical compounds, this burning puts into the atmosphere, are carbon dioxide (CO_2), Carbon monoxide (CO), Nitrogen oxides (NO_x), Sulphur dioxide (SO_2), and tiny solid particles-including lead from gasoline additives-called particulates (Goyal *et al.*, 2003). Deterioration of air quality is a problem that is directly experienced by a majority of the 300 million urban Indians, who constitute about 30 per cent of the India's population. A local survey has indicated that the incidence of respiratory diseases in Delhi is 12 times the national average and that 30 per cent of Delhi's population suffers from respiratory disorders due to air pollution (Kandlikar and Ramachandran, 2000)

Keeping the air quality acceptable has become an important task for decision makers as well as for non-governmental organizations. Particulate matter and gaseous emissions of pollutant emission from industries and auto exhausts are responsible for rising discomfort, increasing airway diseases and deterioration of artistic and cultural patrimony in urban centers (Rao *et al.*, 2003). As many cities around the world become more congested, concerns increase over the level of urban air pollution being generated and in particular its impact on localized human health effects such as asthma or bronchitis. The more this relationship is understood, the better chance there is of controlling and ultimately minimizing such effects. .

Kurukshetra is a religious city and attracts large number of pilgrims from all over India and foreign as well. This city has a large number of daily floating population from near-by extension areas, villages and districts. The increasing tourism trend of the city acts as a contributor of air pollution in the city. The purpose of this study was to analyze the concentration of SPM, SO_2 and NO_2 in ambient air at different locations of the city and to compare the result with National Air Quality Standards.

Materials and Method

Kurukshetra district is located in north-east part of the Haryana state. It is lying between 29°53′ N to 30°15′ N and 76°26′ E to 77°07′ E and covers 3.46 per cent area of the state. In order to monitor the air quality, seven sampling sites of the city namely Kurukshetra University Campus, Kurukshetra University 3rd Gate, Kalyan Nagar Residential Area, Kurukshetra City Market, Kurukshetra Sector-17, New Bus Stand and at Pipli-National Highway-1 were selected. Three important gaseous pollutants namely Suspended Particulate Matter (SPM), Sulphur dioxide (SO_2) and Nitrogen dioxide (NO_2) were identified at all the locations. For this purpose standard methods strictly according to prescribed Indian standard methods for measurements of air pollutants were adopted (ISM, 1969, 1973 and 1975). The concentrations of the pollutants (24 hrs average) were monitored during summer season of 2009 using High Volume Air Sampler (APM 460 PL).

The SO_2 was absorbed from air in a solution of potassium tetrachloride mercurate. The resultant complex was made to react with pararosaniline and formaldehyde to form the coloured pararosaniline methyl sulphuric acid. The absorbance of this solution was measured by spectrophotometer at 560 nm (ISM, 1969). Nitrogen dioxide in ambient air was collected by bubbling it through a solution of sodium hydroxide and sodium arsenate. The resultant nitrite ion concentration was determined by reacting it with sulfanilamide and N-(1-Napthyl 1)- ethyline diamine dihydrochloride, the absorbance was then measured at 540 nm (ISM, 1975). Suspended particulate matter was measured gravimetrically. High volume sampling with Whatmann filter paper was used at an average flow rate being not less than 1.1 cubic metre per minute (ISM, 1973).

Results and Discussion

The monitoring and analysis for three major pollutants SPM, SO_2 and NO_2 was conducted at seven different sites of Kurukshetra City. The results were compared with the ambient air quality standards given by Central Pollution Control Board,

2009. Concentration of NO_2 at all sampling sites was well within the permissible limits, however city market ranked first among the sites with 34.0 μg/m³ of NO_2 followed by Pipli NH-1 (30.0 μg/m³ NO_2) and area of sector-17 (28.4 μg/m³ NO_2)-. The lowest value was observed in KUK campus (13.0 μg/m³) (Figure 3.1).

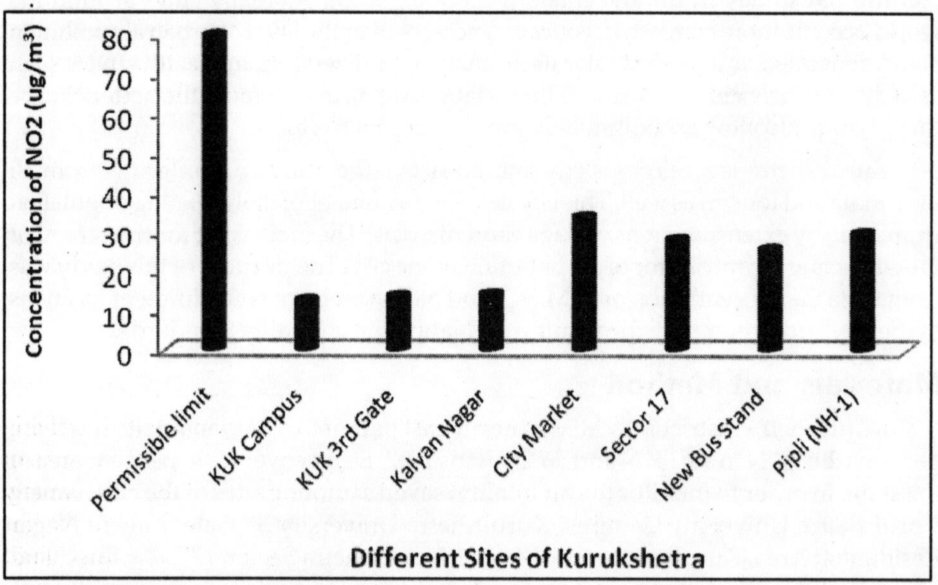

Figure 3.1: Concentration of Nitrogen Dioxide (μg/m³) at different Sites of Kurukshetra

Concentration of sulphur dioxide at all sites was much lower than the permissible limits of ambient air quality standards of CPCB. Among the sites, it was highest in the area of city market (6.0 μg/m³) and minimum in KUK campus (1.5 μg/m³) (Figure 3.2).

The RSPM concentration ranged from 143.0 to169.0 μg/m³. The values were higher than the permissible limit at all the sites. Among the different sites, RSPM content was highest in City market (171.02 μg/m³) followed by New bus stand area (169.81 μg/m³) and Pipli (NH-1) (167.6 μg/m³) (Figure 3.3).

The concentration of Non Respirable Suspended Particulate Matter ranged from 289.2 to 398.01μg/m³. The values were highest for the city market area *i.e.* 398.01μg/m³followed by Pipli NH-1 (360.08 μg/m³) and Sector-17 area (358.18 μg/m³), while the concentration was least at KUK campus (289.02 μg/m³) (Figure 3.4). Permissible limit for Total Suspended Particulate Matter as given by CPCB is 200 μg/m³. The values observed for TSPM was much higher than the permissible limit at all sites. The concentration ranged from 457.8 μg/m³ to 569.03 μg/m³ (Figure 3.5).

Overall, the suspended particulate matter level in all the areas was found to be above the permissible limits and the levels of SO_2 and NO_2 were well within the permissible limits. The air pollution levels were high in the main traffic routes and commercial localities. City market ranked first in SO_2, NO_2 and SPM concentration

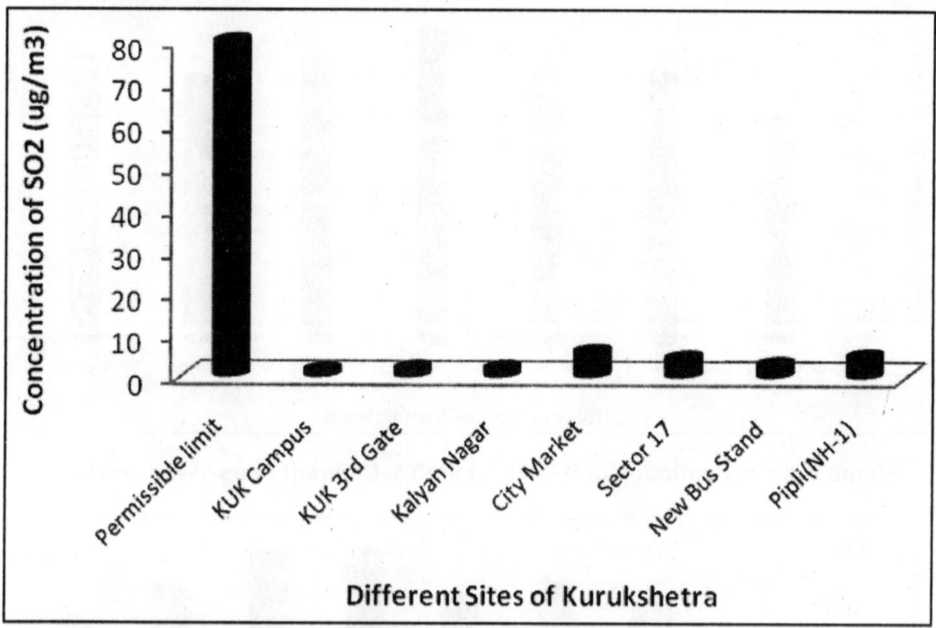

Figure 3.2: Concentration of Sulphur Dioxide (μg/m³) at Different Sites of Kurukshetra

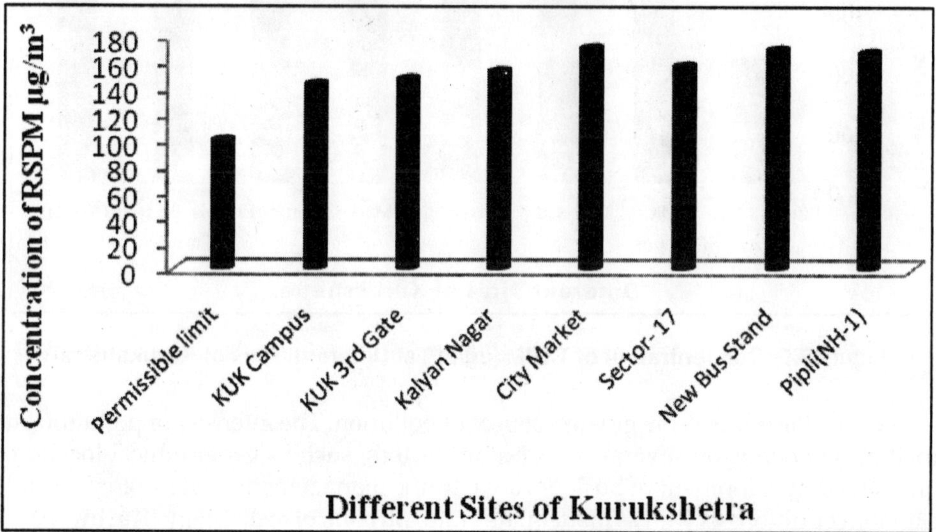

Figure 3.3: Concentration of RSPM (μg/m³) at Different Sites of Kurukshetra

among all the sampling sites.The concentration of all the pollutants was found to be minimum at Kurukshetra University Campus.

A high level of SPM is the most prevalent form of air pollution. This can be attributed to high industrial activities and plying of all kinds of vehicles (Jain *et al.*, 2004; Senthilnathan, 2005).

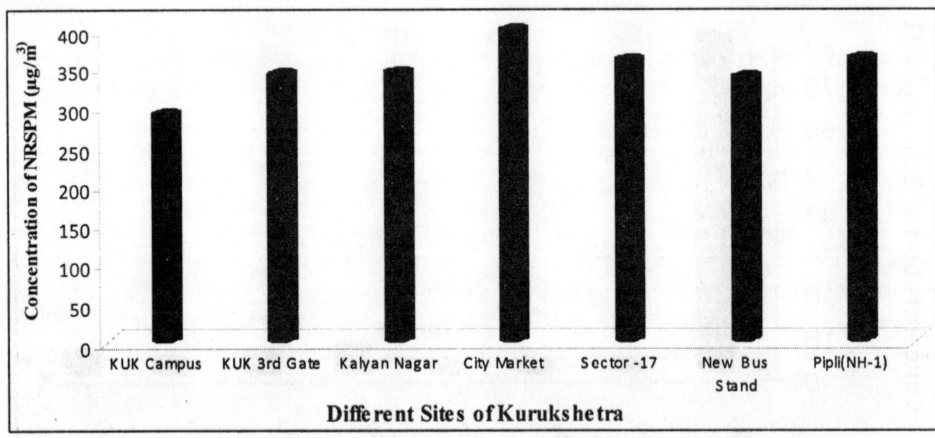

Figure 3.4: Concentration of NRSPM (µg/m³) at Different Sites of Kurukshetra

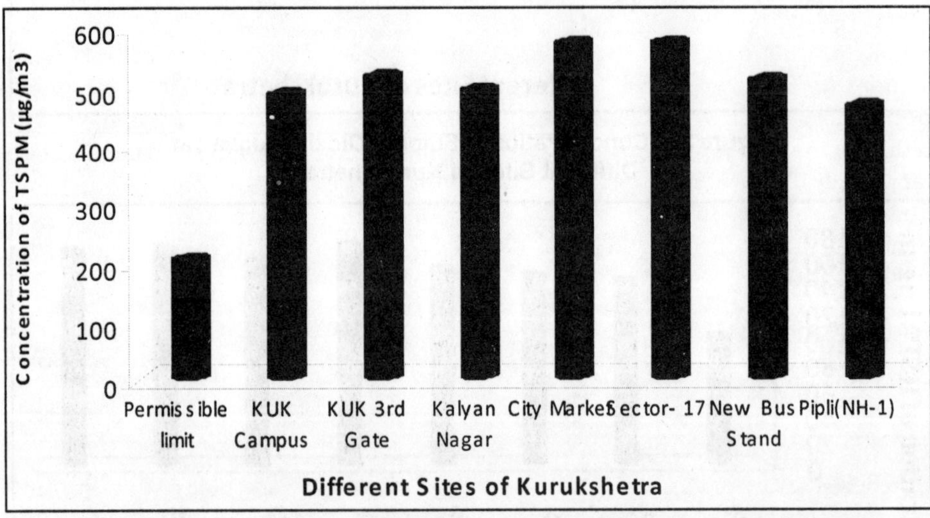

Figure 3.5: Concentration of TSPM (µg/m³) at Different Sites of Kurukshetra

Vehicular traffic is the greatest source of pollution. The intensity of pollution on road also depends on several contributory factors, such as geographical location, unplanned development of business area, traffic management, meteorological and climatic conditions as well as inadequate emission control technology (Kushwaha *et al.*, 2008). Two wheelers are the major transport in the city and these along with three wheelers are the major contributors of air pollution (*e.g.*, RSPM, TSPM). The diesel powered heavy vehicles add to the air pollution problems by emitting out jet black smoke. The problem is further aggravated when some vehicles use adulterated petrol mixed with kerosene oil. The emissions from such vehicles are very high in suspended particulate matter such as carbon and soot particles resulting in air pollution (Sehra *et al.*, 2007). Also, the non-CNG vehicles rejected from Delhi and other metro regions are sold in the markets of nearby small cities and add to the problem of vehicular emissions. The concentration of SO_2 and NO_2 is comparatively lesser than other pollutants. However increasing level of these pollutants, due to industrial activities, urbanization, alarming increase in the number of vehicles, increased inflow of tourists and poor management practices may become a problem in the future with greater health impact. Unless effective strategy is evolved to control the environmental pollution in a time bound manner, the situation may worsen in a few years.

Air quality is strongly dependent on weather and is therefore sensitive to climate change. Urban air pollution control policies are beneficial for human health and ecosystems. Even small contributions to climate change mitigation can be important in economic terms (Prinn *et al.*, 2007). Efficiency-enhancing measures in fossil fuel use may lead to the dual benefits of both local air pollution abatement and GHG emissions abatement. In India, pollution levels are prescribed for vehicles and have been made mandatory. Air pollution regulations result in higher fuel quality and reduced GHG emissions. Estimates indicate the reduction in emission of GHG to be in the range of 10 to 15 per cent up to 2015, if stricter norms are followed (Parik and Parikh, 2002).

Kurukshetra is promoted as religious hub. Therefore, maintenance of air quality standards is important for health of residents and increasing inflow of tourist population. Effective implementation of Euro Norms and Bharat Norms can help in reducing the GHG emissions and improving the quality of air. Bharat 2000 norms are far stricter than the Euro II norms and are one of the most stringent in the world. Bharat Stage IV for all private vehicles, city public service vehicles and city commercial vehicles in indentified cities such as Kurukshetra is also under effective action from April 1, 2010 (CPCB, 2010)

References

Borrego, C., Miranda, A.I., Carvalho, A.C. and Fernandez, C. 2001: Climate change impact on the air quality: The portuguese case. *Global Nest: the Int. J.*, 2(2):199-208.

CPCB, 2010: Central Pollution Control Board. Status of vehicular pollution contol programme in India. Programme Objective Series. PROBES/136/2010. Ministry of Environment and Forest. Government of India.

Goyal, P., Sidhartha. 2003: Present scenario of air quality in Delhi: a case study of CNG implementation., Atmospheric Environment, 37: 5423-5431.

Hamburg, S.P., Harris, N., Jaegar, J., Karl, T.R., McFarland, M., Mitchell, J.F.B., Oppenheimer, M., Santer, S., Schneider, S., Trenberth, K. E., Weigley, T.M.L.1997: Common question about climate change. United Nations Environmental Programme, World Meteorogical Organization.

Houghton, 2001: Climate Change-2001: The Scientific Basis. Cambridge University Press, Cambridge.

IPCC. 2001: Climate Change 2001. The Scientific Basis. Contribution of Working Group I to the Third Assessment Report of the Intergovernmental Panel on Climate Change. Cambridge University Press, New York.

ISM.1969: Indian Standard-Methods for measurement of air pollution, Sulphur dioxide: 5182 (Part-11).

ISM.1973: Indian Standard-Methods for measurement of air pollution, Suspended matter: 5182 (Part-1V).

ISM.1975: Indian Standard-Methods for measurement of air pollution, Nitrogen oxides: 5182 (Part-V1).

Jacob, D.J. and Winner, D.A. 2009: Effects of climate change on air quality. Atmospheric Environment, 43: 51-63.

Jain, R., D.K. and Gupta, A.B. 2004: Status of air quality at selected traffic junctions of Jaipur city. *Nature Environ and Pollution Tech.* 3(4):435-442.

Jayanthi.V. and Krishnamurthy, 2006: Key Airborne Pollutants-Impact on Human Health in Manali, Chennai., *Current Science*, 90(3):405-413.

Kandlikar, M. and Ramachandran, G. 2000: The cause and consequence of particulate air pollution in urban India: A synthesis of the science. Annual Review of Energy and the Environment, 25, 629–684.

Kushwaha., S.K., Singh, Rajesh., Kushwaha, B.P., Pathak, Vandana., Rai. O.P. and Dwivedi. R. 2008: A Study Of Air Quality and Human Health of Certain Villages of Industrial Area, Raigarh. *Indian J. Env. Prot*, 28(8): 709-713.

Mukunda Rao, P.V., Hima Bindu, V., Sagareshwar, G., Jayakumar, Indracanti., Anjaneyulu, Y. 2003. Assessment of ambient air quality in the rapidly industrially growing Hyderabad urban environment.

Parikh, J.K. and Parikh, K. 2002: Climate Change: India's Perceptions, Positions, Policies and Possibilities. OECD.

Prinn, R., Reilly, J.M., Sarofim, M., Wang, C. and Felzer, B. 2007: Effects of air pollution control on climate: results from an integrated global system model. In: Human induced Climate Change: An Interdisciplinary Assessment, eds. M. Schlesinger, H. Khelshgi, J. Smith, F. de la Chesnaye, J.M. Reilly, T. Wilson and C. Kolstad. Cambridge University Press, Cambridge.

Schimmel, D., Enting, I.G., Heimann, M., Wigley, T.M.L., Rayneud, D., Alves, D. and Seigenthler, U. 1995. 'CO2 and the carbon cycle', in J.T. Houghton, L.G. Meira Filho, J., Bruce, H. Lee, B.A. Callander, E. Haites, N. Harris, and K. Maskell (eds), Climate Change 1994.

Senthilnathan, T. 2005: Status of respirable dust particles (RDP) concentration-A case study in Chennai city. *J. Environ. Bio*, 26:425-428.

Sharma, S.K., Bhandari, S. and Bhardwaj, K.R. 2006: Present Status of Air Pollution in Three Towns of Haryana. Eco-Research. *J. Bio-sciences*, 5(1):38-45

Singh, Sehra., Parmjit. 2007: Atmospheric Pollution: A Case study of Degradation Urban Air Quality Over Punjab. *Ind. J. Environmental Protection*, 27(1): 32-37.

2013, Climate Change Effects on Agriculture and Economy *Pages* ***29–36***
Editors: **D.R. Khanna, A.K. Chopra, Vikas Singh, Rakesh Bhutiani & Gagan Matta**
Published by: **BIOTECH BOOKS, NEW DELHI**

Chapter 4

Water Quality and Eutrophication Studies of Some Lakes (Under Ramsar Convention) in the Western Himalayas, India

Omkar Singh, M.K. Sharma, V.K.Choubey and R.D. Singh
National Institute of Hydrology, Roorkee, Uttarakhand

Lakes are important source of freshwater in the Himalayan Region. In view of growing concern about degradation of lake water quality, it is necessary to monitor existing water sources for better conservation and management. In this study, water quality and eutrophication status has been assessed for four Himalayan lakes (Renuka, Mansar, Surinsar, Tsomoriri). DO was observed below limits causing hypolimnetic anoxic condition in the Shiwalik lakes causing fish mortality during winter. Water quality of Tsomoriri Lake, located in Ladakh region (cold desert) was found highly enriched with minerals leading to brackish in nature. The eutrophication status in these lakes was assessed using Carlson's Trophic State Index based on phosphate data, which showed Renuka, Mansar, Surinsar and Tsomoriri lakes under eutrophic condition.

Keywords: *Western Himalayan, Ramsar convention, Water quality, Eutrophication, Fish mortality.*

Introduction

The Himalayas are endowed with hundreds of relatively smaller freshwater lakes, which were used as natural centers of civilization and to cater the civilian needs. Many lakes are unique ecosystems and valuable from the historical, cultural, biological and socio-economic point of view. They serve variety of purposes like drinking, irrigation, recreational, aqua-culture *etc.* However, with the exponential growth of population in India, the land and water resources are facing a tremendous pressure and Himalayan natural resources are no exception to this fact. Consequently, intensive agricultural practices in the lake catchments, encroachments up to the lakeshore, disposal of solid wastes and discharge of domestic and industrial wastes have posed a serious problem of water quality degradation. The felling of trees for energy requirements, cultivation on mountainous catchments coupled with indiscriminate use of chemical fertilizers, insecticides and pesticides has also enhanced problem of water quality degradation and eutrophication of water bodies. Large scale fish mortality has also observed in most of the Himalayan lakes during winter and posing threat to ecological imbalances in the region.

The Ramsar Convention on wetlands has notified 26 lakes/wetlands as international importance in India. Of them, six are located in the Western Himalayan Region namely, Renuka (H.P.), Mansar-Surinsar, Tsomoriri, Wular and Hokera wetlands in Jammu and Kashmir. In this study, water quality and eutrophication status has been assessed for four Himalayan lakes (Renuka Lake of Himachal Pradesh, Mansar, Surinsar, Tsomoriri lakes of Jammu and Kashmir).

Study Area

The present study deals with the lakes located in the State of Jammu and Kashmir (Mansar, Surinsar, Tsomoriri) and Himachal Pradesh (Renuka) in the Western Himalayas. Mansar and Surinsar are located in Jammu region, Tsomoriri in Ladakh Region. All these lakes (Mansar, Surinsar, Tsomoriri and Renuka) are the Ramsar sites. These lakes are of great socio-economic importance and are famous for their picturesque view and most of them are being used for drinking and irrigation purposes.

Table 4.1: General Information about the Lakes under this Study

Sl.No.	*Details*	*Lakes*			
		Mansar	*Surinsar*	*Tsomoriri*	*Renuka*
1	District	Udhampur	Jammu	Leh	Sirmaur
2	Latitude	32° 40' 58.25" N	32° 46' 30" N	32° 48' 16" N	30° 36' N
3	Longitude	75° 5' 11.5" E	75° 2' 30" E	78° 12' 51" E	77° 27' E
4	Altitude (m)	666	604	4527	645
5	Max. depth (m)	38.25	24.05	30.00	13
6	Area (Sq. Km)	0.59	0.32	141.05	0.18

Renuka lake is one of the natural wetlands located in the Sirmour district of the Himachal Pradesh. Renuka lake is a very important tourist and religious place of the H.P. Being situated in the main range of lesser Himalayas, this wetland is of special importance from bio-diversity point of view. The National Wetland Management Committee of the Ministry of Environment and Forests (Govt. of India) has recognized the Renuka lake as one of important wetlands of the country, which requires conservation and management on priority basis (State Council for Science, Technology and Environment, Govt. of H.P.). This lake also finds place in the priority list of lakes, which required immediate attention for restoration under the title "Management of Lakes in India" (Reddy and Char, 2004). The lake rests in a long valley and the surrounding slopes are covered with a variety of vegetation and thick woods. The supply of the lake is through nallahs draining the catchement and probably numerous underlying springs. The information of the lakes under this study is given in Table 4.1.

Review of Literature

Various investigators have reported their studies on various aspects of the Himalayan region lakes (Zutshi, 1989; Omkar and Sharma, 1994-95; Jain *et al.*, 1999; Kumar *et al.*, 1999a, Kumar *et al.*, 1999b; Das and Dhiman, 2003; Shewa, 1998; Rai *et al.*, 2006, 2007). Physico-chemical and biological characteristics of Mansar lake have been studied by Zutshi (1985, 1989), Chandra Mohan (1992) and Gupta (1992). Zutshi *et al.* (1980) have reported that lakes of Jammu and Kashmir are different in their morphology and thermal behaviour and vary from sub-tropical monomictic to dimictic type. Rai *et al.* (2001) performed studies on bathymetry, rate of sedimentation and water quality of Mansar Lake during 1998-99.

Few studies have also been reported for the Renuka Lake, which include mainly morphometry, general water quality and catchment related aspects of the Renuka Lake (Singh *et al.*, 1987; Anonymous, 1996, 2004).

Methodology

In the present study, water quality data of the Mansar and Surinsar (Jammu Region), Tsomoriri (Ladakh Regions) of Jammu and Kashmir and Renuka lake (H.P.) has been used. The data includes mainly: Temperature, pH, EC, TDS, DO, Ca, Mg, Na, K, Alkalinity, HCO_3, Cl, SO_4, NO_3, PO_4, F and Hardness. In few lakes, water quality deterioration due to the bacterial contamination of lakes has also been reported. In-situ measurements of Temperature, pH and EC were also made using portable instruments. Standard Water Sampler was used for depth wise sampling in the Mansar lake. The water quality of the Dal lake has been described based on the published data (Handa *et al.*, 1991). Water quality of the lakes was evaluated for drinking (BIS, 1991). The data of phosphate obtained in different years was used to study eutrophication status of these lakes on the basis of Trophic State Index (Carlson, 1977).

Results and Discussion

The variation of water quality parameters of various studied lakes along with their average values are given in Table 4.2. Table shows a wide range of variation of water quality parameters for the lakes of Ladakh Region in comparison to the lakes of Jammu and Kashmir regions. The study has shown pH values greater than 7.0 for all lakes under this study, which indicate alkaline nature of lake water with varying scale of alkalinity. The average value of pH (8.96) in Tsomoriri is indicative of higher

Table 4.2: Water Quality Status of Mansar, Surinsar and Tsomoriri Lakes of J&K

Parameters	*BIS Limits (Desirable)*	*Mansar*		*Surinsar*	*Tsomoriri*
		Surface	*Bottom*	*Surface*	*Surface*
pH	6.5-8.5	7.95-8.04 (7.99)	7.98-8.12 (8.03)	8.2-8.86 (8.43)	8.9-9.02 (8.96)
EC (μS/cm)	–	184-204 (194)	265-281 (273)	510-600 (545)	3360-3740 (3550)
TDS (mg/l)	500	118-131 (124)	170-180 (175)	310-380 (350)	2150-2393 (2272)
DO (min) (mg/l)	6	3.3-6.6 (5.5)	0.3-0.6 (0.45)	7.2-8.6 (7.8)	–
BOD (max) (mg/l)	2	1.2-6 (2.08)	1.2-14 (7)	0.6-1.2 (0.9)	–
HCO_3 (mg/l)	–	121-129 (125)	173-178 (177)	120-140 (127)	0-2 (1)
TH (mg/l)	300	44-56 (53)	82-94 (88)	100-160 (136)	3161
Cl (mg/l)	250	4-10 (6)	4-8 (7)	7-12 (10)	24 (24)
SO_4 (mg/l)	150	1-4 (2)	1-3 (1.6)	8-18 (11)	32-144 (88)
PO_4 (mg/l)	–	0.01-0.3 (0.1)	0.12-0.18 (0.15)	0.03-0.11 (0.05)	0.03-0.04 (0.03)
NO_3 (mg/l)	45	0.44-2.2 (0.76)	0.5-5 (1.5)	4-6 (5)	–
Ca (mg/l)	75	14-18 (17)	29-32 (30)	32-48 (38)	30-40 (35)
Mg (mg/l)	30	2-4 (2.5)	2-4 (2.8)	5-19 (13)	744-750 (747)
Na (mg/l)	–	15-17 (15.6)	14-16 (15)	10-11 (10)	89-1493 (791)
K (mg/l)	–	3 (2.8)	3 (2.6)	4-6 (4)	98-319 (209)
F (mg/l)	1.5	0.01-0.32 (0.14)	–	–	0.44-0.50 (0.47)

alkaline nature and exceeding the limits for drinking purposes (BIS-10500: 1991). TDS (2272 mg/l), TH (3161 mg/l), Mg (747 mg/l) in Tsomoriri Lake was also found to be exceeding prescribed limits for drinking purposes. It could be primarily due to presence of evaporites in lakes of Ladakh division attributed in cold desert type climates. GSI (1977) has also reported exceptionally very high values of salt containing constituents in the Ladakh division.

The water quality of the Renuka lake was also compared with the BIS norms of drinking water quality (BIS-10500: 1991). The results are given in Table 4.3, which show mean values of few water quality parameters beyond desirable limits *viz.* magnesium (mean 41 mg/l against limit of 30 mg/l), total coliform (mean 61 MPN/100 ml against limit of 10 MPN/100 ml), faecal coliform (mean 50 MPN/100 ml against limit of zero MPN/100 ml)), iron (mean 1.49 mg/l against limit of 0.3 mg/l), lead (mean 0.35 mg/l against limit of 0.05 mg/l), manganese (mean 0.87 mg/l against limit of 0.10 mg/l) and cadmium (mean 0.01 mg/l against limit of 0.01 mg/l), respectively. However, pH, total hardness, TDS, calcium, copper, sulphate, fluoride, alkalinity, DO and BOD were also found to be exceeding the desirable limits in few water samples of the Renuka lake under this study. Therefore, the Renuka lake water requires appropriate water treatment measures before use for drinking purpose.

Table 4.3: Water Quality Status of Renuka Lake (H.P.)

Sl.No.	*Substance or Characteristic*	*BIS Limits (Desirable)*	*Water Quality of Renuka Lake*
1.	pH Value	6.5 to 8.5	5.8-9.1 (7.13)
2.	Total Hardness (as $CaCo_3$) mg/l, Max	300	224-430 (287)
3.	Iron (as Fe) mg/l, Max	0.3	0.0-7.84 (1.49)
4.	Chlorides (as Cl) mg/l, Max.	250	0.0-9.0 (1.8)
5.	Dissolved solids mg/l, Max	500	324-570 (364)
6.	Calcium (as Ca) mg/l, Max	75	32-104 (49)
7.	Magnesium mg/l, Max	30	33-49 (41)
8.	Copper (as Cu) mg/l, Max	0.05	0-0.16
9.	Manganese (as Mn) mg/l, Max	0.10	0.0-7.17 (0.87)
10.	Sulfate (as SO_4) mg/l, Max	200	61-215 (113)
11.	Nitrate (as NO_3) mg/l, Max	45	0.0-42 (4.3)
12.	Fluoride (as F) mg/l, Max	1.0	0-1.6 (0.71)
13.	Cadmium (as Cd) mg/l, Max	0.01	0-0.03 (0.01)
14.	Lead (as Pb) mg/l, Max	0.05	0.17-0.66 (0.35)
15.	Zinc (as Zn) mg/l, Max	5	0.0-1.64 (0.15)
16.	Alkalinity mg/l, Max	200	125-260 (165)
17.	Faecal coliform. MPN, Max	Nil	0-1100 (50)
18.	Total coliform. MPN, Max	Nil	0-1100 (61)
19.	Dissolved Oxygen, mg/l, Min	6.0	0-10.4 (6.7)

Eutrophication Status of lakes in J&K and H.P.

The process of eutrophication is defined as the loading of inorganic and organic dissolved and particulate matter to lakes at rates sufficient to increase the potential for high biological production that leads to a decrease in the capacity of the lake. It is usually measured using one of several Trophic State Index (TSI) of algal weight (biomass): water transparency (Secchi Depth, TSI-SD), algal chlorophyll (TSI-Chl), and total phosphorus (TSI-TP) (Carlson, 1977).

In the present study, eutrophication status of lakes has been studied based on Carlson's Trophic State Index (Carlson, 1977) using phosphate data. The results are given in Table 4.4, which indicates that all lakes (Mansar, Surinsar, Tsomoriri and Renuka) in this study were found under eutrophic condition. As such, more concentrated and integrated efforts are required to lower the present level of eutrophication.

Table 4.4: Eutrophication Status of some Lakes in Western Himalayas

Lakes	*Phosphate (mg/l)*	*TSI (TP)*	*Trophic Status*	*Year*
Mansar (J&K)	14	42.20	Mesotrophic	Zutshi, 1989
	80	67.00	Eutrophic	1999 (Rai *et al.*, 2001)
	100	70.00	Eutrophic	2004
Surinsar (J&K)	50	61.00	Eutrophic	1995
Tsomoriri (J&K)	30	53.00	Eutrophic	2000
Renuka (H.P.)	90	69.00	Eutrophic	2008

Conclusion

In this study, water quality of Tsomoriri lake (Ladakh region) was found to show a distinct characteristics apart from other lakes probably due to climatic condition (cold desert type climate) having a very low rainfall in the order of 100 mm in this region. As such, unusually a very high concentration of some water quality parameters viz. pH, Total Dissolved Solids, Total Hardness, Chloride, Calcium and Magnesium were obtained in Tsomoriri lake of Ladakh region. In Renuka lake, Fe, Mn and Pb were found to be exceeding the desirable limits for drinking water as per BIS (1991). Low Dissolved Oxygen (DO) in the hypolimnion was observed in the Renuka and Mansar lakes, which indicate anoxic conditions during winter. It was also observed that these lakes remain stratified during summer and become mixed during peak winter altogether posing threat to aquatic fauna (fish). Lakes (Mansar, Surinsar, Tsomoriri and Renuka) in this study were found under eutrophic condition. As such, more concentrated and integrated efforts are required to lower the present level of eutrophication.

Acknowledgement

Authors are also thankful to all concerned officers/research staff of NIH Regional Centre, Jammu/EHD (NIH), Roorkee for their contribution during field investigations and laboratory analysis.

References

Anonymous, 1999: Digest of Statistics 1997-98, Directorate of Economics and Statistics, Planning and Development Department, Government of Jammu and Kashmir.

Anonymous, 1999a: Wetlands of India, Space Application Centre (ISRO), June 1998.

Anonymous, 1996: Report on Water Quality Monitoring of Renuka Lake (unpublished), H.P. State Pollution Control Board, Shimla.

Anonymous, 2004: Report on Water Quality Monitoring of Renuka Wetland (unpublished), H.P. State Environment Protection and Pollution Control Board, Shimla.

BIS, 1991: Indian Standard Institute, New Delhi.

Chander Mohan, 1993: Limnology of lake Mansar with particular reference to primary producers, Unpublished Ph.D Thesis, Jammu University, Jammu, 215 pp.

Carlson, R.E. 1977: A trophic state index for lakes. Limnology and Oceanography. 22:361-369.

Das, B.K. and Dhiman, S.C. 2003: Water and sediment chemistry of higher Himalayan lakes in the Spiti Valley: Control on weathering, provenance and tectonic setting of the basin, J. Env. Geol., Vol. 44, No. 6, pp. 717-730.

Handa, B.K., Kumar, A. and Bhardwaj, R.K. 1991: Studies on Dal lake, Srinagar, J&K, eutrophication studies, Bhujal News, CGWB, 6(4).

Jain, A., Rai, S.C. Pal, J. and Sharma, E. 1999: Hydrology and nutrient dynamics of a sacred lake in Sikkim Himalaya, J. Hydrobiologia, 416, pp. 13-22.

Jain, C.K. and Bhatia, K.K.S. 1987: Physico chemical analysis of water and wastewater, UM-26, *National Institute of Hydrology*, Roorkee.

Kumar, Bhishm, Jain S.K., Nachiappan, R.M.P., Rai, S.P., Kumar, Vinod, Dungrakoto, V.C. and Rawat, Y.S., 1999: Hydrological studies of lake Nainital, Kumaun Himalayas, Uttar Pradesh, Final Project Report, National Institute of Hydrology, Roorkee.

Kumar, Vijay, Rai, S.P. and Singh, Omkar, 2006: Water Quantity and Quality of Mansar lake in the Himalayan Foothills, India, Intl J. of Lake and Reservoir Management, Vol. 22 (3), pp. 191-198.

Omkar and Sharma, 1994-95: Water Quality Studies of Surinsar Lake in Jammu Region, Report -CS (AR) 157, NIH, Roorkee, pp. 50

Rai, S.P., Kumar, V., Singh, O., Kumar, B. and Jain, S.K., 2001. "Limnological Study of the Mansar Lake, District Udhampur, J&K", Final Project Report, NIH, WHRC, Jammu Cantt.

Rai, S.P., Kumar, V., Singh, O., Kumar, B. and Jain, S.K., 2006: "Bathymetry, Sedimentation Rate and Physico-Chemical Characteristics of Mansar Lake in the Himalayan Foothills, J&K, *India, J. GSI*, Vol. 67, pp. 211-220.

Rai, S.P., Kumar, V., Singh, O. and Jain, S.K. 2007: "Hydrochemical Characteristics of Mansar Lake, J&K in *India, J. IE (I)-EN*, Vol. 88, pp. 16-22.

Ramsar, 2008: Cited from Proposed Theme of World Wetland Day 2008, Ramsar Convention on Wetlands (www.ramsar.org).

Shewa, W.A. 1998: Eutrophication analysis of lakes of Kumaun region, M.E. Dissertation, Dept. of Civil Engg., IIT, Roorkee, pp. 92.

Singh R., Mishra, S.H., Shyamananda, R.K., Sharma, G., Mahajan, I. and Aggarwal, B.K. 1987. Morphometry and catchment study of Renuka lake, Himachal Pradesh, India, with a note on its flora and fauna, In: Western Himalayas, Vol. II (Eds. Pangtey, Y.P.S. and Joshi, S.C.), Gyanodaya Prakashan, Nainital, pp. 639-649.

Zutshi, D.P., 1985. The Himalayan lake ecosystem, In: Singh J.S. (eds.) Environmental Regeneration in Himalaya: Concept and Strategies, The Central Himalayan Environmental Association and Gyanodaya Prakashan, Nainital, pp. 325-342.

Zutshi, D.P. 1989. 25 Years of ecological research on the lakes of North-Western Himalaya, In: Singh J.S. and Gopal I.B. (eds.) Perspective in Ecology, Jagmaudar Book Agency, New Delhi, pp. 49-65.

Zutshi, D.P., Subha, B.A. and Khan, M.A. 1980: Comparative limnology of nine lakes of Jammu and Kashmir Himalayas. Hydrobiologia, 72, pp. 101-112.

2013, Climate Change Effects on Agriculture and Economy *Pages* **37–45**
Editors: **D.R. Khanna, A.K. Chopra, Vikas Singh, Rakesh Bhutiani & Gagan Matta**
Published by: **BIOTECH BOOKS, NEW DELHI**

Chapter 5

Water Resource Management and its Impact on Public Health

Rashi Gupta and Priyanka Kardam

International Institute of Health Management Research,
Plot No. 3, HAF Pocket, Sector-18A, Dwarka, New Delhi

Potable water is a limited resource for human consumption in developing countries like India. The share of sweet water is only 3 per cent out of total available water on earth which is about 35 million km^3. The share of India is 4 per cent of the total available sweet water whereas its population is 16 per cent of the world population. The major source of water pollution in India is untreated industrial waste, sewage, pesticides. The polluted water is major source of disease like diarrhoea, cholera, typhoid, hepatitis, amoebic dysentery.

The main objective of this study was to understand the water resource management and its effect on the water borne diseases in India. The mixed methodology (both qualitative and quantitative) approach was adapted to carry out this study. Secondary data from various sources like Census of India, National Sample Survey, WHO and SRS bulletin is used. Regression analysis was used to establish relationship between the incidence of water borne diseases and polluted water. The impact of water borne disease due to water pollution is increasing over the past two decades. There is a linear relationship between the water borne disease and water pollution. There is a definite relation between water pollution and population. Cities with more population generated more water pollutants and showed population was at high risk of acquiring water borne disease.

Keywords: *Water borne disease, India, Sewage, Population, Potable water.*

Introduction

Water is essence of life. Water supports all forms of fauna and flora. Water is also essential development of country. It contributes to economic development of country, food production and other general use. The contribution of water to a development of country is indispensable. There is no substitute for water.

Out of total water available on earth only 4 per cent of water is potable. India shares only 4 per cent of this available freshwater where as its population is 16 per cent of the world's population. Global freshwater reserve is rapidly depleting and is expected to leave a significant effect on densely populated areas.

India has been endowed with large freshwater reserves, but increasing population and over exploitation of surface and groundwater has created an alarming situation. The growth in the Indian economy has derived increased water usage by different sectors. Wastewater is increasing significantly and in the absence of treatment and management of wastewater, freshwater bodies are getting polluted.

Although India occupies only 3.29 million km geographical area, which forms 2.4 per cent of the world's land area, it supports over 15 per cent of the world's population. The population of India as on 1 March 2011 stood at 1,210,193,422 (1.21 billion). Thus, India supports about 1/6th of world population, 1/50th of world's land and 1/25th of world's water resources. The total utilizable water resources of the country are assessed as 1086 km.

In the past, several organizations and individuals have estimated water availability for the nation. Recently, the National Commission for Integrated Water Resources Development estimated the basin-wise average annual flow in Indian River systems as 1953 km^3. Utilizable water resource is the quantum of withdrawable water from its place of natural occurrence. Within the limitations of physiographic conditions and socio-political environment, legal and constitutional constraints and the technology of development available at present, utilizable quantity of water from the surface flow has been assessed by various authorities differently. The utilizable annual surface water of the country is 690 km^3.

The annual potential natural groundwater recharge from rainfall in India is about 342.43 km^3, which is 8.56 per cent of total annual rainfall of the country. The annual potential groundwater recharge augmentation from canal irrigation system is about 89.46 km^3. Thus, total replenishable groundwater resource of the country is assessed as 431.89 per cent. After allotting 15 per cent of this quantity for drinking and 6 km^3 for industrial purposes, the remaining can be utilized for irrigation purposes. Thus, the available groundwater resource for irrigation is 361 km^3, of which utilizable quantity (90 per cent) is 325 km^3.

Community water supply is the most important requirement and it is about 5 per cent of the total water use. About 7 km^3 of surface water and 18 km^3 of groundwater is being used for community water supply in urban and rural areas. Along with the increase in population, another important change from the point of view of water supply is higher rate of urbanization. According to the projections, the higher is the economic growth, the higher would be urbanization. It is expected that nearly 61 per cent of the population will be living in urban areas by the year 2050 in high-growth scenario as against 48 per cent in low growth scenario. Different organizations and individuals have given different norms for water supply in cities and rural areas. The figure adopted by the NCIWRD9 was 220 litre per capita per day (lpcd) for class I cities. For the cities other than class I, the norms are 165 for the year 2025 and 220 lpcd for the year 2050. For rural areas, 70 lpcd and 150 lpcd have been recommended for the years 2025 and 2050. Based on these norms and projection of population, it is estimated that by 2050, water requirements per year for domestic use will be 90 km^3 for low demand scenario and 111 km^3 for high demand scenario. It is expected that about 70 per cent of urban water requirement and 30 per cent of rural water requirement will be met by surface water sources and the remaining from groundwater.

Water Resources Management in India

India's water crisis is predominantly a manmade problem. India's climate is not particularly dry, nor is it lacking in rivers and groundwater. Extremely poor management, unclear laws and industrial and human waste have caused this water supply crunch and rendered what water is available practically useless due to the huge quantity of pollution. The tragedy of India's water scarcity is that the crisis could have been largely avoided with better water management practices. There has been a distinct lack of attention to water legislation, water conservation, efficiency in water use, water recycling and infrastructure. Historically water has been viewed as an unlimited resource that did not need to be managed as a scarce commodity or provided as a basic human right. These attitudes are changing in India; there is a growing desire for decentralized management developing, which would allow local municipalities to control water as best needed for their particular region.

Since independence, India's primary goals have been economic growth and food security, completely disregarding water conservation. This has caused serious ramifications being felt today, as many citizens still operate under these principles. Unlike many other developing countries, especially those with acute water scarcity issues such as China, Indian law has virtually no legislation on groundwater. Anyone can extract water: homeowner, farmer or industry as long as the water lies underneath their plot of land. The development and distribution of cheap electricity and electric pumps have triggered rapid pumping of groundwater and subsequent depletion of aquifers. There are approximately 20 million individual wells in India that are contributing to groundwater depletion. The owners of these wells do not have to pay for this water, so there is no incentive to conserve or recycle it; infact they are incentivized to overdraw resources. Generally, the more water they use, the more they can produce. Industry applies the same logic, and rather than reusing the water used for cooling machines, they dump it back into rivers and canals, along with the pollution it has accumulated. India needs to keep boosting agricultural production in order to feed its growing population, but to do so without jeopardizing the amount of water available, farmers must switch to less water intensive crops.

The central government in India also lacks the ability to store and deliver potable water to its citizens, especially as supply shrinks. There is currently a water storage crunch, because means for storage, such as temple tanks and steep wells, have fallen

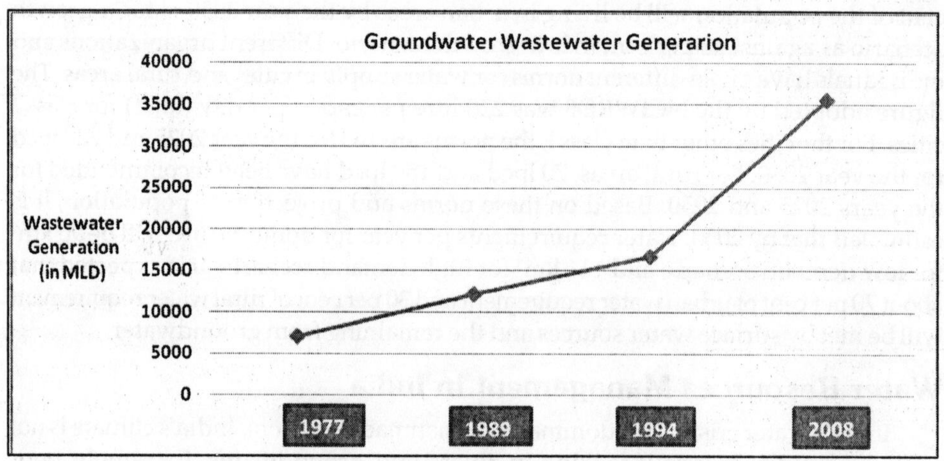

Figure 5.1: Groundwater Wastewater Generation

apart. China is able to store 5 times as much water per person as India, making it blatantly clear how poor India's water management is, in fact that almost of all of that water is too contaminated to use. None of the 35 Indian cities with a population of more than one million distribute water for more than a few hours per day.

Meanwhile, the quantity of sewage is constantly increasing due to population growth. Those not connected to sewage lines end up dumping their waste into canals, which empty into a storm drain that runs into the Yamuna, dumping all of the waste into the river. When the water reaches downstream cities they have to heavily treat it, which subsequently drives up the cost.

Every river in India is polluted to some degree. The water quality in underground wells violates the desired levels of dissolved oxygen and coliform, the presence of which is one measure of filth, in addition to having high concentrations of toxic metals, fluoride and nitrates. India's rivers also have high fluoride content, beyond the permissible limit of 1.5 ppm, which affects 66 million people. The polluted water then seeps into the groundwater and contaminates agricultural products when used for irrigation. Over 21 per cent of transmissible diseases in India are related to unsafe water. Millions of the poorest are affected by preventable diseases caused by inadequate water supply and sanitation.

Water related diseases are the most common cause of deaths. The paucity of clean water for domestic use has led to the increase in the number of deaths in both the urban and rural parts of developing economies and India is no different. Deaths due to water related diseases in India are in the range of nearly 80 percent. Here is a list of the most deadly water related diseases that occur in India.

Diarrhoea

Diarrhoea remains the most prevalent water related disease in India. It mostly affects children under the age of 5 and often leads to death. Diarrhoeal infection is spread through food and drinking water that has been contaminated. A diarrhoeal attack can last up to 2 weeks and leave the person completely dehydrated. Diarrhoea can spread through multiple viruses that is found in contaminated water. The poorer sections of the society come in daily contact with this water and that is the why the rate of diarrhoea is highest amongst them (Figure 5.2).

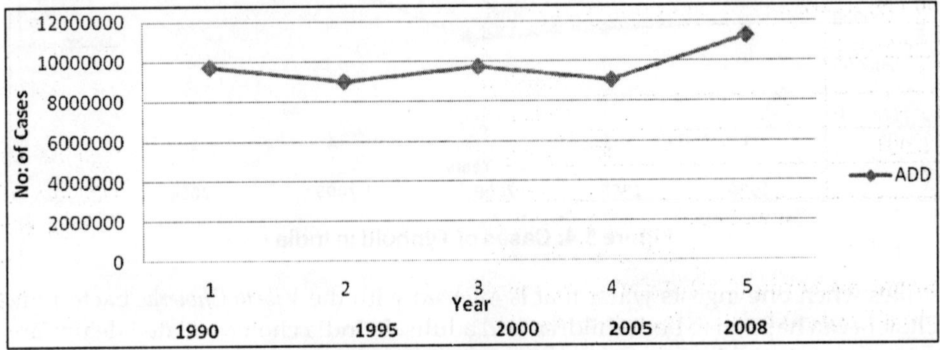

Figure 5.2: Cases of Diarrhoea in India

Viral Hepatitis (Jaundice)

Hepatitis A and E viruses are the most common cause of jaundice in our country. Common treatments with alternative medicines are of questionable benefit and may cause harm in some patients (Figure 5.3).

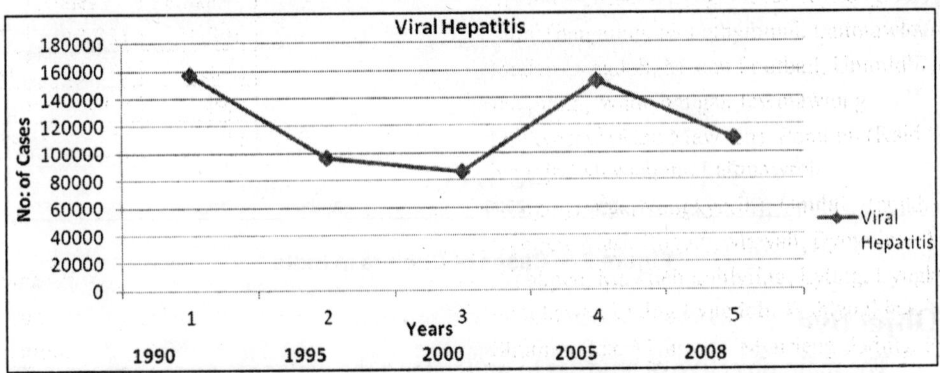

Figure 5.3: Cases of Viral Hepatitis in India

Typhoid Fever (Enteric fever)

Infection by certain bacteria causes this fever. It begins as high grade fever, loss of appetite and lethargy. Blood tests are useful in the diagnosis of typhoid fever (Figure 5.4).

Cholera

A person may get cholera by drinking water or eating food contaminated with the cholera bacterium. The disease can spread rapidly in areas with inadequate treatment of sewage and drinking water. It can kill in hours if left unattended. Cholera

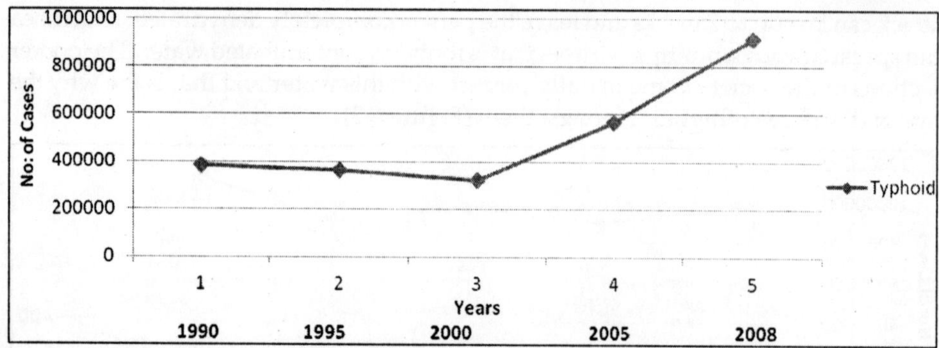

Figure 5.4: Cases of Typhoid in India

strikes when one ingests water that is infested with the *Vibrio Cholerae* bacterium. Cholera can happen to both children and adults. In India cholera related deaths are most common in places with shortage of good quality water. In 2010, nearly 140 people died of cholera in Odisha (formerly known as Orissa) (Figure 5.5).

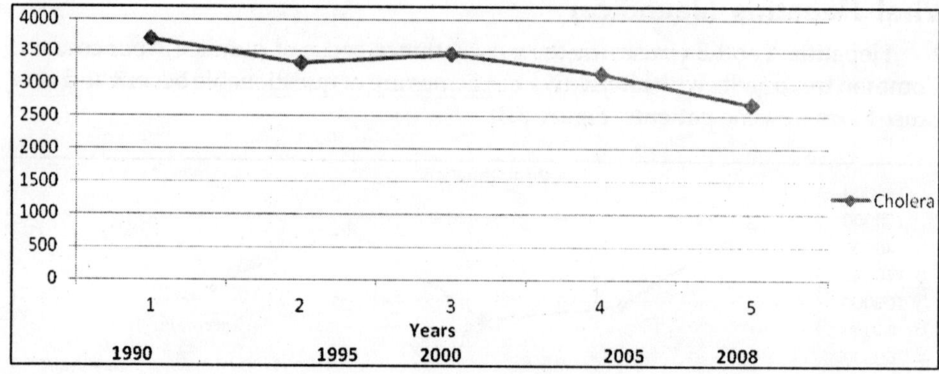

Figure 5.5: Cases of Cholera in India

Objective

The main objective of study is to understand the water resource management and its effect on the water borne disease in India.

Methodology

The mixed methodology (both qualitative and quantitative) approach was adapted to carry out this study. Secondary data from various sources like census of India, National Sample Survey. WHO and SRS bulletin is used. Regression analysis is used to establish relationship between the incidence of water borne diseases and polluted water. It is found that at 95 per cent CI 99 per cent chances are there is link between unimproved water supply and sanitation with water borne diseases (Table 5.1).

Table 5.1: Regression Statistics

Multiple R	0.999986508
R Square	0.999973016
Adjusted R Square	0.999892064
Standard Error	0.075848543
Observations	5

Statistical Analysis

Table 5.2: ANOVA

	df	*SS*	*MS*	*F*	*Significance F*			
Regression	3	213.194247	71.064749	12352.63868	0.006613964			
Residual	1	0.005753002	0.005753002					
Total	**4**	**213.2**						
	Coefficients	*Standard Error*	*t Stat*	*P-value*	*Lower 95 per cent*	*Upper 95 per cent*	*Lower 95 per cent*	*Upper 95 per cent*
Intercept	1959.932697	4.323799075	453.2894946	0.001404442	1904.993621	2014.871773	1904.993621	2014.871773
X Variable 1	–0.000102987	5.01276E-06	–20.54493994	0.030962258	–0.00016668	–3.92937E-05	–0.00016668	–3.92937E-05
X Variable 2	7.65653E-05	3.63927E-06	21.03861082	0.030236834	3.03239E-05	0.000122807	3.03239E-05	0.000122807
X Variable 3	4.76643E-08	6.57542E-08	0.724886264	0.600690665	–7.87822E-07	8.83151E-07	–7.87822E-07	8.83151E-07

Results and Conclusion

The impact of water borne disease due to water pollution is increasing over the past two decades. There is a linear relationship between the water borne disease and water pollution. There is a definite relation between water pollution and population. Cities with more population generated more water pollutants and showed population was at high risk of acquiring water borne disease.

The evidence cited in this review shows that water borne disease is a significant problem worldwide, and is a concern in India. A significant part of this disease relates to the consumption of contaminated drinking water and could be prevented by provision of adequate water and sanitation and promotion of household hygiene practices such as hand washing and point-of-use water treatment and safe storage.

The detrimental environmental consequences of over-exploitation of groundwater need to be effectively prevented by the Central and State Governments. Overexploitation of groundwater should be avoided, especially near the coasts to prevent ingress of seawater into freshwater aquifers. Clearly, a joint management approach combining government administration with active people participation is a promising solution. In critically overexploited areas, bore-well drilling should be regulated till the water table attains the desired elevation. Artificial recharge measures need to be urgently implemented in these areas. Amongst the various recharge techniques, percolation tanks are least expensive in terms of initial construction costs. Many such tanks already exist but a vast majority of these structures have silted up. In such cases, cleaning of the bed of the tank will make them reusable. Promotion of participatory action in rehabilitating tanks for recharging would go a long way in augmenting groundwater supply. Due to declining water table, the cost of extraction of groundwater has been increasing over time and wells often go dry. This poses serious financial burden on farmers. Hence, special programmes need to be designed to support these farmers. Finally, the role of government will have to switch from that of a controller of groundwater development to that of a facilitator of equitable and sustainable development. However, because people have triggered this crisis, by changing their actions they have the power to prevent water scarcity from devastating India's population, agriculture, and economy.

In view of the existing status of water resources and increasing demands of water for meeting the requirements of the rapidly growing population of the country as well as the problems that are likely to arise in future, a holistic, well planned long-term strategy is needed for sustainable water resources management in India. The water resources management practices may be based on increasing the water supply and managing the water demand under the stressed water availability conditions. Data monitoring, processing, storage, retrieval and dissemination constitute the very important aspects of the water resources management. These data may be utilized not only for management but also for the planning and design of the water resources structures. In addition to these, now a days decision support systems are being developed for providing the necessary inputs to the decision makers for water resources management. Also, knowledge sharing, people's participation, mass communication and capacity building are essential for effective water resources management

References

Nath, K.J. 2003: Home hygiene and environmental sanitation: a situation analysis for India. *International Journal of Environmental Health Research*. 13, suppl: 19-28.

Nath KJ,Bloomfield Sally. Household water storage, handling and point-of-use treatment. A review commissioned by the International Scientific Forum on Home Hygiene.

National Water Policy, 2002:India.Ministry of Water Resources.

Diererich, B.H., Hendreson, J.M. Urban Water Supply Conditions and Needs in seventy five developing countries, World Health Organization.

Kumar, Rakesh, Singh, R.D. and Sharma, K.D. 2005: Water resources of India, Current Science, Vol. 89, No. 5, 10 September 2005 National Institute of Hydrology, Roorkee.

www.imd.ernet.in/section/climate/annual-rainfall.htm, IMD website, 2004.

Groundwater Resources of India, Central Groundwater Board, New Delhi, 1995.

National Health Profile, Central Bureau of Health Intelligence, New Delhi- 2000.

National Health Profile, Central Bureau of Health Intelligence, New Delhi- 2005.

National Health Profile, Central Bureau of Health Intelligence, New Delhi- 2008.

2013, Climate Change Effects on Agriculture and Economy *Pages* **47–53**
Editors: **D.R. Khanna, A.K. Chopra, Vikas Singh, Rakesh Bhutiani & Gagan Matta**
Published by: **BIOTECH BOOKS, NEW DELHI**

Chapter 6

Hydrobiology Assessment of the Freshwater Reservoir Nyari–II in Relation to Fisheries

***A.P. Goswami*[1] *and P.C. Mankodi*[2]**

[1]*Biology Department, M.V.M. Science and Home Science College, Rajkot, Gujarat*

[2]*Division of Fishery and Aquatic Biology, Department of Zoology, Faculty of Science, M.S. University of Baroda, Varodara, Gujarat*

Nyari–II reservoir is located in Rajkot district of Gujarat. The reservoir provides water for irrigation in surrounding agriculture fields and drinking water to the urban area of Rajkot. This reservoir is also used for stocking of freshwater fishes and is good source of inland fisheries of the Rajkot district. To evaluate sustained fisheries and estimate potential of fisheries, the hydrobiology of the reservoir was studied. The present study deals with the seasonal variation in the water quality of the reservoir during June, 2005 to May, 2006. The water samples were collected from the reservoirs for each month in the morning hours. Various physico-chemical parameters like pH, temperature, alkalinity, hardness, chloride, dissolved oxygen, phosphate, nitrate *etc.* were analysed. The plankton dynamics and productivity were estimated. Significant seasonal variation was observed during the study for various parameters. Further, considerably good diversity and density with seasonal variations of plankton was recorded from the waters of the reservoir. The good ratio of gross and net productivity was recorded leads to consider the reservoir as good potential area for fisheries. The hydrological status of the reservoir is correlated with its plankton status and fisheries potential.

Keywords: *Hydrobiology, Plankton dynamics, Fisheries potential.*

Introduction

Gujarat state, broadly falls within the ecological term of semi arid zone (BES-GOG, 1986; GEC, 2000), is naturally water scarcity prone area. The water is one of the essential constituents of all living organism. Reservoir is a low line area of a river with surrounding earthen depression of a river with constructed dam to develop surface water sheet. Depending upon the water stocking capacity broadly reservoirs can be classified as major, medium or minor reservoirs (WRD-GOG, 1989). As principally the water of the reservoir is used for irrigating agricultural fields and for the potable water supply to nearby inhabitants, they are termed as irrigation tanks also. Many of such reservoirs due to their good water holding capacity during summer termed as long seasonal or perennial reservoirs. Such reservoirs are used for natural capture fisheries or planned fish culture activities. Fishing activity always results into economical gain from the reservoirs, as primarily they are designed for other purposes. The rain water is pure but the reservoir holds impurities, turbid water with variation in its hydrological parameters. In the present study various changes have been observed for the water quality; parameters may be attributed to disposal of garbage, sewage and agricultural runoff. The hydrological parameters such as pH, Dissolved Oxygen, Biological Oxygen Demand, Chemical Oxygen Demand, Total hardness, phosphates and nitrates were assessed by recommended international standard methods in APHA (1998). To overcome this long lasting major problem of water scarcity several varied capacity reservoirs have been created. Many of such reservoirs are exploited for fisheries activities though truly under exploited. Exploration of reservoirs for economical gain – fisheries – requires to be estimated for standing fish stock and its potential efficiency to increase the reward by fishing. Good fisheries sustenance is possible only if the water and soil quality is maintained. The ideal approach in such cases is to select a sample reservoir and thoroughly analyse it for ecological conditions and fisheries status. Inland reservoir in this case was treated as the unit, an ecosystem. The quality estimation was done for physico–chemical properties of water and soil, the biological conditions and finally fisheries aspects. The variation in above mention parameters influences each other as well as in total to the hydrological state of the reservoir. This may also influence on biological nature of the reservoir. Any reservoir requires to the assessed for its water budget, water quality. The water quality assessment helps in understanding its biological nature as well as pollution status.

Study site

An Inland reservoir, Nyari–II, is located in the Rajkot district of Gujarat state. This perennial reservoir is rain fed as well as receives flood water through Nyari River. Primarily the water resource is identified to be utilized for irrigation and community water supply to RUDA (Rajkot Urban Development Area). This long seasonal reservoir has the catchment area of 314 sq km and water storage capacity of 88.94 FRL, which is used for capture fisheries. Fisheries department of government of Gujarat has under taken this reservoir for stocking of Carp seeds due to its good quality and quantity of water. Study site freshwater reservoir Nyari – II is located in Rajkot district Latitude: 22° – 21′ – 45″ N and Longitude: 70° – 40′ – 15″ E.

Methodology

Water samples were collected monthly for the entire year 2005–06, from immediate sub – surface of the freshwater reservoir. Three samples were collected randomly from various zones of the reservoir and were pulled together and final sample was drawn for the analysis on the same day. These samples were processed, preserved *etc.* for further detailed quality estimations in the laboratory. All the quality parameters were analysed as per standard methods mentioned in APHA (1998) (Table 6.1). The estimation of dissolved oxygen (DO) was carried out by fixing the same in the sample on site by adding Winkler's reagents and then brought to the laboratory for further analysis in different container.

Table 6.1: Methods Used for Analyzing Different Parameters

Sl.No.	*Parameter*	*Method*	*Instrument*
1.	Temperature	Laboratory or Field method	Mercury Thermometer
2.	Electrical Conductivity	Instrumental method	Conductivity meter
3.	Turbidity	Nephlometric method	Nephlo meter, Turbidity meter
4.	pH	Electrometric method	pH meter
5.	DO	Winkler's method - Azide modification method	BOD bottle, Titrationassembly
6.	BOD	Winkler's method - Azide modification method	BOD bottle, Titration assembly
7.	COD	Titrimetric Open Reflux Method	COD digestion apparatus
8.	Alkalinity	Titrimetric	Titration Assembly
9.	Nitrogen	Colorimetric method (PDA)	Spectrophotometer
10.	Phosphate	Stannous Chloride Method	Spectrophotometer
11.	Sulphate	Turbidometric Method	Spectrophotometer
12.	Free CO_2	Titrimetric Method	Titration assembly
13.	Primary Productivity NPP and GPP	Titrimetric Method,DO – Difference	BOD bottle, Light and Dark bottle

Results and Discussion

Hydrological assessment of subjected reservoir was carried out during year 2005 to 2006 on monthly basis. Water quality was determined for some physical and chemical parameters that were closely related to fisheries status of the reservoir. The average surface temperature was 28.33 °C and ranged from 21.0 °C to 33.0 °C and usually minimum was recorded in January and maximum in the month of September-October. Similar observation was made by Kumar and Kapoor (2006) for such temperature condition is due to close relationship established between atmosphere and water. The turbidity of the water plays a vital role in interaction of penetrating light and photosynthetic pigments. Here, comparatively water is quite transparent and turbidity ranged from 2.0 to 6.0 NTU during entire period of observation. The chemistry of water interacts with the planktonic forms and generates organic carbon as well as plays important role in maintaining the balance between chemical and biochemical state of reservoir. In general pH of water remained alkaline with average pH 7.85, only in July the fluctuation was recorded more. According to Venkateshwarlu (1983) such reservoirs were classified under alkaliphilous state which is found to be good for fisheries activity. Salts in water are the indicators of active chemical nature weather suspended or dissolved. Hence, the estimation of total solids as well as dissolved and suspended solids is required. In present investigation total solids were observed to be minimum during monsoon season and maximum usually in May and June months. The average 422 mg/l of total dissolved solids was recorded during the complete period of study as range from 218 to 680 mg/l.

Electrical conductivity (EC) is closely related to the nature of salts and solids. It ranged from 0.5 to 2.0 mho/cm and recorded minimum in September-October and maximum during the summer period. The organic status of a water body is sustained always if the status of dissolved gases and nutrients remains within normal required ranges (Water standards IS 10500, 1991). In this present study by and large the status of dissolved gases like dissolved oxygen, free carbon dioxide and chloride remained within the optimum ranges required. Dissolved oxygen value was ranged from 3.77 to 10.9 mg/l with an average of 7.19 mg/l for entire year; minimum dissolved oxygen is recorded during monsoon and maximum during winter season. The biological oxygen demand (BOD) and chemical oxygen demand (COD) for the water samples were estimated to qualify the water qualities suitability for fisheries purpose. The data indicates that the reservoir is not having either organic or inorganic pollution threats, therefore is suitable for fish stocking. Minimum value of BOD is around 3.0 mg/l may be due to retarded microbial activity and high value (8.0 mg/l <) of BOD is observed during post monsoon may be attributed to high organic decomposition (Ara *et al.*, 2003).

Nutrients like Ammonical Nitrogen, Nitrite and Phosphate ranged within the normal limit which is the indication of balanced state between chemical nature of water and its productive potential. Organic decomposition produces Sulphate which ranged from 30 to 100 mg/l and do not represent any significant relationship of its quantity during different season. Estimated values of total hardness ranged from 210 to 540 mg/l with general average at 330 mg/l and minimum during October and maximum during June. Such moderate state of hardness in the water body might be due to the contribution through the runoff from the agricultural fields from the command area (Forsyth and McColl, 1975; Pandarkar and Mane, 2006). In the water of this reservoir generally high total alkalinity values were recorded ($_{\text{þþ}}$200 mg/l). It is indicator of the higher trophic status (Sarwar and Wazir, 1991), which is favourable condition to stocking of major carps seeds. Suitability of water for pisciculture or positive potential for fisheries requires good status of physico–chemical condition of the various above mentioned abiotic parameters. On assessment of the ecological status of Nyari-II reservoir most of all the parametric results were within the stated quality criteria hence, may be considered that the reservoir is good for fisheries activity (Subbamma and Ramsharma, 1992; Chandraprakash, 2001).

Primary productivity was estimated as mgc/m^3/d and reported as gross primary productivity (GPP) and net primary productivity (NPP). GPP ranged between 425 to

Table 6.2: Monthly Change in Different Physico-Chemical Parameters in Water of Nyari–II Reservoir

Sr.No.	*Month*	*Temp. (°C)*	*pH*	*TDS (mg/l)*	*DO (mg/l)*	*BOD (mg/l)*	*COD (mg/l)*	*NH_3-N*	*PO_4-P*	*NO_3-N*	*Salinity (ppt)*	*Turbidity (N.T.U.)*	*NPP*	*GPP*
1.	June	30	7.47	218	6.37	0.71	22	0.56	0.060	0.090	23.00	4.0	265.63	425.00
2.	July	31	8.02	262	6.50	2.88	08	0.56	0.210	0.080	15.50	4.5	409.38	1025.00
3.	August	31	7.96	238	3.77	2.74	13	0.56	0.010	0.070	11.75	4.8	446.88	1231.25
4.	September	32	7.40	416	8.67	7.77	20	0.84	0.035	0.150	9.38	3.1	40.63	1100.00
5.	Octtober	33	7.69	392	7.68	1.03	23	0.56	0.050	0.680	8.38	2.8	68.75	1009.38
6.	November	32	7.04	580	6.82	0.84	22	0.56	0.210	2.047	10.25	4.5	221.88	903.13
7.	December	22	8.20	402	8.15	1.81	28	0.28	0.330	0.018	10.38	4.9	87.50	578.13
8.	January	21	8.31	510	10.90	3.65	14	0.84	0.230	0.093	15.00	2.6	237.50	493.75
9.	Febryary	23	8.34	580	6.14	3.11	26	0.28	0.310	0.023	14.75	4.8	106.25	687.50
10.	March	26	8.06	620	6.94	1.87	17	0.56	0.180	0.041	16.25	3.8	125.00	1068.75
11.	April	29	7.40	474	4.77	0.79	21	1.12	0.008	0.135	17.75	5.8	265.63	956.25
12.	May	30	7.44	680	4.72	9.71	37	0.56	0.007	0.130	19.88	5.1	118.75	868.75
	Average	28.33	7.85	422	7.19	2.64	19	0.56	0.163	0.329	14.36	4.2	199.48	862.24
	Maximum	33	8.34	680	10.9	7.77	28	0.84	0.33	2.047	23.0	5.8	446.88	1231.25
	Minimum	21	7.04	218	3.77	0.71	8	0.28	0.01	0.018	8.38	2.6	40.63	425.00

1350 mgc/m^3/d and was reported high during post monsoon. Present study revealed distinct seasonal and bimodal pattern of variation in GPP having peaks in winter and summer. Solar radiation perhaps is the prime factor for such variation in GPP (Kumar and Chaudhary, 2007). NPP value showed similar pattern like GPP with a slight sift in the monthly variation. This may be due to reduction in productive zone, water level and nutrient recycling (Saran *et al.*, 1985).

Table 6.3: Fish Catch Fresh and Dehydrated

Sl.No.	*Year*	*Stocking Fish Seeds*	*Fresh Fish (kg)*	*Dehydrated Fish (kg)*	*Total (kg)*
1.	1998-99	239000	9380	3450	12830
2.	1999-00	40,000	5335	1100	6435
3.	2000-01	600000	14396	6000	20396
4.	2001-02	1205000	24192	6250	30442
5.	2002-03	6,00,000	14890	6250	21140
6.	2003-04	60,000	12920	100	13020
7.	2004-05	120000	73990	5600	79590
8.	2005-06	360000	78360	100	78460
9.	2006-07	252000	8170	3450	12830
10.	2007-08	328500	2340	1100	6435

It has been observed that several Cypriniformes (fishes) were present as native fish of this reservoir and the Indian Major Carps were also introduced as a part of pisciculture programme, feed on this high amount of zooplankton. On the analysis of various parameters we may consider that Nyari–II reservoir is ecologically very well balanced ecosystem and it can sustain good biodiversity as well as fisheries. Several macrofauna observed in this reservoir was having direct or indirect relationship with ecosystem as well as fisheries; therefore, biodiversity analysis was carried out as one of the significant component. Native fishes were represented by major families like Cyprinidae, Siluridae, Gobidae, Channidae and Cichlidae. The regular and organized fishing activities are going on in this reservoir and good amount of inland fish catch have been regularly reported from Nyari–II reservoir. Regular stocking of Indian Major Carps like Rohu, Catla and Mrigal is done by the government fisheries department considering the water budget of the reservoir. Total fish yield return of this group has been reported well and since 2004 – 05 the yield has increased, the catch of Catla and Rohu was above 10,000 kg in the year 2005-06. The past records revealed good state of fisheries in Nyari–II reservoir was one of the important key factors to select the site for ecological analysis. Data collected for ecological analysis and fish catch were subjected to appropriate statistical analysis to arrive to necessary conclusion.

Acknowledgement

Authors are thankful to The Head, Department of Zoology, The M. S. University of Baroda, Vadodara to provide necessary laboratory facilities to carry out this work in the department. One of the authors Goswami A.P. is thankful to the Principal, Shree M.V. Mahila Science and Home Science College, Rajkot for the permission to carry out this research work.

References

Anonymous. 1989: Water Resources Department: major, midium and minor river valley projects, Government of Gujarat.

Anonymous. 2000: State environmental action programme. Gujarat Ecological Commission, Vadodara.

APHA. 1998: Standard Methods for the Examination of Water and Wastewater. 20th Edn. American Public Health Association, Washington D. C.

Ara, S., Khan, M.A. and Zargar, M.Y. 2003: Physico–chemical characteristics of Dal lake water of Kashmir valley, India. *Indian J. Environ. and Ecoplan.* 7 (1): 47 – 50.

Chandra, Prakash. 2001: Status of soil and water quality parameters in blood stalk management. Course manual CAS training programme on brood stock management and genetic selection in fish seed production (Feb.-March, 2001). CIFE, Mumbai.

Forsyth, D.H. and McColl, R.H.S. 1975: Limnology of lake Nagahewa, North Island. New Zeland. *N. Z. J. Freshwat. Res.* 9: 311 – 322.

IS-10500. 1991: Indian standard drinking water – specification (1st revision). Bureau of Indian Standard. Manak Bhavan, New Delhi.

Kumar, Rajesh and Kapoor, Kamal. 2006: Water quality monitoring in respect to physic-chemical characteristics of a tropical lake of Udaipur city of Rajasthan. *Indian J. Environ. and Ecoplan.* 12 (1): 775-782.

Kumar, B.N. and Chaudhary, S.K. 2007: Assessment of primary productivity of phytoplankton of Jagarpur wetland, Bhagalpur, Bihar. *Indian J. Environ. and Ecoplan.* 14 (3): 531 – 534.

Pandarkar, A.K. and Mane, U.H. 2006: Hydrography of a Visapur Dam near Ahmednagar, Maharashtra. *J. Curr. Sci.* 10(2): 687-692.

Sarwar, S.G. and Wazir, M.A. 1991: Dynamics of aquatic ecosystem in relation to fish growth exposed to ammonium sulphates. *J. Environ. Biol.* 12 : 37 - 43.

Saran, H.M. and Adoni, A.D. 1985: Limnological studies on seasonal variation in phytoplanktonic primary productivity in Sagar Lake. *Proc. Nat. Symp. Pure and Appl. Limnology (ed. A. D. Adoni) Bull. Bot. Soc.* Sagar. 32: 71- 76.

Subbamma, D.V. and Ramasarma, D.V. 1992: Studies on the water quality characteristics of a temple pond near Machilipathnam, Andhrapradesh. *J. Aqua. Biol.* 7: 22- 27.

Venkateshwarlu, V. 1983: Taxonomy and ecology of algae in river Moosi Hyderabad, India II. Basillariophyceae, Biboliotheca, Phycologica. *Indian J. Cramer.* 66: 1 – 44.

2013, Climate Change Effects on Agriculture and Economy *Pages* **55–64**
Editors: **D.R. Khanna, A.K. Chopra, Vikas Singh, Rakesh Bhutiani & Gagan Matta**
Published by: **BIOTECH BOOKS, NEW DELHI**

Chapter 7

Lethal Effect of Metal on Animals with Special Reference to their Clinical Applications, Particularly of Heavy Metals

Deepa Sethi Singh

Department of Zoology,
Hindu College, Moradabad, M.J.P. Rohilkhand University,
Bareilly, U.P.

Among the metals of different nature there are pairs of metals that are linked antagonistically and are called antagonistic pair of metals. The metals in antagonistic pair counter balance adverse effects of each other on body and are present in specific ratio of their concentration. When their ratio of concentration changes under any circumstances in the body, it adversely affects health status by reducing immune response in specific type of cell systems.

Heavy metals at the tolerable concentration in animal body exercise toxic effects in controlling environmental infections. These metals cure various types of infections at their specific, particular concentration related to the nature of infection. These concentrations of heavy metals catalyse particular type of bio-chemical reactions that are important for health keeping activities. They also eliminate toxic material from body through complex salt formation. These metals actively participate in energy storing and energy releasing process of the body through change their oxidation state.

Keywords: *Metal, Lethal effect, Disease management.*

Introduction

Metals are backbone of life on earth and are distributed in water, soil, air, plants and animals *etc*. In specific concentrations in the body it not only regulates usual function and structure but also protect from invasion of environmental infections. The nature and relative concentration of metals present in a living body depends on the nature of environmental profile. A significant change in concentration level of a metal or in its relative concentration with other metals in the body leads to the onset of specific type of health problem that is related to the nature and level of imbalanced (deficient metals). A metal atom travels from soil, through water, air, plants to animals and back in due course of time.

A cell is smallest living unit of life and organise with other cells during formation of organ/tissues. The organised cells regulate particular function such as maintenance of temperature, pressure cellular communication renewal process *etc*. A discrepancy developed in metallic content of a cell adversely affects its structure and function besides during abnormality in functional character of regulatory mechanism of specific nature that starts passing on sign and symptoms of a disease before onsets.

The metallic contents of a cell under go change under the influence of

(*a*) Environmental pollution

(*b*) Mol/ore nutrition

(*c*) Stressful condition

(*d*) Emotional injuries

(*e*) Adverse effects of medication and adversely affects

1. Operation level of blood-brain-barrier
2. Immune response (cell mediated immunity)
3. Production of free radicals
4. Level of electro-kinetic potential of body fluid pH
5. Level of nutrition and metabolism

The concentration of a metal cellular fluid of a cell depends upon pH level *i.e.* with the change in pH levels; concentration of a metal is liable to change. The concentration of metal of different chemical nature at specific level of cellular fluid pH is different *i.e.* concentration level of various types of metal present in a cell regulates pH level of its cellular fluid. With change in their relative concentrations, the levels of cellular fluid pH change. The level of cellular fluid pH in different types of cell system is different and is responsible for different concentration level of various type of metal present in them *i.e.* related concentration of various types of metal present in different type of cell system are different and are responsible for their different structure and functions with different level of deficiency of a metal in animal body that affects cell system of different nature related to each level of deficiency. The affected cell system became prone to the attack of specific type of infection *i.e.* in different level of deficiency a metal involves different type of cell system and infection during onset of health problem of various nature. The cell in an organ/tissue mutually communicates through single ionic-channel. The nature of single ionic-channel depends upon the nature of metal that is present in both the cells. The nature of single ionic-channel operating among similar and dissimilar cell system are different and depends upon the nature of common metals present in them.

The metal present in animal body may be classified on the basis of their.

(*a*) Requisite normal concentration levels *viz.* macro, micro/trace and non essential trace metals. Macro metals are present relatively in higher concentration than micro (essential and non essential). Essential trace metals are present in higher concentration than non essential trace metals.

macro >>> micro essential trace >>> non essential trace metals.

(*b*) Energy consuming and energy releasing behaviour during metabolism *viz.* hyper and hypo caloric metals. The hypo caloric metals consume certain amount of energy from the body that not only depends upon the chemical nature of metal but also on its metabolized amount. The hyper caloric metal releases specific amount of energy and enhances energy level of the body. The amount of energy depends upon its chemical nature as well as on the amount metabolized. The related concentration levels of these metals in the body changes with the change in environmental profile/season and adjust it in such away that body may with-stand against atmospheric changes *i.e.* relative concentration of hypo and hyper caloric metals are different and are precipitated in the expression of their gene.

Among the metals of different nature there are pairs of metals that are linked antagonistically and are called antagonistic pair of metals. The metals in antagonistic pair counter balance adverse effects of each other on body and are present in specific ratio of their concentration. When their ratio of concentration changes under any circumstances in the body, it adversely affects health status by reducing immune response in specific type of cell systems. A metal during its metabolism requires specific level of carbohydrates in the body. The inadequate amount of carbohydrates available in the body leaves certain amount of metal un-metabolized and raised its concentration in the body. The required level of carbohydrates in the body for different types of metals is different and depends upon the chemical nature of the metal and its amount to be metabolized. Carbohydrates also require specific level of concentration of particular types of metals for its proper metabolism in the body. The non-availability of requisite amount of metal leaves certain amount of carbohydrates un-metabolized and raises its concentration level in the body fluid. The raised level of the carbohydrates or related metals reduce its immune response and leads to the onset on the health problems by invasion of infection related to the level and nature of un- metabolized metals or carbohydrates level in the body.

The body fluid contains acids and metals of various natures in specific concentration that regulate its pH level. A significant change in their level of body fluid pH, render body nutritionally deficient. At different levels of change in their concentration, affects cell system of different nature that is related to their chemical nature. The normal pH level of water is 7.0 and is neutral but normal pH level of blood is 7.4 and is slightly alkaline. The body fluid pH level when become 7.0 it is acidic and exercise adverse affects on body. The hyper acidic level of body fluid becomes prone to the invasion of environmental effects. The nature of infection depends upon the level of hyper acidity. The concentration level of an acid is in dynamic equilibrium with particular concentration of a metal in body fluid and contributes to its pH level. The concentration level of an acid and metals in dynamic equilibrium fluctuates with in specific limits without affecting significantly level of body fluid pH. A metal or acid may be in dynamic equilibrium with other acids or metals of different nature but with different level of their concentration. A significant change in concentration of acid and metal linked in dynamic equilibrium adversely affect level of cellular fluid pH in particular type of cell system related to their chemical nature and level of change in concentration.

The developed hyper acidity affects only those cells system that undergo change in level of cellular fluid pH. The nature of affected cell system not only depends on the chemical nature of an acid (in hyper state) but also on its magnitude of change in concentration. An acid (in hyper state) deplete metal that is co-related with it in dynamic pair in cellular fluid. At different level of hyper state of an acid, deplete different type of metals from different type of cell system. The change in concentration of a metal in dynamic pair also exhibits similar effects on concentration level of the acid in specific type of cell system. The relative concentration levels of acids and metals in dynamic pair in body fluid/cellular fluid of cell systems of different nature are different and are responsible for variation in their pH level so long as their concentration remains unchanged the pH level of body fluid/cellular fluid remains constant *i.e.* significant change in their concentration adversely affects pH level in related fluid. With the change in level of body fluid pH, the relative concentration of acids and metals in it changes and modifies level of temperature pressure electro-kinetic-potential, immune response, communication in specific type of cell systems by depleting particular type of metal/metals related to the chemical nature of acids (in hyper state).

The level of body fluid pH is index of health status *i.e.* abnormality developed in it leads to the onset of a health problem. Different magnitude of change in pH level is associated with health problem of different nature. A significant change in the concentration of a metal in body fluid:

1. Alters the level of cellular fluid pH by developing hyper acidity in particular cell system related to its nature and level of change in its concentration
2. Degrades level of nutrition, metabolism, enzyme functions and immune response of body
3. Alters level of cellular characteristics of the cell.

The developed hyper acidity in the body may be due to hyper level of any of the acids present in it but affected cell system is one that acquires hyper concentration level of the acid.

Only this cell system initiates disease process. The acid of different chemical nature when individually present in hyper state, affect different type of cell system related to the chemical nature and the level of hyper state. Equal level of acids in hyper state alters pH level of body fluid to different extents and develops abnormality of different nature in body. Normal pH level supervises growth, development, maintenance of body health by keeping various types of acids and metals in their requisite concentrations. Various regulatory requisite concentrations of specific type of metals in particular type of cell system in the body. Environmental infection invade body only when it gets specific level of body fluid for its survival that is infections of particular character require specific level of body fluid pH for its localization. This specific level for different type of infections is different and is available only in particular type of cell system in the body.

The management of body fluid pH within normal limit is one of the effective way/means to keep the body healthy. It is equally effective for clinical management of health problems irrespective of any nature of disease causing factors. The other way requires development of universal anta-acid that may be effective against hyper level of acids of different chemical nature present in body/cellular fluid.

The abnormality developed in pH level of body fluid adversely affects specific type of cell system by developing hyper acidity and depleting particular type of metal. The nature and level of depleted metal decides the nature of health problem to be onset *i.e.* a metal with its different levels of depletion may be involved in the onset of different type of health problems using different type of cell system for each health problem. The nature of infection involved in disease process not only depends upon the nature of depleted metal but also on its level of depletion (deficiency). The depletion of a metal from body fluid also develops hyper acidity due to raised concentration level of the related acid in dynamic equilibrium. The chemical nature of acid in hyper state not only depends upon the nature of depleted metal but also on its level of depletion *i.e.* at various levels of depletion a metal cause raise in concentration level of acids of different nature related to each level of depletion in different type of cell systems.

Thus (a) development of abnormality (Hyper acidity in body fluid is an essential condition for onset of a health problem) (b) Onset of a health problem is associated with the depletion of (deficiency of specific type of metal/metals) (c) The nature of onset health problem not only depends upon the chemical nature of depleted metal but also on its level of depletion that attracts different type of disease causing infections related to each level of its deficiency in the body. Heavy metals at the tolerable concentration in animal body exercise toxic effects in controlling environmental infections. These metals cure various types of infections at their specific, particular concentration related to the nature of infection. These concentrations of heavy metals catalyse particular type of bio-chemical reactions that are important for health keeping activities. They also eliminate toxic material from body through complex salt formation. These metals actively participate in energy storing and energy releasing process of the body through change their oxidation state.

The accumulation of heavy metals proceeds with the change in energy level of body *i.e.* certain amount of energy is released or absorbed during their accumulation. The amount of energy change in the body depends upon the chemical nature of heavy metal as well as on its accumulated amount.

The total absence of a heavy metal in animal body is associated with the degradation in health status due to slow run of specific type of bio-chemical reaction that produce important health aiding bio-chemicals such as hormones.

Conclusion

1. The total absence of a heavy metal in the body leads to the degradation.
   (*a*) In the level of physiological functions *i.e.* quality and quantity of synthesized bio-chemicals are adversely affected and leads to the degradation in health status *e.g.* **As** cause infertility due to inadequate production and sub quality of sperm characteristics.
   (*b*) In level of immunity of a cell/organ/body.
   (*c*) Provide favourable ground for development of environmental infection in the body.
   (*d*) Alters level of seasonal adjustment of body energy.
2. The heavy metals though not actively involved in human physiology but play a key-roles in supporting physiological process by.
   (*a*) Eliminating invasion of routine infection.
   (*b*) Maintaining energy level of body.
   (*c*) Scavenging ROS of specific nature.
   (*d*) Drain out toxicants through salt/complex formation.
   (*e*) Supporting specific type of regulatory mechanism that operate for maintaining adequate level of renewal process, temperature, digestion, reproduction *etc.*
3. The heavy metals are dietary anti-oxidants and offer best weapon to boost health of sperm in male. The highly reactive free-radicals damage these cells. The sperm are venerable to the attack of free radical but can be protected by adequate amounts of specific type of heavy metals in the cellular fluid.
4. A heavy metal exercise toxic effect at and above its particular concentrations level called lethal concentration for different type of heavy metals in the body. This concentration level is different and depends upon its oxidation state and density. It is also different for different types of animals and depends upon their tolerance limits (level of body fluid pH).
5. Toxic effects of heavy metals are concentration related and are influenced by level of body fluid pH that regulates its accumulation rate in the body.

Discussion

1. The requisite level of body fluid pH, Keep various types of metal at concentration that support body health by participating in physiological processes of various nature. At normal level of body fluid pH various acids and metals are in specific concentration of dynamic equilibrium the concentration of an acid is in dynamic equilibrium with number of metals in their specific concentrations and concentrations of a metal also forms

dynamic equilibrium with specific concentration of number of acids. The significant change in concentrations of an acid or a metal in dynamic equilibrium in body fluid changes the level of body fluid pH. The different levels of cellular fluid in different type of cell systems is different because chemical nature and relative concentration of acids and metals present in dynamic equilibrium are different.

2. The hyper acidity developed in the body, increases accumulation of heavy metals and reduce level of essential metal.
3. The altered pH level of body fluid, imbalances, concentration level ratios of metals in specific pair of antagonistic metals. In antagonistic pair metals counter balance their adverse effects and regulate bio-chemical effects on mutual basis during their action.
4. Level of a metal in a body depends upon the pH level of body fluid, the pH level of cellular fluid, in different type of cell systems in different so concentrations level of a metal in them is also different.
5. The pH level of body fluid, depends upon the concentration levels of different types of acids and metals present in it. Variation in pH level of cellular fluid, among cells reveal variation in chemical nature and relative concentration level of different type of acids and metals present in them.
6. The altered pH level of body fluid adversely affects structures and functions of particular cells system (related to the chemical nature of acid in hyper state) by changing the level of cellular fluid, pH that modifies normal level of nutrition, metabolism, enzyme, activities, immune response, cellular communication and renders the cell prone to the invasion of specific type of infections.
7. Essential metals are required in specific level of concentrations related to their chemical nature in the body. The concentration of a metal fluctuates within certain limits. When a metal acquires hyper level by crossing upper limit of fluctuation, it starts exercising lethal effects by changing course of certain bio-chemical reaction and produce toxic chemicals besides reducing the level of routine chemicals.
8. The hypo-concentration level of an essential metal (below lower limit of fluctuation of concentration) in the body degrades normal level of nutrition's metabolism, enzyme functions, immune response cellular communication.
9. The nature of affected cell system, during developed hyper acidity in the body depends on the chemical nature and is magnitude of hyper level *i.e.* acidity of different chemical nature when present in their hyper states, affects cells system of different nature.
10. Ayurvedic, unani, homeopathic systems of medicine widely uses heavy metals in the clinical management of human health problems of various natures. In "Metallomics" journal international team of scientist recently reported that certain compounds of gold are selectively toxic to cancer cells but are harmless on normal cells.

11. Normal pH level of body fluid keep heavy metals below their lethal concentration levels for getting their support in health keeping process. Heavy metals catalyse specific type of bio-chemical reactions that are involved in the synthesis of important chemicals of specific nature. These reactions in the absence of related heavy metal proceed so slowly that produced chemicals are in insufficient amount that is specific amount of heavy metals are essential in the body for production of particular type of bio-chemical.
12. Heavy metals below lethal concentration in the body exercise toxic effects against specific type of microbes/infections and eliminate toxic substances through complex formation.
13. Accumulation of heavy metal in body occurs at particular level of body fluid pH related to its chemical nature, oxidation state and density. Lower oxidation state of a metal that accumulated at lower level of acidity while for its higher oxidation state, higher level of acidity of specific order is required *i.e.* for accumulation of heavy metal in a cell/organ favourable pH level of body fluid is essential.
14. The developed hyper acidity in body fluid tends to increase (a) accumulation rate of heavy metal and (b) depletion rate of essential metal. At specific level of body fluid pH, heavy metal competes with essential metals for their assimilation into tissue cells. The reduced level of essential metals leads to the onset of health problems related to various levels of its deficiencies in the body.
15. Restoration of normalcy in abnormal level of body fluid pH depletes accumulated heavy metals and reduces them to harmless level in the body.
16. Accumulation of heavy metals proceeds with the generation or release of certain amount of energy *i.e.* alters energy level of body. The amount of energy change in body not only depends on the chemical nature of heavy metals but also on its accumulated amount.
17. Total absence of a heavy metal adversely affects production rate, quality and quantity of particular type of bio-chemicals *e.g.* animal infertility is associated with the absence of As (Arsenic).
18. Arsenic drug is prepared by interaction of arsenious-oxide with warm milk through curd formation followed by extraction of butter that is converted on warming into fat (Ghee) after vitro-vivo study it is used in treating animal infertility of various natures.
19. Clinical potential of a heavy metal is based on its (a) catalytic effects (b) accumulations related change is energy level of body (c) complex salt formation character that drain-out toxicants (d) accumulations that modifies the profile of physiological environment (through change in level of body fluid pH) and threaten survival of localized infections of specific nature.
20. Animal body suffers with a health problem due to change in the normal level of following factors that are inter-related (a) body fluid pH (b) concentration level of specific type of metal/metals (c) 'Nitrogen balance'

These factors modifies (i) rate of production of free radicals (ii) Activities of enzymes (iii) rate of synthesis and decomposition of proteins.

21. Animal health problems are clinically managed by restoration of normalcy in abnormal concentration level of a disease related metal/metals through oral in-take of its suitable salt, irrespective of any nature of disease causing factor.
22. Lethal effects of a metal on animal body are managed by (a) administration suitable salt of the metal that is related to lethal effect causing essential matter in an antagonistic pair of metals. This not only reduce absorption rate of lethal effect causing metal but also improve absorption rate of administrable metal that brings both metals within their requisite limits (b) depleting the accumulated amount of heavy metal by restoring normalcy in abnormal level of body fluid pH through oral administration of suitable salt of essentials metals of that is associated with the body in deficient state.
23. The compound of essential metals of clinical importance are prepared by inter- action of metal oxide with ammonium-chloride in vapour-phase in a sealed container and are used after vitro-vivo study in resolving number of health problems related to different level of its deficiencies in the body. Metals 'imides' not only look after metal related aspect of the health problem but also execute anta-acidic effects and correct abnormal level of 'Nitrogen balance' in the body.
24. Therapeutic potential of a heavy metal depends upon their following characteristic (i) variable oxidation state (ii) complex salt forming character (iii) catalytic property (iv) Low level of bio-availability of its salt (v) hyper caloric nature (vi) concentration related toxic effects (vii) energy generating and energy storing characteristic.

Therapeutically important compound of a heavy metal should contain (a) metal in lower oxidation state (b) H-atom in its molecule that assists its assimilation into tissue cells system.

References

Bourai, A.A.: 2011: 'Health and environmental effect of radium' abstract National conference on emerging trends in chemistry- biology interface, Nainital, November 03-05, 2011 page 52.

Gupta, S. 2011: '*Invitro* studies of effects of various heavy metals on haematological parameters' abstract National conference on emerging trends in chemistry-Biology interface, Nainital, November 03-05, page 73.

Sethi, D. 2010: 'Occupational health problem causing effects of industrial effluents on body fluid pH' abstract National seminar on management of environmental pollution for sustainable energy production, Bareilly, 92.

Sharma, D.K. 2011: 'Analysis of trace elements in spinach (*Spinacea oleracea*) and its biological and environmental importance' abstract National conference on emerging trends in chemistry- biology interface, Nainital, November 03-05, 2011 page 80.

Singh D.S. 2011: 'Role of cellular matter in prevention and control of animal health problems with special reference to liver disorder (jaundice) abstract National conference on emerging trends in chemistry- biology interface, Nainital, November 03-05, page 31.

Singh, D.S- 'Clinical potential of mollusca shell with special reference to the management of pollution related health hazards' abstract National seminar on recent advancement in life sciences: potential benefits and Ricks, Meerut 05th 06th 2011.

Singh, D.S. 2011: 'Antibiotic' behaviour of cellular metals during management of infectious 'diseases in human' abstract International conference on new horizons in microbial bio technology and pharmaceutical sciences Pounta Sahib (India) April 24th -25th, page 80.

Singh, D.S. 2011: 'Health hazards related to industrial waste treatment (E-waste) for extraction of noble metals', abstract national seminar on New innovation in parasitological research, Dehradun, 21-30th March, Abs.no.63.

Singh, D.S. 2011: 'Prevention of adverse effect of industrial pollution on human health through maintenance of body fluid pH within normal limits' abstract national conference on biodiversity *vis-à-vis* environmental degradation in hilly terrains, Dehradun, 15-16th May, page 66.

Singh, D.S. 2011: 'Role of heavy metals in biological system be redefined with particular reference to their clinical potential'; abstract National conference on emerging trends in chemistry- biology interface, Nainital, November 03-05, page 32.

Vidushi, 2011: 'Investigations of chromium, lead, nickel, zinc, cobalt, copper and cadmium in hair samples of auto drivers and bricks furnace workers by atomic absorption spectra photo meter technique' abstract National conference on emerging trends in chemistry- biology interface, Nainital, November 03-05, 2011 page 47.

2013, Climate Change Effects on Agriculture and Economy *Pages* **65–72**
Editors: **D.R. Khanna, A.K. Chopra, Vikas Singh, Rakesh Bhutiani & Gagan Matta**
Published by: **BIOTECH BOOKS, NEW DELHI**

Chapter 8

Ecological Studies of Zooplankton found in Freshwater Lake of Khekara–Nullah Dam, Nagpur, Maharashtra

***Bharati Tapase*[1], *J.L. Tarar*[2] *and Pravin Charde*[1]**
[1]*Sevadal Mahila Mahavidyalaya and Research Academy, Nagpur, Maharashtra*
[2]*Department of Environmental Science, Institute of Science, Nagpur, Maharashtra*

A group of zooplanktons is the characteristic indicator of water quality, eutrophication and pollution status and also form an important link in the aquatic food chain. They play an important ecological role as grazers, suspension feeders and predators within the aquatic community. Zooplankton community of freshwater bodies consists an extremely diverse assemblage of organisms represented by most of the over invertebrate phyla. Extensive works on ecology of zooplankton of freshwater bodies of India have been carried out by many workers. But knowledge about zooplankton present in Khekara-Nullah Dam in Nagpur district of Maharashtra is not yet known. Such studies are focused on the ecology of zooplankton and their importance as a fish food. It also indicates the pollution and the trophic status of water body. An attempt has been made to investigate seasonal variations in zooplankton along with physico-chemical parameters. The present work was carried out for two years (June, 2007–May, 2009) to collect data of abiotic and biotic components of Khekara-Nullah Dam. The biological indicator *i.e.*, zooplankton attains the lowest and the highest population 167.67 ± 68.18 in 2007-2008 and 170.00 ± 68.36 in 2008-2009. Zooplanktonic communities in these years were represented by *Rotifer*, *Ostracoda*,

Cladocera and *Copepoda*. The abundance of *Cypris*, *Ostracoda* of total zooplankton was significantly low during the winter season when compared to monsoon and summer seasons, during which the abundance of all these variables was significantly higher. In all 1013.0 zooplanktonic forms were recorded. During the entire study period seventeen positive and three negatively significant correlations were observed. The regression analysis revealed that *Rotifers* are positively correlated with potassium and T.D.S. *Ostracodas* are positively correlated with hardness. *Cladocera* with potassium and *Copepoda* positively correlated with temperature. In general, groupwise composition of zooplankton communities indicated following ranking: *Rotifers* > *Copepoda* > *Cladocera* > *Ostracoda*

Keywords: *Zooplankton, Ecology, Khekara-Nullah Dam, Regression.*

Introduction

Zooplankton is microscopic organisms that formulate the base of food chains and food webs in all aquatic ecosystems. They are mostly small, many of them are minute and their structure can only be seen clearly with the aid of a binocular or compound microscope, with the exception of some large animals. Although they belong to different taxa, they have one thing in common *i.e.* lack of strong locomotory organs like fish, they are weak in locomotion and can only drift about in water at mercy of waves and currents, being incapable of moving anywhere as fish (Zheng, 1984).

Zooplankton plays an important food item of omnivorous and carnivorous fishes (Alam *et al.*, 1987). The larva of carps feed mostly on zooplankton because zooplankton provide the necessary amount of protein requires for the rapid growth and development of different organs in fishes and also brood fishes productivity (Bardach *et al.*, 1972). Many researchers worked on the physico-chemical conditions and seasonal variations of zooplankton (Ali and Islam, 1981; Bhuiyan, 1981; Bhuiyan and Nessa, 1998 and Cottenie, 2001).

The present study has been undertaken to know the role of zooplankton on the seasonal variations and occurrence of some zooplankton in respect to physico-chemical parameters in Khekara Nullah dam near Nagpur city.

Materials and Methods

Morphology of Pond

This investigation has been conducted in Khekara Nullah dam site at Khapa range forest near Nagpur city of Vidarbha region (21°32' N latitude and 78°56' E longitude) 55 kms away from Nagpur. The dam was constructed in 1988. The water of dam was clear and bloom was sighted during the study period. The dam water receives direct sunlight throughout the day.

Sample Collection

Water sample were collected seasonally (June, 2007–May, 2009) from surface layer. Water temperature was recorded by a centigrade thermometer and pH was noted by digital pH meter. DO was determined immediately by Winkler's method (APHA, 2005) and other chemical parameters such as alkalinity were estimated according to the procedure given by Welch (1952). Zooplankton were collected by plankton net of no. 20 count bolting cloth and the numerical assessment was done by Sedgwick-Rafter counting cell. They are identified with the help of compound microscope and by the following keys given by Ward and Whipple (1959).

Results and Discussion

Some selected physico-chemical parameters of water were studied and these are shown in Table 8.1. Table 8.2 represents the seasonal values and diversity of zooplankton variables. Interrelation between zooplankton and environmental variables are shown in Table 8.3. Percentage population of plankton variables are shown by graphical method in Figure 8.1.

Rotifers

Season wise grouped data revealed that, population were almost similar for both the years during the study period of monsoon and summer seasons. The abundance of rotifers was dominant in winter season. They show positive correlation with 17 physico-chemical variables (Table 8.3).

Rotifers are one of the most important organisms which determine the degree of water pollution. Certain species and genera of rotifers are used as indicators of water quality, eutrophic status and productivity of a water ecosystem (Sladeek, 1983). In the present study rotifers showed the strongest positive correlation (r = 1.0) with potassium, (r = 0.995) with TDS, (r = 0.992) with calcium hardness and negatively correlated with DO (r = -0.947) and fluoride (r = -0.812) (Joseph, 2010).

Ostracodas

Season wise grouped data revealed that, abundance of ostracodas was almost similar in monsoon, winter and summer season (Table 8.2). Surana *et al.* (2005) studied the community structure of zooplankton group of Chimdi lake Sunsari, Nepal.

In the present study, Ostracoda show significant positive correlation (r=0.999) with total hardness, calcium hardness (r=0.988) and magnesium hardness (r=1.00) and negative correlation with DO (r=-0.820) and fluoride (r=-0.618).

The regression analysis of Ostracoda in all seasons revealed that 99 per cent of its variation was due to calcium and magnesium. These are positively correlated with 17 physico-chemical variables (Table 8.3).

Cladocera

The abundance of cladocera showed similar findings for both years and showed significant positive correlation with potassium (r=1.0), TS (r=0.995), TDS (r=0.990), Ca-hardness (r= 0.986), pH (r=0.916) and negative correlation with turbidity (r = -0.291) Tables 8.2 and 8.3.

Towhid *et al.* (2003) studied relationship between zooplankton abundance and physico-chemical parameter in Lake Sunderban (Bangladesh) and reported that

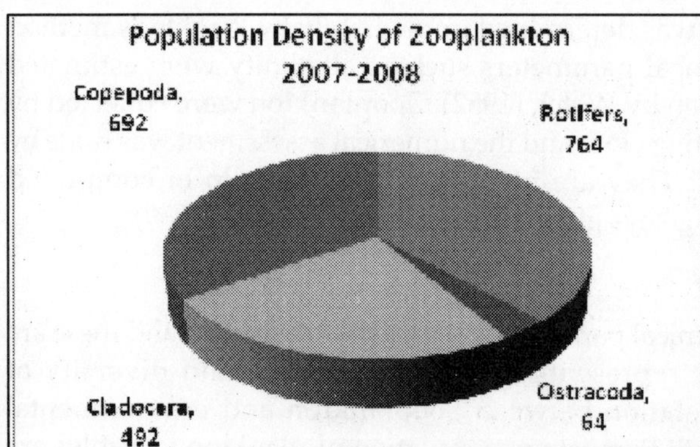

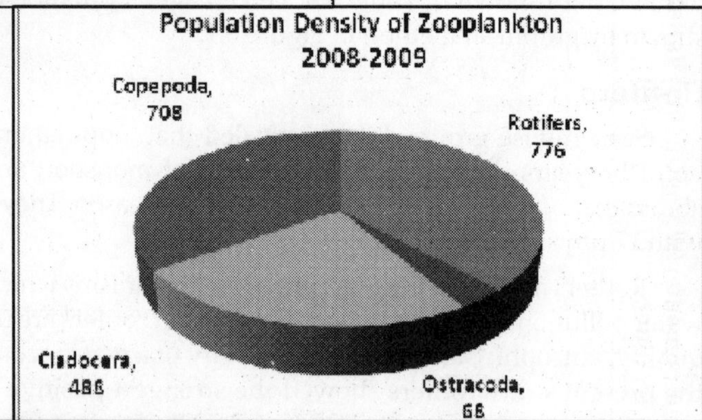

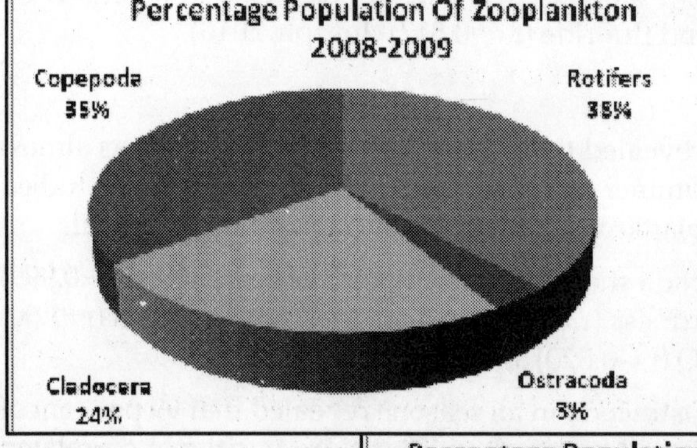

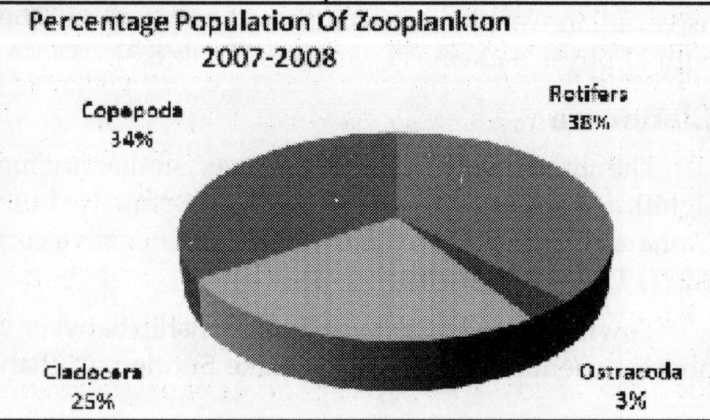

Figure 8.1: Population and Percentage of Species for the year 2007-2008 and 2008-2009

Table 8.1: Seasonal Changes in the Environmental (Water Quality) Variables in Khekara Nullah Dam, Nagpur (June, 2007–May, 2009)

Parameter	Season							
	Year 2007-08				Year 2008-09			
	Monsoon	Winter	Summer	Average	Monsoon	Winter	Summer	Average
Temperature (°C)	28.01	26.0	30.8	28.27±1.97	29.0	27.0	30.2	28.73±1.32
pH	8.3	7.14	8.6	8.01±0.63	8.4	7.13	8.7	8.08±0.68
Conductivity (mS/cm)	0.398	0.281	0.512	0.40±0.09	0.431	0.291	0.521	0.41±0.09
Turbidity (NTU)	9.0	3.1	2.0	4.70±3.07	8.0	3.4	2.0	4.47±2.56
Alkalinity (mg/l)	217	301	320	279.33±44.75	220	265	280	255.00±25.50
Hardness (mg/l)	102	107	172	127.00±31.89	110	96	170	125.33±32.10
Ca-Hardness (mg/l)	63	60	69	64.00±3.74	80	73	97	83.33±10.08
Mg-Hardness (mg/l)	39	47	103	63.00±28.47	30	23	73	42.00±22.11
T.S. (mg/l)	618	418	712	582.67±122.60	720	617	912	749.67±122.25
T.D.S. (mg/l)	138	123	149	136.67±10.66	157	138	198	164.33±25.04
T.S.S. (mg/l)	490	295	563	449.33±113.13	563	577	814	651.33±115.16
D.O. (mg/l)	6.8	7.7	5.9	6.80±0.73	6.9	7.8	6.5	7.07±0.54
Fluoride (mg/l)	0.212	0.182	0.291	0.23±0.05	0.121	0.21	0.12	0.15±0.04
BOD (mg/l)	9.31	6.31	12.8	9.47±2.65	9.43	7.81	13.8	10.35±2.53
COD (mg/l)	28.21	38.7	40.21	35.71±5.34	30.2	28.1	31.1	29.80±1.26
Chloride (mg/l)	87.08	81.0	98.0	88.69±7.03	98.21	87.0	99.2	94.80±5.53
Sulphate (mg/l)	18.7	10.8	20.1	16.53±4.09	19.7	10.8	22	17.50±4.83
Phosphate (mg/l)	0.917	0.81	1.1	0.94±0.12	1.1	0.98	1.17	1.08±0.08
Sodium (mg/l)	12.08	9.2	13.1	11.46±1.65	13.08	12.1	13.08	12.75±0.46
Potassium (mg/l)	3.9	2.8	5.65	4.12±1.17	4.2	3.9	4.6	4.23±0.29

Table 8.2: Average Seasonal Values and Diversity of Zooplankton in Water of Khekara Nullah Dam during June, 2007–May, 2009

Zooplankton Variables	*2007–2008*					*2008–2009*				
	Monsoon	*Winter*	*Summer*	*Average*	*Diversity*	*Monsoon*	*Winter*	*Summer*	*Average*	*Diversity*
Rotifers	64	26	101	63.67 ± 30.62	764	60	30	104	64.67 ± 30.39	776
Ostracoda	5	1	10	5.33 ± 3.68	64	4	3	10	5.67 ± 3.09	68
Cladocera	42	21	60	41.0 ± 15.94	492	39	21	62	40.67 ± 16.78	488
Copepoda	57	36	80	57.67 ± 17.92	692	60	36	81	59.0 ± 18.38	708
Total Plankton	168	84	251	167.67 ± 68.18	2012	163	90	257	170.0 ± 68.36	2040

Table 8.3: Coefficient of Correlation of Physico-chemical Variables with Biological Variables of Khekara Nullah Dam

	Parameters																			
	Temperature	*pH*	*Conductivity*	*Turbidity*	*Alkalinity*	*Hardness*	*Ca-Hard*	*Mg-Hard*	*T.S.*	*T.D.S.*	*T.S.S.*	*D.O.*	*Fluoride*	*BOD*	*COD*	*Chlorude*	*Sulphate*	*Phosphate*	*Sodium*	*Pottasium*
Rotifers	0.968	0.900	0.973	-0.328	0.344	0.972	0.992	0.960	0.998	0.995	0.894	-0.947	-0.812	0.989	0.944	0.848	0.906	0.966	0.807	1.000
Ostracoda	0.861	0.743	0.870	-0.578	0.592	0.999	0.988	1.000	0.976	0.983	0.983	-0.820	-0.618	0.991	0.815	0.666	0.752	0.857	0.610	0.953
Cladocera	0.977	0.916	0.981	-0.291	0.308	0.963	0.986	0.948	0.995	0.990	0.876	-0.959	-0.834	0.982	0.956	0.867	0.922	0.976	0.829	1.000
Copepoda	0.995	0.954	0.996	-0.185	0.203	0.928	0.963	0.908	0.978	0.970	0.819	-0.984	-0.889	0.956	0.982	0.916	0.958	0.994	0.885	0.993
Total Zooplankton	0.977	0.915	0.981	-0.293	0.310	0.963	0.987	0.949	0.995	0.991	0.877	-0.958	-0.833	0.983	0.956	0.866	0.921	0.975	0.828	1.000

Cladocerans were positively correlated with calcium. In the present investigation, the regression analysis for all seasons was almost 100 per cent due to potassium and hardness.

Copepoda

The copepods constitute dominant plankton group of both fresh and marine habitats. It includes Cyclops male and female copepods were significantly low during monsoon and winter season as compared to summer season (Table 8.2).

However, the abundance of copepods showed positive correlation with calcium hardness (r = 0.963), pH (r = 0.954) and negative correlation with DO and fluoride (Table 8.3).

The regression analysis in all season showed that 99 per cent of the variation was due to phosphate, temperature (99.5 per cent) and pH (95.4 per cent) (Table 8.3). Munawar (1970), studied limnology of freshwater pond in Hyderabad, India and reported that the abundance of cyclops show positive correlation with pH of the water.

Abundance of Total Zooplankton

The sum of all zooplankton is significantly less during monsoon and winter seasons (Table 8.2). A wide diversity of zooplankton was found in the dam water during the investigating period. The data of two years with reference to number of zooplankton per season and average percentage of total population of zooplankton species are given in Figure 8.1. The dominant group of zooplankton was the rotifer during the investigating years, constituting 38 per cent of total zooplankton population. The ostracoda found in the dam water was few in numbers and their population percentage during two years was 3 per cent respectively (Tijare, 2010).

Environmental (Water Quality) Variables

About 14 physico-chemical variables *i.e.* temp., pH, conductivity, turbidity, COD, chloride, sulphate, phosphate, sodium and potassium did not show any significant differences in both the years (Table 8.1). Total solids and hardness were significantly more during summer season, when compared to monsoon and winter seasons. These results are in good agreement with those of Khan (2005) who observed significant fluctuation in the physico-chemical analysis. In general, group wise composition of zooplankton communities indicated following ranking: *Rotifers* > *Copepoda* > *Cladocera* > *Ostracoda.*

Conclusion

It is evident from our study that several (water quality) variables *viz.* temp, DO, pH, alkalinity, turbidity, phosphate, BOD, show a little variation, but all the parameters are within normal range and showed positive correlation with zooplankton variables. It is good indication for fish culture and ecology.

References

APHA, 2005: *Standard methods for the examination of water and wastewater.* American Public Health Association, 21st edition, Washington, U.S.A.

Ali, M. and Islam, M.A. 1981: *Studies on the plankton of a lake in Bangladesh* Agricultural University Campus. *Bangladesh Journal of Fisheries*, 10(10):82-88.

Alam, A.K.M.N., Islam, M.A., Mollah, M.F.A. and Haque, M.S. 1987: *State of zooplankton newly constructed pond and their relation to some metrological and limnology factors. Bangladesh Journal of Fisheries*, 10(1): 83-88.

Bardach, E.J., Ryther, H.J. and Melamy, O.W. 1972: *The farming and husbandry of freshwater and marine organisms*, John-Willey and Sons, Inc. New York.

Bhuiyan, A.S. and Nessa, Q. 1998: *Seasonal variations in the Occurrence of some Zooplankton in a fish pond. Bangladesh Journal of Fisheries Research*, 2(2): 201-203.

Bhuiyan, A.S., Hafizuddin, A.K.M. and Pasha, M.K. 1981: *Diurnal movement of plankton in a pond.* Chittagong University Studies, part II (5): 41-49.

Cottenie, K.N. Nuytten, E. Michels and Messter, L.D. 2001: *Zooplankton community structure and environmental conditions in a set of interconnected ponds. Hydrobiol.*, 442: 339-350.

Joseph, Beenamma and Sadanand M.Yamakanamardi 2010: *Winter, Summer and Rainy seasonal changes in the abundance and biomass of zooplankton in relation to environmental variables in Kukkarahalli lake of Mysore, Karnataka State*, India. J. Aqua. Biol, Vol. 25 pp 8-17

Khan, T., Kumar,A.D. Singh, V.V. and Trivedi, R.C. 2005: *Physico-chemical analysis of Drains in Delhi. Pollution Research* 24(4): 737-744.

Munawar, M. 1970: *Limnology studies on freshwater ponds of Hyderabad*. India 11. The Biotope Hydrobiological, 35: 17-165

Sladecek, Vladimir. 1983: *Rotifers as indicators of water quality*, Hydrobiologia 100: 169-201.

Towhid, I.A. and Kabir, A.M. 2003: *Relationship between zooplankton abundance and physico-chemical parameter in Sunderban lake, Khulna, Bangladesh*. Pak, *J. Bio.Sci.* 6(8): 762-765.

Tijare, R.V. and Thosar, M.R. 2010: *Studies on ecology of zooplankton found in freshwater lake of Gadchiroli city, Maharashtra. J. Aqua. Biol.*, 25 (1), 22-23.

Ward, G.B. and Whipple, G.C. 1959: *Freshwater Biology*. John Willy and Sons, New York, 1248.

Welch, P.S. 1952: *Limnology*. Second Edition Mc Graw Hill Book Co. New York and London.

Zeng, Z. 1984: *Marine Planktology*, China Ocean Press. Beijing.

2013, Climate Change Effects on Agriculture and Economy *Pages* ***73–80***
Editors: **D.R. Khanna, A.K. Chopra, Vikas Singh, Rakesh Bhutiani & Gagan Matta**
Published by: **BIOTECH BOOKS, NEW DELHI**

Chapter 9
Contamination of Heavy Metals in Vegetables Irrigated with Textile Effluent at Haridwar, Uttarakhand

Vinod Kumar and A.K. Chopra
Department of Zoology and Environmental Science, Gurukula Kangri University, Haridwar, Uttarakhand

The heavy metals content was estimated in textile effluent irrigated soil and vegetables *viz. Phaseolus vulgaris, Vigna radiata, Vica faba* and *Linum usitatissimum.* The results revealed that textile effluent irrigation showed significant accumulation of various heavy metals *viz.* Fe, Zn, Cd, Cu, Cr, Co and Ni in the soil and *P. vulgaris, V. radiata, V. faba* and *L. usitatissimum* crops after irrigation. The contamination of heavy metals in soil were in order of Cd>Ni>Cr>Co>Zn>Cu>Fe for *P. vulgaris, V. faba* and Cd>Ni>Cr>Co>Zn>Fe>Cu for *V. radiata, L. usitatissimum* after irrigation with textile effluent. In crop plants the contamination of heavy metals were in order of Ni>Co>Cu>Cd>Cr>Zn>Fe for *P. vulgaris,* Co>Ni>Cr>Cu>Cd>Fe>Zn for *V. radiata,* Cr>Co>Ni>Cu>Cd>Fe>Zn for *V. faba* and Co>Cr>Cd>Cu>Zn>Fe>Ni for *L. usitatissimum* after irrigation with textile effluent.

Keywords: *Textile effluent, Heavy metals, Vegetables, Contamination.*

Introduction

One of the most important environmental problems faced by the world is management of wastes. Different industries create a variety of wastewater pollutants which are difficult and costly to treat. Wastewater characteristics and levels of pollutants vary significantly from industry to industry. The presence of dyes in industrial effluents is a major problem due their adverse effect to many forms of life. Dyes are natural or synthetic colourants used in various industries such as textiles, tanneries, paints, pulp and paper. The urban agricultural soils are often irrigated with city effluents for growing vegetables (Ibrahim and Salmon, 1992; Qadir *et al.*, 1998; Chopra *et al.*, 2009). Farmers use it as a source of irrigation and nutrients (Ghafoor *et al.*, 1994; Vinod and Chopra, 2011a,b,c). The wastewater may constitute of hazardous toxic materials, chemical compounds and organic material. Extensive irrigation by the effluents released from household and industrial sectors have led to the accumulation of heavy metals in the soil and different parts of the vegetable crops (Vinod and Chopra, 2011a).

Trace elements are the essential part of rock and soil, natural waters and living matters. It is well known that the biomass uptakes trace metals naturally available to them through soil by accepting them as components vital to their normal functions and have even stored them in their tissues (Godbold and Hütterman, 1985; Breckle, 1991; Chopra *et al.*, 2009). Some plant species have developed strategies for avoiding toxic concentrations build up of heavy metals at sensitive sites within the cell. If heavy metals are absorbed by the plant, they may be stored away from metabolically active compartments, thus preventing phytotoxic effects (Memon *et al.*, 2001; Vinod and Chopra, 2010; 2011a,c).

Urban agricultural lands in the metropolitan centres are probably potentially toxic due to heavy metals deposits from diverse sources (Voutsa *et al.*, 1996; Chopra *et al.*, 2009). The farming activities and land utilization especially for industrial purposes are regulated *via* fertilizer application, sewage sludge or irrigation with wastewater (Devkota and Schmidt, 2000; Frost and Ketchum, 2000; Chopra *et al.*, 2009; Chakresh *et al.*, 2011). Such effluents and wastes hardly decompose completely but leave undegradable contaminants in the environment (Voutsa *et al.*, 1996). The heavy metals in these soils are taken up by plants *via* roots and later accumulated in the stem and other edible portions of plants which eventually lead to poisoning when consumed (Nieboer and Yassi, 1998; Vinod and Chopra, 2011a).

The nature of soil is considered one of the most important factors that determine heavy metal content of food plants (Itanna, 2002; Madyiwa *et al.*, 2002a; Vinod and Chopra, 2010, 2011b, c) probably because this is the binding and retention site for the toxicants. Heavy metals accumulation in agricultural soils pose long-term problem of uptake accumulation with its inherent negative human and animal health implications (Chumbley, 1982; Lester, 1987; Ferguson, 1990; Sauve *et al.*, 1996). Keeping above in view, the present investigation was undertaken to assess the contamination of heavy metals in vegetables irrigated with textile effluent at Haridwar (Uttarakhand), India.

Materials and Method

Experimental Design

The present study was conducted in Green House of the Department of Zoology and Environmental Sciences, Gurukula Kangri University Haridwar, for studying the effect of textile effluent on *Phaseolus vulgaris, Vigna radiata, Vica faba* and *Linum usitatissimum*. Pots were used for growing the crop plants. The experiment was replicated by six times.

Effluent Collection and Analysis

The textile effluent was collected from the outlet of Alps Industries, SIDCUL, Haridwar, using plastic container and brought to the laboratory and was analyzed for various physico-chemical parameters *viz*. pH, electrical conductivity, biochemical oxygen demand, chemical oxygen demand, total kjeldahl nitrogen, phosphate, sulphate iron, zinc, cadmium, copper, chromium, cobalt and nickel content following standard methods (APHA, 2005; Chaturvedi and Sankar, 2006).

Sowing of Seeds, Irrigation Pattern and Collection of Crop Parameters Data

The seeds of *P. vulgaris, V. radiata, V. faba* and *L. usitatissimum* were procured from ICAR, Pusa, New Delhi and sterilized with 0.01 mercuric chloride and was soaked for 12 hrs. Seven seeds were initially sown in each pot at equal distance between plant to plant (7.5 cm.). Five plants were maintained in each pot by thinning out of the seven and watered with 500 ml of textile effluent twice in a week and no drainage was allowed. The plant materials were collected to determine the heavy metals content after harvesting the crops (Chandrasekar *et al.*, 1998).

Heavy Metals Analysis

The content of heavy metals in textile effluent, soil and crop plants were analyzed after digestion the samples. For digestion, 5-10 ml sample of textile effluent, 0.5-1.0 gm sample of air dried soil/plant was taken in digestion tube and add 3 ml conc. HNO_3 digested on electrically heated block for1 hour at 145.0 °C. Then 4 ml of $HClO_4$ was added and heated up to 240.0 °C for an additional hour. Cooled and filtered through Whatman # 42 filter paper and volume upto 50 ml was made and analyzed following the standard methods (APHA, 2005; Chaturvedi and Sankar, 2006).

Results and Discussion

Characteristics of Textile Effluent

The values of various physico-chemical parameters of textile effluent *viz*. EC, pH, BOD, COD, TKN, PO_4^{3-}, SO_4^{2-}, Fe^{2+}, Zn, Cd, Cu, Pb, and Cr are presented in Table 9.1.

The results revealed that the textile effluent was slightly acidic in nature pH (6.87). Among various parameters, BOD (2462.50 mg L^{-1}), COD (6495.50 mg L^{-1}), TKN (456.00 mg L^{-1}), SO_4^{2-} (954.50 mg L^{-1}), Fe^{2+} (52.75 mg L^{-1}), Zn (29.24 mg Kg^{-1}), Cd (35.99

mg Kg $^{-1}$), Cu (29.50 mg Kg $^{-1}$), Co (36.30 mg Kg $^{-1}$), Cr (47.54 mg Kg $^{-1}$), Ni (25.64 mg Kg $^{-1}$) were not found in the prescribed limit of Indian Irrigation Standards (BIS, 1991).

Table 9.1: Physico-chemical and Heavy Metal Characteristics of Textile Effluent

Parameters	*Textile Effluent*	*BIS for Irrigation Water*
EC (dS m^{-1})	6.87 ± 1.42	–
pH	6.87 ± 0.31	5.5-9.0
BOD(mg l^{-1})	2462.50 ± 11.82	100
COD(mg l^{-1})	6495.50 ± 11.82	250
TKN (mg l^{-1})	456.00 ± 10.77	100
PO_4^{3-} (mg l^{-1})	476.50 ± 9.71	–
SO_4^{2-} (mg l^{-1})	954.50 ± 5.97	1000
Fe (mg l^{-1})	52.75 ± 4.27	1.0
Zn (mg kg^{-1})	29.24 ± 2.41	15
Cd (mg kg^{-1})	35.99 ± 3.76	2.00
Cu (mg kg^{-1})	29.50 ± 2.60	3.00
Co (mg kg^{-1})	36.30 ± 2.88	1.00
Cr (mg kg^{-1})	47.54 ± 4.37	2.00
Ni (mg kg^{-1})	25.64 ± 2.11	1.00

Mean ±SD of six values.

Heavy Metals in Soil

The values of various heavy metals *viz.* Fe, Zn, Cd, Cu, Ni, Co and Cr of the soil after irrigation with textile effluent are given in Table-2. In the present study the content of different heavy metals *viz.* Fe, Zn, Cd, Cu, Cr, Ni and Co of the soil were recorded to be significantly higher after irrigation with textile effluent. The contamination of heavy metals in the soil were observed in order of Cd>Ni>Cr>Co>Zn>Cu>Fe for *P. vulgaris, V. faba* and Cd>Ni>Cr>Co>Zn>Fe>Cu for *V. radiata* and *L. usitatissimum* after irrigation with textile effluent (Table 9.2).

Table 9.2: Contamination of Various Heavy Metals (mg kg^{-1}) in the Soil after Irrigation of Different Crops Plants with Textile Effluent

Crop Plants	*Fe*	*Zn*	*Cu*	*Cd*	*Cr*	*Co*	*Ni*
P. vulgaris	7.06	16.16	7.31	538.00	71.36	62.50	171.83
V. radiata	9.11	24.28	8.07	708.00	81.63	77.50	225.16
V. faba	4.81	11.08	5.07	417.00	46.81	34.50	127.50
L. usitatissimum	6.30	13.47	4.40	321.50	61.81	52.50	136.83

The content of heavy metals was recorded higher in effluent irrigated soil in comparison to control soil. Among heavy metals the maximum contamination factor was found of Cd (538.00, 708.00, 417.00 and 321.50) for the soil used for *P. vulgaris, V.*

radiata, V. faba and *L. usitatissimum* cultivation with textile effluent (Table-2). Vinod and Chopra (2011a) observed earlier that distillery effluent irrigation increased the Fe, Zn, Cd, Cu, Pb and Cr of the soil and it was in order of Cd>Cr>Pb>Zn>Cu>Fe after irrigation with distillery effluent. The findings were also supported by Chopra *et al.* (2011). Similar findings have also been reported by (Patterson *et al.*, 2008; Vinod *et al.*, 2010; Chopra *et al.*, 2011; Vinod and Chopra, 2011a).

Heavy Metals in Vegetables

The values of various heavy metals Fe, Zn, Cd, Cu, Ni, Co and Cr in *P. vulgaris, V. radiata, V. faba* and *L. usitatissimum* after irrigation with textile effluent are given in Table-3. In the present study, the textile effluent irrigation resulted in the accumulation of various heavy metals in the entire crop plants *viz. P. vulgaris, V. radiata, V. faba* and *L. usitatissimum* differently. The concentration of Cd, Cu, Cr, Co and Ni were found to be significantly higher after irrigation with textile effluent. The accumulation of heavy metals in plant parts of the vegetable crops were observed to be crop specific, differ from crop to crop it may be due to the uptake efficiency and tolerance limit of these crop plants. The contamination of heavy metals were recorded in order of Ni>Co>Cu>Cd>Cr>Zn>Fe for *P. vulgaris*, Co>Ni>Cr>Cu>Cd>Fe>Zn for *V. radiata*, Cr>Co>Ni>Cu>Cd>Fe>Zn for *V. faba* and Co>Cr>Cd>Cu>Zn>Fe>Ni for *L. usitatissimum* after irrigation with textile effluent (Table 9.3). Though, content of heavy metals in *P. vulgaris, V. radiata, V. faba* and *L. usitatissimum* were slightly higher in the textile effluent irrigated plants as compared to control, but these were under the permissible limits of FAO/WHO-Codex alimentarius commission (2001).

Table 9.3: Contamination of Various Heavy Metals (mg kg^{-1}) in different Crop Plants after Irrigation with Textile Effluent

Crop Plants	*Fe*	*Zn*	*Cu*	*Cd*	*Cr*	*Co*	*Ni*
P. vulgaris	1.34	1.48	2.21	3.25	1.85	4.89	6.57
V. radiata	2.02	1.5	2.35	3.82	6.66	8.41	8.18
V. faba	1.75	1.53	3.42	3.6	6.99	5.67	4.64
L. usitatissimum	3.3	4.16	5.19	4.78	6.96	7.83	3.28

Chandra *et al.* (2009) reported the accumulation and distribution of toxic metals (Cu, Cd, Cr, Zn, Fe, Ni, Mn, and Pb) in wheat and mustard plants irrigated with mixed distillery and tannery effluents. Analysis of effluents and soil samples has shown high metals content than the permissible limit except Pb. Pandey *et al.* (2008) also observed that the distillery effluent contains potentially hazardous toxic heavy metals (Cd, Cr, Ni and Zn) which affect the seed germination and seedling growth of maize (*Zea mays* L.) and rice (*Oryza sativa* L.). Vinod and Chopra (2011a) reported the maximum accumulation of heavy metals Fe, Zn, Cd, Cu, Pb and Cr in soil and *Trigonella foenum graecum* with 100 per cent concentration of distillery effluent and it was in order of Pb>Cr>Cd>Cu>Zn>Fe after irrigation with distillery effluent.

Conclusion

The present study concluded that the textile effluent increased the heavy metals *viz.* Fe, Zn, Cd, Cu, Cr, Co and Ni of the soil used in the cultivation of *P. vulgaris, V. radiata, V. faba* and *L. usitatissimum*. The textile effluent irrigation showed significant contamination of different heavy metals in the soil as well as cultivated vegetable crops *viz. P. vulgaris, V. radiata, V. faba* and *L. usitatissimum*. The contamination of heavy metals in soil were in order of Cd>Ni>Cr>Co>Zn>Cu>Fe for *P. vulgaris, V. faba* and Cd>Ni>Cr>Co>Zn>Fe>Cu for *V. radiata, L. usitatissimum* after irrigation with textile effluent. The contamination of different heavy metals were recorded in order of Ni>Co>Cu>Cd>Cr>Zn>Fe for *P. vulgaris*, Co>Ni>Cr>Cu>Cd>Fe>Zn for *V. radiata*, Cr>Co>Ni>Cu>Cd>Fe>Zn for *V. faba* and Co>Cr>Cd>Cu>Zn>Fe>Ni for *L. usitatissimum* after irrigation with textile effluent. Thus the accumulation of various heavy metals was observed to be crop specific and it may be due to their uptake efficiency and tolerance limit against different heavy metals.

Acknowledgement

The University Grants Commission, New Delhi, India is acknowledged for providing the financial support in the form of UGC research fellowship (F.7-70/ 2007-2009 BSR) to Dr. Vinod Kumar.

References

Abida Rehman, Haq Nawaz Bhatti and Habib-ur-Rehman Athar. 2009: Textile effluents affected seed germination and early growth of some winter vegetable crops: A case study. *Water Air Soil Pollut.*, 198:155–163 (DOI 10.1007/s11270-008-9834-5).

APHA. 2005: In: *Standard Methods for the Examination of Water and Wastewater*. American Public Health Association, 21st edition, Washington, DC. pp 1368.

Breckle, C.W. 1991: Growth under heavy metals. In Y. Waisel, A. Eshel, and U. Kafkafi (Eds.) Plant roots: The hidden half (pp. 351–373). New York, NY: Marcel Dekker.

Buurman, P.B., Van Langer and Velthrost, E.J. 1996: In: *Manual of Soil and water analysis*. Backhuys Publisher, Leiden, Netherland.

Carter, M.R. 1993: In: *Soil sampling and method of analysis*, Lewis Publishers, Boca Raton, FL.

Chandra, R., Bhargava, R.N., Yadav, S. and Mohan, D. 2009: Accumulation and distribution of toxic metals in wheat (*Triticum aestivum* L.) and Indian mustard (*Brassica campestris* L.) irrigated with distillery and tannery effluents. *Journal of Hazardous Materials*, 162(2-3): 1514-1521.

Chandrasekar, N., Subramani A. and Saravana, S. 1998: Effect of sugar mill effluent on germination and early seeding growth of black gram (*Vigna mungo* L.) Hepper Var. ADT-3). *Indust. Poll. Cont.*, 14: 73-78.

Chaturvedi, R.K. and Sankar, K. 2006: In: *Laboratory manual for the physico-chemical analysis of soil, water and plant*. Wildlife Institute of India, Dehradun. pp 97.

Chopra, A.K., Chakresh Pathak and Parasad, G. 2009: Scenario of heavy metal contamination in agricultural soil and its management. *J. Appl. and Nat. Sci.*, 1(1): 99-108.

Chopra, A. K., Srivastava, Sachin and Kumar, Vinod. 2011: Comparative study on agro-potentiality of paper mill effluent and synthetic nutrient (DAP) on *Vigna unguiculata* L. (Walp) Cowpea. *J. Chem. Pharm. Res.*, 3(5): 151-165.

Chumbley, C.G. 1982: Cadmium and lead content of vegetable crop grown on land with a history of sewage sludge application. *Environ. Pollut.*, 4: 231-237.

Codex Alimentarius Commission (FAO/WHO) 2001: Food additives and contaminants. Joint FAO/WHO. Food standards programme; ALINORM 01/ 12A, 1-300.

Devkota, R. and Schmidt, G.H. 2000: Accumulation of heavy metals in food plants and grasshoppers from the Taigetos Mountains, Greece. *Agric. Ecosyst. Environ.*, 78: 85-91.

Ferguson, E.J. 1990: *Heavy metals in plants*. In: Ferguson, E.J. (Ed.), The heavy elements chemistry, environmental impact and health effects. Pergamon Press, Oxford, UK, pp. 7-30.

Frost, H.L. and Ketchum, L.H. 2000: Trace metal concentration in durum wheat from application of sewage sludge and commercial fertilizer. *Adv. Environ. Res.*, 4: 347-355.

Godbold, D.L. and Hüttermann, A. 1985: Effect of zinc, cadmium and mercury on root elongation of *Picea abies* (Karst) seedlings, and the significance of these metals to forest die-back. *Environmental Pollution*, 38: 375–381.

Ghafoor, A., Rauf, A., Arif, M. and Muzaffar, V. 1994: Chemical composition of effluents from different industries of the Faisalabad city. *Pakistan J. Agri. Sci.*, 31: 367–9.

Ibrahim, M. and Salmon, S. 1992: Chemical composition of Faisalabad city sewage effluent. Nitrogen, Phosphorus and Potassium contents. *J. Agri. Res.*, 30: 381–90.

Itanna, F. 2002: Metals in leafy vegetables grown in Addis Ababa and toxicological implications. *Ethiopian Journal of Health Development*, 6, 295-302.

Kaushik, A. Nisha, R., Jagjeeta, K. and Kaushik, C.P. 2005: Impact of long and short term irrigation of a sodic soil with distillery effluent in combination with bioamendments. *Bioresource Technology*, 96 (17): 1860-1866.

Lester, J.N. 1987: In: *Heavy metals in wastewater and sludge treatment processes*. CRC Press Inc., Boca Raton, USA, 208 pp.

Madyiwa, S., Chimbari, M., Nyamangara and Bangira, C. 2002a: Cumulative effects of sewage sludge and effluent mixture application on soil properties of a sandy soil under a mixture of star and kikuyu grasses in Zimbabwe. Phys. *Chem. Earth Parts*, 27: 747-753.

Memon, A.R., Aktoprakligül, D., Zdemur, A. and Verti, A. 2001: Heavy metal accumulation and detoxification mechanisms in plants. *Turkish Journal of Botany*, 25: 111–121.

Nieboer, E. and Yassi, A. 1998: Other effects of chromium compound. *In: Chromium in the natural and Human Environment; Advances in Environmental Science and Technology*. J.O. Nriagu and E. Nieboer (Eds). John Wiley and Sons. pp 553 – 546.

Pandey, S.N., Nautiyal, B.D. and Sharma, C.P. 2008: Pollution level in distillery effluent and its phytotoxic effect on seed germination and early growth of maize and rice. *Journal of Environmental Biology*, 29(2): 267-270.

Patterson, S.J., Chanasyk, D.S., Mapfumo, E. and Naeth, M.A. 2008: Effects of diluted Kraft pulp mill effluent on hybrid poplar and soil chemical properties. *Irrig. Sci.*, 26: 547–560.

Qadir, M., Ghafoor, A., Hussain, S.I., Murtaza G. and Mahmood, T. 1998: *Metal ion Contamination in Vegetables and Soils Irrigated with city Effluent*. pp. 89–92. In Proc. Environmental Pollution: third Nat. Symp. Modern Trends in Contemporary Chemistry, Islamabad.

Sauve, S., Cook, N., Hendershot, H.W. and Mcbridge, B.M. 1996: Linking tissue copper concentrations and soil copper pools in urban contaminated soils. *Environ. Pollut.*, 94: 153-157.

Kumar, Vinod, Chopra, A.K., Pathak, Chakresh and Pathak, Sachin. 2010: Agro-potentiality of paper mill effluent on the characteristics of *Trigonella foenum-graecum* L. (Fenugreek). *New York Science Journal*, 3(5): 68-77.

Kumar, Vinod and Chopra, A.K. 2010: Influence of sugar mill effluent on physico-chemical characteristics of soil at Haridwar (Uttarakhand), India. *J. Appl. and Nat. Sci.*, 2 (2): 269-279.

Kumar, Vinod and Chopra, A.K. 2011a: Fertigation effect of distillery effluent on agronomical practices of *Trigonella foenum-graecum* L. (Fenugreek). *Environ Monit Assess*, DOI 10.1007/s10661-011-2033-7.

Kumar, Vinod and Chopra, A.K. 2011b: Impact on physico-chemical characteristics of soil after irrigation with distillery effluent. *Arch. Appl. Sci. Res.*, 3 (4): 63-77.

Kumar, Vinod and Chopra, A.K. 2011c: Alterations in physico-chemical characteristics of soil after irrigation with paper mill effluent. *J. Chem. Pharm. Res.*, 3(6): 7-22.

Voutsa, D., Grimanis, A. and Samara, C. 1996: Trace elements in vegetables grown in an industrial area in relation to soil and air particulate matter. *Environ. Pollut.*, 94: 325-335.

2013, Climate Change Effects on Agriculture and Economy *Pages* **81–84**
Editors: **D.R. Khanna, A.K. Chopra, Vikas Singh, Rakesh Bhutiani & Gagan Matta**
Published by: **BIOTECH BOOKS, NEW DELHI**

Chapter 10

The Natural Radiation Pollution in Garhwal Himalaya: An Overview

A.A. Bourai, G.S. Gusain and R.C. Ramola
Department of Physics, H.N.B.Garhwal University Campus, Badshahithaul, Tehri Garhwal, Uttarakhand

The natural radionuclides are widely distributed in different geological formation such as soil, rocks and water. In environment there are mainly two source of natural radiation. High energy cosmic ray particles incident on earths atmosphere and radionuclide that originated in the earth's crust and are present every where including human environment. Cosmic rays may be divided into primary and secondary rays. The cosmic rays before they interact with atmosphere are termed as primary cosmic rays and consists two components Galactic and Solar. The galactic component comes from outer space and origin could be infinite numbers of star of the universe. The solar component originating from the sun is a minor component and is observed during solar flares. The galactic component has the particles with enormous energy up to 10^9 eV. It consists of 79 per cent high energy protons, about 20 per cent alpha particles and rest being electrons, protons and neutrons.

The secondary cosmic rays produced by the interaction of primaries nitrogen, oxygen and organ of the earth's atmosphere resulting in the production of neutrons, protons, mesons and electrons. The cosmic ray interactions also produce a number of radioactive nuclide known as cosmogenic radionuclide such as ^{3}H, ^{14}C.

The radionuclides are mostly isotopes of heavy elements which have been in existence since the origin of the earth and distributed in different geological formation such as soil, rocks and groundwater. Weathering of bed rocks, the main reservoir of the radionuclides releases the radionuclide to the soil. From there they are taken by the plants and ultimately reach the human system through the animal grazing on these plants. Leaching of bed rocks and soil may lead to concentration of these radionuclides in groundwater. In some selected areas this could reach to vary high values. Radium isotopes are important in such process due to their solubility in the water. Radionuclides from ^{238}U series, ^{232}Th series and ^{40}K series are the main sources of the terrestrial radiation on earth surfaces. One of the elements of each series (^{238}U, ^{232}Th and ^{40}K) is a rare gas being the isotopes of radon (^{222}Rn, ^{220}Rn, ^{219}Rn). Radon and its decay products can build up to high values and in enclosed spaces with poor ventilation. These include mines, houses with poor ventilation and basement areas.

The present paper includes the assessment of natural radionuclide pollution due to exposure of radon, thoron and their progeny. Simultaneously, the radionuclide pollution due to terrestrial radionuclides (^{238}U, ^{232}Th, ^{40}K) is also estimated in the form of Gamma radiation dose in the environment.

Keywords: *Natural radiation, Gamma radiation, Cosmic particles, Galactic and Solar component.*

Introduction

The knowledge of radionuclides distribution and radiation level in the environment is important for assessing the effects of radiation exposure from both terrestrial and extraterrestrial sources. Terrestrial radiation is due to radioactive nuclides present in varying amounts in soil, building materials, water, rocks and atmosphere (Singh *et al.*, 2005; Ramola *et al.*, 2008).The world is naturally radioactive and around 90 per cent of human radiation exposure arises from natural sources such as cosmic radiations, exposure to radon gas and terrestrial radiations. The naturally occurring radionuclides present in soil include ^{226}Ra, ^{232}Th and ^{40}K. Some of these radionuclides are transferred to man through food chain or inhalations, while the extraterrestrial radiations originate from outer space as primary cosmic rays (UNSCEAR, 2000). External exposures will occur as a result of irradiation and internal exposures will occur as a result of inhalations. Furthermore, because of extensive quantities of radionuclide in soil and in vegetation, exposures also will occur as a result of ingestion. Radon is a naturally occurring gas that seeps out of rocks and soil. Radon comes from uranium that has been in the ground since the time the earth was formed and the rate of radon seepage is very variable, partly because the amounts of uranium in the soil vary considerably. Radon flows from the soil into outdoor air and also into the air in homes from the movement of gases in the soil beneath homes. Outside air typically contains very low levels of radon, but it builds up to higher concentrations indoor when it is unable to disperse. Some underground mines, especially uranium mines, contain much higher levels of radon. Although radon is chemically inert and electrically uncharged, it is radioactive, which means that radon atoms in the air can spontaneously decay, or change to other atoms.

Experimental Technique

For the assessment of natural radiation level different techniques have been used. For the assessment of indoor radon concentration a Twin cup dosimetry technique is used in which the concentration of ^{222}Rn, ^{220}Rn and their progeny are separately measured (Ramola *et al.*, 1996; Eappen and Mayya, 2004). The concentration of naturally occurring radioactive element such as ^{226}Ra, ^{232}Th and ^{40}K ^{226}Ra, ^{232}Th, ^{40}K were assessed in soil samples using high efficiency and low resolving gamma ray spectrometer. The prepared soil sample was placed in the protection unit of gamma ray spectrometry for the counting time of three hours. The estimation of natural radionuclide in soil sample was carried out using NaI (Tl) gamma radiation detector of size 63 × 63 mm with multi-channel analyzer. The radionuclides in soil were identified from energy peak 1460 keV for ^{40}K, the activity of ^{226}Ra from the 1764 keV gamma line of ^{214}Bi and that of ^{232}Th from 2620 keV gamma line of ^{208}Tl. This spectral analysis was performed with the aid of computer software. The particular peak of gamma spectra was measured taking the reference of 661 keV of ^{137}Cs. The activity concentration of these soil samples were calculated from the intensity of each line, taking into account the mass as well as geometry of the samples, counting time and efficiency of the detector (Ramola *et al.*, 2008).

Table 10.1: Radioactivity Concentration from the Soil Samples of Study Area

Study Area	*Radionuclide*	*Activity (Bq/Kg)*		
		Min	*Max*	*Average*
Thauldhar	^{226}Ra	BDL	131.5	64
	^{232}Th	0.88	177.8	41.2
	^{40}K	195.4	1659.3	913.5
Budhakedar	^{226}Ra	BDL	97.1	64.5
	^{232}Th	28.0	384.4	117.5
	^{40}K	124.0	1406.2	663.2
Kumaun	^{226}Ra	36.4	166.6	83.8
	^{232}Th	17.0	94.7	50.4
	^{40}K	394.8	137.8	826.4
Ukhimafh	^{226}Ra	38.4	141.7	80.5
	^{232}Th	57	155.9	118.9
	^{40}K	3.0	672.8	341.0

Results and Discussion

The ^{226}Ra, ^{232}Th and ^{40}K contents in the soil samples from Garhwal Himalaya region were found to vary from BDL to 131.5 ± 11.5 Bq/kg with an average of 64 Bq/Kg, 8.8 ± 3.0 Bq/kg to 177.8 ± 13.3 Bq/kg with an average of 41.2 Bq/kg and 195.4 ± 14.0 Bq/kg to 1659.3 ± 40.7 Bq/kg with an average of 913.5 Bq/kg, respectively. The radium equivalent activity in this area was found to vary from 46.4 Bq/kg to 441.7 Bq/kg with an average of 190.5 Bq/kg. The total absorbed gamma dose rate in this area was found to vary from 21.9 nGy/h to 201.3 nGy/h with an average of 87.7 nGy/h. The contribution to total radiation level due to ^{226}Ra series, ^{232}Th series and ^{40}K was found to be 30.5 per cent, 26.6 per cent and 42.9 per cent, respectively. The total indoor effective dose from the study area was found to vary from 0.11 mSv/y to

0.99 mSv/y with an average of 0.45 mSv/y. The total outdoor effective dose was found to vary from 0.03 mSv/y to 0.25 mSv/y with an average of 0.11 mSv/y. The concentrations of indoor radon and thoron were found to vary from 10.3 Bq/m^3 to 190.5 Bq/m^3 and 1 Bq/m^3 to 144.6 Bq/m^3, respectively. The levels of radon and thoron daughters vary from 0.1 mWL to 30.9 mWL and 0.1 mWL to 143.7 mWL (Ganesh *et al.*, 2008). The radon concentration in soil-gas was found to vary from 1098 ± 293 Bq/m^3 to 31,776 ± 1591 Bq/m^3, with an average of 7456 Bq/m^3 in summer and from 3501 ± 527 Bq/m^3 to 42,883 ± 1847 Bq/m^3, with an average of 17,148 Bq/m^3 in winter. In groundwater, it was found to vary from 8 ± 1 Bq/l to 3047 ± 25 Bq/l with an average of 510 Bq/l in summer and from 26 ± 1 Bq/l to 2311 ± 25 Bq/l with an average of 433 Bq/l in winter (Ramola *et al.*, 2007). During this study Budhakedar, Rudraparyag and Ukhimath region of Garhwal Himalaya have found high radiation background area.

Conclusion

Investigation of the radionuclides characteristics in Garhwal Himalayas and its radiological impact on the neighboring environment are presented in details. A large variation in radionuclide contents and gamma dose rate shows that the radionuclides are not uniformly distributed in the soil. The distribution of radionuclides was found to depend widely upon the rock formation and geology of the area. Based on the results obtained from the Garhwal Himalaya a considerable variation in the activity of concentrations of radon, thoron and their progeny were found in the houses of Garhwal Himalaya.

References

Eappen, K.P., Mayya, Y.S. 2004: Calibration factors for LR-115 (type-II) based radon thoron discriminating dosimeter. Radiat. Meas. 38, 5–17. George, A.C., 1984. Passive, integrated measurements of indoor.

Prasad, Ganesh, Prasad, Yogesh, Gusain, G.S. and Ramola, R.C. 2008: Measurement of radon and thoron levels in soil, water and indoor atmosphere of Budhakedar in Garhwal Himalaya,India. Radiation Measurements 43 (2008) S375–S379.

Ramola, R.C., Choubey, V.M., Negi, M.S., Prasad, Yogesh and Prasad, Ganesh. 2007: Radon occurrence in soil–gas and groundwater around an active landslide. Radiation Measurements 43 (2008) 98 – 101.

Ramola, R.C., Gusain, G.S., Badoni, Manjari, Prasad, Ganesh, Prasad, Yogesh and Ramachandran, T.V. 2008: ^{226}Ra, ^{232}Th and ^{40}K contents in soil samples from Garhwal Himalaya, India, and its radiological implications. J. Radiol. Prot. 28: 379–385.

Ramola, R.C., Rawat, R.B.S., Kandari, M.S., Ramachandran, T.V., Eappen, K.P., Subba Ramu, M.C. 1996: Calibration of LR-115 plastic track detectors for environmental radon measurements. Indoor Built Environment 5, 364–366.

UNSCEAR. 2000: United Nations Scientific Committee on the Effects of Atomic Radiation. Sources and effects of ionizing radiation. Report to General Assembly with annexes. United Nations sales Publications, United Nations New York; Assembly B, 97-105.

2013, Climate Change Effects on Agriculture and Economy Pages 85–89
Editors: D.R. Khanna, A.K. Chopra, Vikas Singh, Rakesh Bhutiani & Gagan Matta
Published by: BIOTECH BOOKS, NEW DELHI

Chapter 11

Separation of Pesticides as Pollutants from Water by Molecularly Imprinted Polymers (MIPs) Membranes

H.V. Pant[1], *Suresh Kr*[2], *Prashant Singh*[3] *and K.P. Singh*[4]

[1]*Department of Chemistry, S.G.R.R. (PG) College, Pathri Bagh, Dehradun, Uttarakhand*

[2]*Department of Chemistry, Government College, Talwari, Chamoli, Uttarakhand*

[3]*Department of Chemistry, D.A.V. (P.G.) College, Karanpur, Dehradun, Uttarakhand*

[4]*Department of Biophysics and Nanotechnology, CBSH, G.B. Pant University of Agriculture and Technology, Pantnagar, Uttarakhand*

Chemists, chemical engineers and biologists are engaged in research in the field of preparation and characterization of synthetic and biological membranes. Nowadays curiosity to get the membrane of desired pore size has forced the researcher to modify the surface of polymer membrane. The same is done by molecular imprinting of desired template molecules. Molecularly imprinted polymers offer potential for the removal of pesticides specifically from waste and drinking water. Currently, purification is carried out using nano filtration and reverse-osmosis membranes. Unfortunately, these membranes remove all ions, which can make the purification process less effective.

Molecular imprinting of polymer membranes is recent technology being used in separation and sensing technology. Stability of such membrane in polar and non polar solvent enhances the applicability of such MIP membranes. Molecularly imprinted polymers (MIPs) represent a new class of materials that have artificially created receptor structures. First of all membrane is characterized and their applicability and utility in sensing technology is tested. The MIP matrix are used both for preparation of membranes for separation and specific binding sites for sensors. It is followed by imprinting of selected pesticide on a suitable polymers matrix and the imprinted polymers are used for separation technology and sensing device in biosensors.

In the present paper, an overview of application of MIP's in separation of pesticides have been discussed and reviewed.

Keywords: *Molecularly imprinted polymer Membranes, Pesticides, Water.*

Introduction

Molecularly imprinted polymers (MIPs) represent a new class of materials that have artificially created receptor structures (Wulff, and Sarhan, 1972; Mosbach, 1998; Piletsky, 1998). Since their discovery in 1972, MIPs have attracted considerable interest from scientists and engineers involved with the development of chromatographic adsorbents, membranes, sensors and enzyme and receptor mimics. A great variety of chemical compounds have been imprinted successfully. The resulting polymers are robust, inexpensive and, in many cases, possess an affinity and specificity that is suitable for industrial application.

Widespread interest in MIP technology is evident from the increasing number of publications and groups working in this area (which has doubled between 1996 and 2006) and the steady stream of new patents. Two areas of research continue to provide the focus for scientific activity: materials science and separation technology. Initially, MIP research was aimed at the design of specific adsorbents and at understanding the nature of their specificity. The broad structure of research activity has remained unchanged over the last ten years. Comparatively new areas in which MIPs provide opportunities for advancement that have appeared in the last decade include the development of assays and sensors, membranes, capillary electrophoresis (CE), and the production of polymers with special functions such as drug-release matrices.

MIP technology is now in the process of maturing after a slow start because of difficulties in resolving problems associated with poor performance of MIPs in aqueous environments and because of the absence of any general procedure for MIP design. The three most promising areas in which MIPs could be commercially applied are: (i) separation; (ii) assays and sensors; and (iii) catalysis.

Indiscriminate use of pesticide/herbicides can leave environmentally unsafe residues in soil and water. Traditional testing methods for herbicides, radiation, humidity, chemicals and microorganisms are relatively costly and time-consuming. The most common testing method can take several days. Other methods may provide faster detection but require complicated procedures, expensive reagents and equipment and extensively trained personnel. Nearly all of these methods require samples to be manually collected on the plant floor and taken to a laboratory for subsequent analysis. Use of molecularly imprinted conducting polymer membranes in identifying heat, moisture, chemicals impurities and morphological changes has open a new era of environmental sensing technology. In the present work, a plan of preparation of molecularly imprinted polymer membrane is given. Characterization of such membranes will be done with their applicability and utility in sensing technology. The MIP matrix are used both for preparation of membranes for separation and specific binding sites for sensors. Under the plan of work, two pesticides namely Deltamethrin and Chlorpyrifos will be imprinted on a suitable polymers matrix and the Imprinted polymers will be used for separation technology and sensing device in biosensors.

The proposed studies fall under the interdisciplinary area of pure chemistry, Soil science, Agronomy, Horticulture, Plant pathology, Animal Science, Ecology, Applied Chemistry, Medical Science, Biotechnology, Climate Change Science, Environmental Science and Analytical Chemistry. Through this multidisciplinary approach, the water and soil samples analyzed for Deltamethrin and Chlorpyrifos pesticides will be analyzed and the outcome would help in better availability of pure water and soil to society for good health and improved economy. The outcome of the study is expected to be beneficial for decreasing the pesticides present in water/soil. Molecularly imprinted polymers offer potential for the removal of pesticides, endocrine-disrupting compounds and heavy metals from waste and drinking water. Currently, purification is carried out using nanofiltration and reverse-osmosis membranes. Unfortunately, these membranes remove all ions, which can make the purification process less effective. MIP polymers can also be used to remove heavy metals, rare metals and radioisotopes with high specificity (Ching, 2011; Maeda and Bartsch, 1998). They are expected to be effective in extreme environments, such as wastewater from caustic cleaners and acid rinses that are used to clean pasteurizers. Molecular imprinting is a method for creating polymer matrices that display a selective molecular recognition behavior (Wulff, 1986; Takeuchi *et al.*, 1999). Such materials enable the separation based on affinity, particularly when used as stationary phases in HPLC, in capillary electrophoresis and as selective membranes (Yoshikawa *et al.*, 1995, 1999). Several other applications of the molecularly imprinted polymers (MIPs) have also been reported, *e.g.*, shifting the reaction equilibrium by a product removal (Cormack, and Mosbach, 1999; Brüggemann, 2001).

Membranes are important tools in separation science *e.g.* for water purification from toxic metals (Eyal, A.M. and Bressler, 1993) and other substances (Yoshidome *et al.*, 1998). Oscillatory liquid membrane systems allow also the identification and quantitative determination of certain water pollutants (Yoshidome, 2000; Larter, 1990). Artificial systems can be used for modeling the oscillatory behavior of biological membranes (Toko, 1998). Very interesting is the behavior of cells in human body responsible for taste. In estimation of food quality mainly the organoleptic methods are used. However, these methods are not objective enough. In recent years a lot of efforts are made for designing a taste sensor based on lipid membranes (Toko, 1996; Yoshikawa, and Matsubara, 1986). On the account of their unique feature of oscillation (amplitude, frequency, phase) it may be suggested that they would be more sensitive to different molecules than lipid membranes.

Objectives

The objectives of the study are:

1. Preparation of MIP membranes
2. Characterization of MIP membranes and matrix
3. Thermo dynamical and transport studies
4. Testing of MIP membrane in separation device

Methodology

The different steps of methodology involve:

1. Preparation of polymer with suitable templates of pesticides.
2. Preparation of polymer membrane without template
3. Removal of template form polymer and formation of MIP
4. Identical treatment just same as template removal
5. UV-VIS spectrometric, FTIR and SEM study
6. Irreversible thermodynamic and transport study of both types of membranes.
7. Separation study with both types of membranes
8. To test the utility of such Membranes in sensing (Biosensor) devices.

The tools to be used for experimental work:

1. Membrane casting device for the preparation of polymer membrane by phase inversion technique with constant stirring magnetic stirrer attached to water bath,
2. Digital micro conductivity bridge for the measurement of conductance with connectivity with glass cell,
3. Digital potentiometer with the connectivity of a transducer and PC for recording and analysis of data,
4. Membrane cell assembly (i) for saturated calomel electrode [SCE] (ii) for Ag/AgCl electrode (iii) for conductivity cell of micro conductivity bridge,
5. Millipore filtration kit for separation and
6. QCM biosensor setup for detection experiment.

Significance of the Study

It is obvious from the literature and the experiments that MIP membranes play an important role in separation and sensing technology. These membranes are being used extensively for the separation of hazardous chemicals. It is useful to prepare cheaper and effective membranes for such techniques. Thermodynamic properties and characterization play a role to give a desired perfection to polymer membranes which are going to be used in various technologies. Molecular imprinting of polymer membranes are recent technology to get a membrane of desired pore size and are being used in separation and sensing technology. Stability of such membrane in polar and non polar solvent enhances the applicability of such MIP membranes. Thus this research work deals a very important phenomenon of great importance.

References

Brüggemann, O. 2001: *Anal. Chim. Acta.*, 435, 197-207.

Ching-Chwang Huang *et al.*, Ind. J. Chem. Tech. Volume 18, 7-12, 2011.

Cormack, P.A.G. and Mosbach, K. 1999: React. Funct. Polym. 41, 115-124.

Eyal, A.M. and Bressler, E. 1993: *Biotechnol. Bioeng.*, 41, 287 –295.

Larter, R. 1990: *Chem. Rev.*, 90, 355 – 381.

Maeda M. and Bartsch R.A., *ACS Symp. Ser*, 703, 1–8, 1998.

Mosbach, K. *et al.*, *ACS Symp. Ser.* 703, 29–48, 1998.

Piletsky, S.A. *et al.*, *J. Membr. Sci.* 157, 263–278, 1999.

Takeuchi, T., Haginaka, J., Chromatogr. J. 1999: B 728, 1-20.

Toko, K. 1996: Mat. Sci. Eng. C, 4, 69-92,

Toko, K. 1998: Meas. Sci. Technol., 9, 1919 –1936.

Wulff, G. ACS Symp. Ser. 308, 186-198, 1986

Wulff, G. and Sarhan, A., *Angew. Chem. Int. Ed. Engl.* 11, 341–344, 1972.

Yoshidome, T., Higashi, T. and Kamata, S. 2000: Chem. Letters, 550 – 551.

Yoshidome, T., Higashi, T., Mitsushio, M. and Kamata, S. 1998: Chem. Letters, 855 – 856.

Yoshikawa, Izumi, M.J., Kitao, T., Koya, S., Sakamoto, S. 1995: *J. Membr. Sci.*108, 171-175.

Yoshikawa, K. and Matsubara, Y. 1986: Biophysical Chemistry, 23, 211-214.

Yoshikawa, M., Ooi, T., Izumi, J. 1999: *J. Appl. Polym. Sci.* 72, 493-499.

2013, Climate Change Effects on Agriculture and Economy *Pages 91–102*
Editors: **D.R. Khanna, A.K. Chopra, Vikas Singh, Rakesh Bhutiani & Gagan Matta**
Published by: **BIOTECH BOOKS, NEW DELHI**

Chapter 12

Studies on Physico-chemical, Microbiological Properties of Rhizospheric and Non-Rhizospheric Soils in Compartment No. 792 of Gorewada Forest

Kirti V. Dubey [1], ***Pravin N. Charde*** [1], ***Ashish Lambat*** [1] ***and Gagan Matta*** [2]

[1]*Sevadal Mahila Mahavidyalaya, Sakkardara Square, Nagpur, Maharashtra*

[2]*Department of Environmental Science, Gurukula Kangri University, Haridwar, Uttarakhand*

The study of Gorewada forest was carried out to find out physicochemical as well as microbiological properties of soil collected from the rhizosphere of different plant species. Studies have shown that the forest present in compartment No. 792 is of dry deciduous type having trees, shrubs, herbs, climbers and grasses and some barren patches of land. Soil of the different plant species had clayey texture, porosity and water holding capacity was high. The electrical conductivity, organic carbon and cation exchange capacity of pit soil was somewhat high compared to barren soil; however levels of these chemical properties were lower as compared to the rhizospheric soil samples of dominant plant species. Total and available nutrient content was high in rhizospheric soil as compared to the pit soil and barren. The bacteria, fungi, actinomycetes and nitrogen fixer namely *Rhizobium* and *Azotobacter* and Vesicular Arbuscular Mycorrhizae (VAM) spores were found to be high in pit soil as compared to barren soil. However, the rhizospheric soil samples had comparatively high counts of tested microorganisms. Findings of this particular study will formed a basis for developing future plans required for development of forest cover with a diverse variety of plants of high economic value like timber, medicinal and ornamental including even the exotic and rare species by using the biotechnological approach of bioaugmentation.

Keywords: *Gorewada forest, Rhizospheric soils, Rhizobium, Azotobacter, VAM, Bioaugmentation.*

Introduction

Many parts of country are facing problem of forest resources depletion due to lack of proper physico-chemical and microbiological status of forest soils. This problem can be alleviated by site-specific soil augmentation process which has now been accepted as a tool for the sustainable management of forest resources. In forest management practices, high scientific skills are required. However, important steps are proposed herein as preliminary study is based on following aims and objectives. One of the main aims of the proposed study is to study native plant species in and around Gorewada forest. Another aim is to characterize rhizospheric soil of native plant species with respect to its physico-chemical and microbiological properties. This study will help to generate the basic data required for developing an appropriate bioaugmentation process that can be used to resolve the low nutrient status and less counts of useful micro flora present in forest soil for proper functioning of biogeochemical cycle. Project also aims to isolate site- specific biofertilizers *i.e.* strains of *Rhizobium, Azotobacter* and Vesicular Arbuscular Mycorrhiza (VAM) and to select a suitable organic amendment for improving the physico-chemical and microbiological properties of soil. Results of above mentioned study will form a platform for future studies in order to define the most supportive and nutritive rhizosphere for the sustainable development of the forest canopy by using bioaugmentation process.

Materials and Method

The study was carried out on the project entitled, "Studies on different Flora, physico-chemical, microbiological properties of Rhizospheric and Non Rhizospheric Soils in Compartment No. 792 of Gorewada Forest" by Sevadal Mahila Mahavidyalaya and Research Academy, Nagpur. During the study it was observed that the parent material dominant in this area is basalt rock. The climate of the area is mainly characterized as dry sub humid climate with hot summer and mild winter. The Gorewada forest is located in Nagpur at 21°45′N latitude and 78° 15′E longitude. The annual rainfall is 750 to 1000 mm. Site surveys were conducted at Gorewada forest during December-2007 to March-2008 by the college team of microbiologist, botanist and soil scientist. During every visit, as many specimens as possible were collected were brought to the college laboratory to identify all plants by using reference floras by Hooker (1872-1897), Ugemuge (1986), Almeida (1990), Joshi (2000) and Singh *et al.*

(2001). Beside this representative soil samples from minimum five rhizospheres of same plant species were collected from each compartment with the help of hand auger, and were mixed in equal proportions to get a composite sample on dry weight basis, air dried, homogenized, sieved through 2.0 mm mesh and stored in polythene bags. Analysis of composite barren, pit and rhizospheric soil samples of dominant plant species available in experimental site were conducted for different physico – chemical and microbiological properties to know the fertility status of the soil. Particle size analysis of soil samples was done by international pipette method, (Piper, 1966), Bulk density was determined by using clod method as described by Black *et al.* (1965), Particle density was determined by a pycnometer method as described by Black *et al.* (1965), Water holding capacity was determined by using Keen- Roozkowski brass boxes as described by Piper (1965a), The pH of soil was determined by using 1:2.5 soil water suspension ratio as described by (Jackson, 1967), Electrical conductivity was determined with the help of conductivity meter using 1: 2.5 soil water suspension ratios, (Jackson, 1967), Organic carbon was determined by following Walkly and Black wet digestion method, (Black; 1965b) Available N (was determined by following alkaline permanganate (0.31 per cent $KMnO_4$) method (Subbaih and Asija, 1956), Available P_2O_5 determined colorimetrically by following Bray's II method at 660 nm wavelength, (Black; 1965b), Available K_2O determined air dried soil with 50.0 ml neutral N NH_4OAc (Jackson, 1967). Total N: Samples were digested using conc. H_2SO_4 (wet digestion) and taken for nitrogen distillation by micro Kjeldahl method using Kjel Plus instrument (Tondon, 1993). Total phosphorous is estimated by digesting the soil with mixture of nitric acid and perchloric acid (9:4) and determines the phosphate by ammonium vanadomolybdate as described by Jackson (1967). Total potassium is estimated by digesting the soil with mixture of nitric acid and perchloric acid (9:4) and determines the potassium by using flame photometer. (Hesse, 1971). Microbes such as bacteria, fungi, actinomycetes and nitrogen fixing strains of *Rhizobium* and *Azotobacter* were analyzed by following standard procedure for soil microbial populations and were expressed in terms of colony forming units cfu/g (Black *et al.*, 1965). Vesicular Arbuscular Mycorrhizal (VAM) spores were enumerated by Wet Sieving and decanting technique (Gerdmann and Nicolson, 1963).

Results and Discussion

Type of Flora Present

There are about 42 different plant species classified as trees, shrubs, herbs, climbers and grasses in compartment number 792 (Table 12.1). Among the tree species dominant type were *Acacia catechu, A. leucophloea. A. nilotica, Albizia odoratissima, Butea monosperma, Cassia siamea, Mitragyna perbifolia and Ziziphus mauritiana.*

Textural Classification

Determination of the texture of forest soil is also important as it plays an major impact on forest soil productivity. The proportion of sand, silt and clay in the topsoil and subsoil layers is called texture. Sandy soils are normally very well drained and often lack nutrients due to constant leaching loss. At the other end of the spectrum is the pure clay soils comprised of very small, fine soil particles. Consistency of the

Table 12.1: List of Various Plant Species Present in Compartment Number 792

Sl.No.	Botanical Name	Family	Local Name
		Trees	
1.	*Acacia catechu*	Leguminosae	Khair
2.	*Acacia leucophloea*	Leguminosae	Hivar
3.	*Acacia nilotica*	Leguminosae	Babhul
4.	*Albizia odoratissima*	Leguminosae	Shiras
5.	*Andropogon pumilus*	Poaceae/Gramineae	Diwartan
6.	*Azadirachta indica*	Meliaceae	Kaduneem
7.	*Bauhinia racemosa*	Leguminosae	Apta
8.	*Butea monosperma*	Leguminosae	Palas
9.	*Cassia siamea*	Leguminosae	Cassia siamea
10.	*Dalbergia sissoo*	Leguminosae	Shisham
11.	*Dandrocalamus strictus*	Poaceae	Bamboo
12.	*Dolichandrone falcata*	Bignoniaceae	Mersing
13.	*Maytenus emarginata*	Celastraceae	Bharati
14.	*Mimosa hamata*	Leguminosae	Chilati
15.	*Mitragyna perbifolia*	Rubiaceae	Kalamb
16.	*Phoenix sylvestris*	Palmae	Sindi
17.	*Semicarpus anacardium*	Anacardiaceae	Biba
18.	*Tamarindus indica*	Leguminosae	Imli/Chinch
19.	*Tectona grandis*	Verbenaceae	Teak
20.	*Terminalia bellirica*	Combretaceae	Bihada
21.	*Ziziphus glaberrima*	Rhamnaceae	Guti
22.	*Ziziphus mauritiana*	Rhamnaceae	Bor
		Shrubs	
23.	*Lantana camera*	Verbenaceae	Ghaneri
24.	*Xanthium strumarium*	Asteraceae	Gokharu
		Climbers	
25.	*Cocculus hirsutus*	Menispermaceae	Vasanvel
26.	*Coix lacryma-jobi*	Poaceae/Gramineae	Kasai/Garu
27.	*Hemidesmus indicus*	Periplocaceae	Khobarvel/Anantvel
28.	*Ipomea quamoclit*	Convolvulaceae	Ganeshvell
		Herbs	
29.	*Ageratum conyzoides*	Asteraceae	Osadi
30.	*Alternanthera pungens*	Amaranthaceae	Galighosh
31.	*Blumea eriantha*	Asteraceae	Nirmudi
32.	*Cassia tora*	Leguminosae	Tarota

Contd...

Table 12.1–*Contd...*

Sl.No.	*Botanical Name*	*Family*	*Local Name*
33.	*Hyptis suaveolens*	Lamiaceae/Labiatea	Rantulas
34.	*Parthenium*	Asteraceae	Gajarghas
35.	*Sida acuta*	Malvaceae	Chikana/Tupkadi
36.	*Sida cordata*	Malvaceae	Bhuichikana
37.	*Tridax procumbens*	Asteraceae	Kambarmodi
38.	*Vernonia cinerea*	Asteraceae	Sahadevi
39.	*Vicoa indica*	Asteraceae	Sankuli (Sonuli)
		Grasses	
40.	*Apluda mutica*	Poaceae/Gramineae	Ponai
41.	*Hackelochloa granularis*	Poaceae/Graminaeae	Phulwa
42.	*Iseilema laxum*	Poaceae/Gramineae	Mushan

subsoil layer is another important factor in forest soil productivity. The combination of soil-sized particles and the physical and chemical properties of each individual particle type in a given soil determine the soil's consistence class. The textural classification *i.e.* sand, silt and clay percentages of soils collected from 792 compartment of Gorewada forest is presented in the Table 12.2 and results show that the textural class of all the soil samples analyzed were clayey in texture. The rhizospheric soil of the *Albizia odoratissima* had highest clay content of 62.12 percent and very low content of sand (17.57 per cent) and silts (20.31 per cent). Due to clayey textural class the clay content was high in all the rhizospheric soils of different species as compared to the sand and silt percentages. It was in this compartment reported that the majority of the soils have high amount of clay as compared to sand and silt. As basalt (rock) being the parent material for formation of these soils, therefore, it is known to impart higher amount of clay to the soil.

Table 12.2: Textural Classification of Barren and Rhizospheric Soils of Dominant Plant Species available from Compartment Number 792 of Gorewada Forest

Sl.No.	*Sample Name*	*Sand (%)*	*Silt (%)*	*Clay (%)*	*Textural Class*
	Compartment number 792				
1.	Barren patch	19.29	28.61	52.1	Clayey
	Rhizospheric soils of dominant plant species				
1.	*Ziziphus mauritiana* (Bor)	22.34	26.29	51.37	Clayey
2.	*Acacia catechu* (Khair)	28.39	27.45	44.16	Clayey
3.	*Bauhinia racemosa* (Apta)	20.71	23.53	55.76	Clayey
4.	*Acacia nilotica* (Babhul)	20.39	27.68	51.93	Clayey
5.	*Albizia odoratissima* (Shiras)	17.57	20.31	62.12	Clayey
6.	*Tamarindus indica* (Imli/ Chinch)	24.02	23.84	52.14	Clayey
7.	*Mitragyna perbifolia* (Kalamb)	23.87	25.69	50.44	Clayey

Physical Properties

The physical properties of soils are dominant factors affecting the fertility of forest soil which determine the availability of O_2 in the rhizospheric area of the plant species for efficient nutrient turn over by microorganisms. Results of different physical properties of the soils collected from compartment number 792 of Gorewada forest are presented in the Table 12.3 and results show that the barren soil was although clayey in texture the porosity and WHC was very low (51.32 per cent and 50.45 per cent, respectively) and the bulk density was very high (1.29 mg m^{-3}). Due presence of vegetation in the soil the rhizospheric soil showed increase in porosity and WHC percentages. This phenomenon was observed in case of rhizospheric soil of *Albizia odoratissima* which reported the highest WHC (61.45 per cent) and porosity (65.34 per cent) with the lower BD of 1.10 mg m^{-3}. Similar findings were observed in case of the rhizospheric soils of other plant species namely *Acacia nilotica, Mitragyna perbifolia, Acacia catechu, Ziziphus mauritiana* and *Bauhinia racemosa* where the WHC ranged from 52.65 percent to 61.45 percent.

Table 12.3: Physical Properties of Barren and Rhizospheric Soils of Dominant Plant Species Available from Compartment Number 792 of Gorewada Forest

Sl.No.	*Sample Name*	*Bulk Density (Mg m^{-3})*	*Porosity (%)*	*Water Holding Capacity (%)*
	Compartment number 792			
1.	Barren patch	1.29	51.32	50.45
	Rhizospheric soils of dominant plant species			
1.	*Ziziphus mauritiana* (Bor)	1.14	58.92	56.39
2.	*Acacia catechu* (Khair)	1.15	56.6	52.65
3.	*Bauhinia racemosa* (Apta)	1.09	58.87	56.34
4.	*Acacia nilotica* (Babhul)	1.17	60.74	57.78
5.	*Albizia odoratissima* (Shiras)	1.1	65.34	61.45
6.	*Tamarindus indica* (Imli/Chinch)	1.13	57.36	54.45
7.	*Mitragyna perbifolia* (Kalamb)	1.1	58.59	55.34

Chemical Properties

Forest soil quality with respect to nutrient status can be ascertained by studying soil chemical properties. It has been found that leaf litter layer plays an important role in forest growth on soils (Perala and Alban, 1982) and leaf litter accumulation enhances soil chemicals (Singh and Mudgal, 2000). The status and release of nutrients in soil is largely depending on nature of vegetation; polyculture leads to release more nutrients in soil than monoculture (Mishra, 2010). Release of nutrient from litter through decomposition process is recognized as an important part of the nutrient cycle whereby essential mineral elements tied up in the plant biomass are made available for further plant growth (MacLean and Wein, 1978). The chemical properties of soil collected from the rhizosphere of different plant species found in compartment number 792 are presented in the Table 12.4. The result shows that pH and the EC of this site were observed to be high as compared to the other compartments. Soil organic matter (SOM) or soil organic carbon (SOC) is commonly recognized as one of the key chemical parameters of soil quality. The organic carbon status varied from 1.78 percent in barren soil to the 2.34 percent in the *Mitragyna perbifolia*. The leguminous plant species like *Acacia catechu*, *Acacia nilotica*, *Mitragyna perbifolia* and *Tamarindus indica* were reported high organic carbon, *viz.* 2.90, 2.22, 2.34 and 2.24 percent respectively. The CEC was obviously high in the rhizospheric soil of all the species of the compartment as compare to the barren soil. The rhizospheric soil exhibits biological nitrogen fixation in leguminous plants; hence rhizosphere contains the higher level of nutrient availability and exchange of the cations and thus CEC is high in the rhizosphere. In this compartment the CEC was very high in the rhizospheric soil of the *Mitragyna perbifolia* (60.61CEC cmol (p^{+}) kg^{-1}) which was followed by the *Tamarindus indica* (58.67 cmol (p^{+}) kg^{-1}) and 58.20 cmol (p^{+}) kg^{-1} in the *Acacia catechu*.

Table 12.4: The Chemical Properties of the Barren and Rhizospheric Soils of Dominant Plant Species available from Compartment Number 792 of Gorewada forest

Sl.No.	*Sample Name*	*pH*	*Electrical Conductivity dS m^{-1}*	*Organic Carbon (%)*	*Cation Exchange capacity cmol (p+) kg^{-1}*
	Compartment number 792				
1.	Barren soil	7.8	0.3	1.78	53.14
	Rhizospheric soils of dominant plant species				
1.	*Ziziphus mauritiana* (Bor)	7.44	0.32	1.89	56.57
2.	*Acacia catechu* (Khair)	7.76	0.32	2.9	58.2
3.	*Bauhinia racemosa* (Apta)	7.67	0.22	2.21	56.66
4.	*Acacia nilotica* (Babhul)	7.12	0.29	2.22	55.34
5.	*Albizia odoratissima* (Shiras)	7.32	0.26	1.9	57.59
6.	*Tamarindus indica* (Imli/Chinch)	7.35	0.28	2.24	58.67
7.	*Mitragyna perbifolia* (Kalamb)	7.65	0.2	2.34	60.61

Total Nutrient Status

The status and release of nutrients in forest soil are highly linked with microenvironment, litter dynamics, altitude and type of vegetation (Mishra, 2011). The total nutrient status of the 792 compartment is given in the Table 12.5. The highest N (0.26 per cent), K (0.57 per cent) and S (0.06 7 per cent) content was found in the rhizospheric soil of *Acacia catechu*, because the *Acacia catechu* is the leguminous plant which harbors *Rhizobium* in its root nodules and its growth is preferentially stimulated in the rhizosphere of legume than in that of non legumes. Moreover, legumes excrete a large number of substances into the rhizosphere principally sugars, amino acids and vitamins such as biotin and pantothenic acid. Therefore the total N was high in the rhizosphere of this plant. The K and S content are high due to the rhizospheric effect of soil microorganisms residing in the root zone of *Acacia catechu*.

In rhizospheric soils of other plant species the total N ranged from 0.1 to 0.25, P: 0.045 to 0.068; K: 0.27 to 0.45 and S: 0.043 to 0.063 percent.

Table 12.5: Total Nutrient Status of the Barren and Rhizospheric Soils of Dominant Plant Species available from Compartment Number 792 of Gorewada Forest

Sl.No.	Sample Name	Total Nutrient Status			
		Total N (%)	Total P (%)	Total K (%)	Total S (%)
	Compartment number 792				
1.	Barren patch	0.22	0.045	0.35	0.05
	Rhizospheric soils of dominant plant species				
1.	*Ziziphus mauritiana* (Bor)	0.12	0.067	0.45	0.055
2.	*Acacia catechu* (Khair)	0.26	0.056	0.57	0.067
3.	*Bauhinia racemosa* (Apta)	0.19	0.068	0.42	0.056
4.	*Acacia nilotica* (Babhul)	0.12	0.045	0.3	0.056
5.	*Albizia odoratissima* (Shiras)	0.25	0.05	0.27	0.047
6.	*Tamarindus indica* (Imli/Chinch)	0.12	0.054	0.32	0.063
7.	*Mitragyna perbifolia* (Kalamb)	0.24	0.057	0.34	0.043

Available Nutrient Status

The available nutrient status of the compartment number 792 is given in the Table 12.6. The nitrogen content of this site was low as compared to the other compartment. In all the rhizospheric soil of the different plant species of this compartment the highest N content was observed in the *Ziziphus mauritiana* (0.022 per cent) followed by the *Mitragyna perbifolia* (0.021 per cent). The P_2O_5 and K_2O was

Table 12.6: The Available Nutrient Status of Barren and Rhizospheric Soils of Dominant Plant Species available from Compartment Number 792 of Gorewada Forest

Sl.No.	Sample Name	Available Nutrient Status		
		N (%)	P_2O_5 (%)	K_2O_5 (%)
	Compartment number 792			
1.	Barren soil	0.012	0.00111	0.0224
	Rhizospheric soils of dominant plant species			
1.	*Ziziphus mauritiana* (Bor)	0.022	0.00112	0.0345
2.	*Acacia catechu* (Khair)	0.016	0.00129	0.0435
3.	*Bauhinia racemosa* (Apta)	0.012	0.00063	0.0234
4.	*Acacia nilotica* (Babhul)	0.018	0.0011	0.0378
5.	*Albizia odoratissima* (Shiras)	0.016	0.00129	0.0443
6.	*Tamarindus indica* (Imli/ Chinch)	0.019	0.00122	0.0238
7.	*Mitragyna perbifolia* (Kalamb)	0.021	0.0012	0.0367

Table 12.7: The Microbiological Count of the Barren and Rhizospheric Soil of Dominant Plant Species Available in the Compartment Number 792 of Gorewada Forest

Sl.No.	*Name of Sample*	*Counts of Different Types of Useful Microorganisms*					
		Bacteria cfu g^{-1}	*Fungi* cfu g^{-1}	*Actinomycetes* cfu g^{-1}	*Rhizobium* cfu g^{-1}	*Azotobacter* cfu g^{-1}	*VAM Spores* cfu 10^{-1} g
	Compartment number 792						
1.	Barren patch	14×10^5	11×10^3	7×10^3	12×10^3	17×10^3	4
	Rhizospheric soils of dominant plant species						
1.	*Ziziphus mauritiana* (Bor)	49×10^5	34×10^3	40×10^3	36×10^3	43×10^3	9
2.	*Acacia catechu* (Khair)	53×10^5	42×10^3	36×10^3	46×10^3	34×10^3	10
3.	*Bauhinia racemosa* (Apta)	52×10^5	37×10^3	41×10^3	39×10^3	24×10^3	12
4.	*Acacia nilotica* (Babhul)	43×10^5	40×10^3	39×10^3	48×10^3	27×10^3	10
5.	*Albizia odoratissima* (Shiras)	41×10^5	32×10^3	34×10^3	29×10^3	26×10^3	9
6.	*Tamarindus indica* (Imli/Chinch)	32×10^5	41×10^3	27×10^3	26×10^3	21×10^3	10
7.	*Mitragyna perbifolia* (Kalamb)	40×10^5	35×10^3	23×10^3	29×10^3	20×10^3	3

content was high in the *Albizia odoratissima* 0.00129 and 0.00443 percent, respectively. The *Acacia catechu* also showed the high P_2O_5 and K_2O *viz.* 0.00129 and 0.0435 percent, respectively.

Microbiological Counts

Soil bacteria are an essential component of the biotic community in natural forests and they are largely responsible for ecosystem functioning, because they participate in most nutrient transformations. Moreover, soil micro flora plays a pivotal role in improving the physico-chemical and nutrient conditions of soil and in stimulating plant growth (Singh *et al.*, 1999). Microorganisms are beneficial in increasing the soil fertility and plant growth as they are involved in several biochemical transformation and mineralization activities in soils. Therefore, in the present study the microbiological analysis of soil samples collected from the rhizosphere of the different plant species available in compartment number 792 was carried out. Results presented in the Table 12.7 highlights that highest bacterial and fungal count was reported in the rhizospheric soil of *Acacia catechu* (53 X 10^5, 42 X 10^3 cfu g^{-1}respectively) due to high organic carbon and CEC of the soil. The actinomycetes were highest in the *Bauhinia racemosa* (41 X 10^3 cfu g^{-1}) followed *Ziziphus mauritiana*. The nitrogen fixing bacteria *i.e. Rhizobium* was high in the *Acacia nilotica* (48 X 10^3 cfu g^{-1}) closely followed by the *Acacia catechu* (46 X 10^3 cfu g^{-1}) being a leguminous plant. However, the non symbiotic nitrogen fixing bacteria *i.e. Azotobacter* was highest in the *Ziziphus mauritiana*. The count of VAM spores was high in the rhizosphere of *Bauhinia racemosa* (12 cfu 10^{-1} g). In the barren soil, the microbiological counts of useful microorganisms was very low owing to low total and available nutrient status of barren soil in terms of contents of carbon, N, P and K levels. It has been found that association occurs between vegetation and microbes, to release plant root exudates, macro and micro elements into the soil rhizospere to create a new environment (niche) for the growth of microorganisms (Stafford *et al.*, 2005). The connection between vegetation types and microbes develops in the forest soil ecosystem, where each type of microbe filled a special niche play a different role in the nutrient cycle. Microbes, which were potential as bio-fertilizers were often found in rhizosphere. The following microbes were identified from compartment No. 792 were namely, *Azospirillum, Rhizpbium, Azotobacter, Pseudomonas, Aspergillus* and *Streptomyces* which have potential to become bio-fertilizers or bio-controls. Rhizosphere is rich with biological activity because microbes feed on the carbon compounds exuded by root, while organic and inorganic materials released by the plants into the areas (in the form of exudates), will be useful for life continuity of soil microbes (Rosch *et al.*, 2002).

Conclusion

Development of richness in the vegetation wealth and biodiversity of soil is of prime importance in the forest cover as well as on the non - forest area to conserve the bio-resources and tree wealth. Many parts of country are facing problem of forest resources depletion, due to lack of proper physico-chemical and microbiological status of forests soils. Therefore, sustainable management of forest soil is required. This can be achieved by virtue of performing the proposed preliminary study of native plant species in and around Gorewada forest, characterization of rhizospheric soil of native plant species with respect to its physico-chemical and microbiological properties, isolation of site specific bio-fertilizer strains of *Rhizobium*, *Azotobacter* and Vesicular Arbuscular Mycorrhiza (VAM) and selection of a suitable organic amendment to improve the physico-chemical and microbiological properties of soil from Gorewada forest. Findings of this particular study will formed a basis for developing future plans required for development of forest cover with a diversed variety of plants of high economic value like timber, medicinal and ornamental including even the exotic and rare species by using the biotechnological approach of bioaugmentation. The bioaugmentation process can be used to resolve the low nutrient status and less counts of useful micro flora present in forest soil.

References

Almeida, S.M., 1990: *The Flora of Savantwadi.* Vol. I–II, *Scientific Publishers, Jodhpur - India.*

Black, C.A. 1965b: Methods of Soil Analysis: Part – II Am. Soc. Agron., Madison, Wisconsin, U.S.A.

Black, C.A., Evans, D.D., White, J.L., Ensminger, L.E. and Clark, F.E. 1965a: *Methods of Soil 9,* Part II, Medison, Wisconsin, USA.

Gerdmann, J.W. and Nicolson, T.H. 1963: Spores of mycorrhizal endogone species extracted from soil by wet sieving and decanting. *Trans. Br. Mycol. Soc.* 46: 235-244.

Hesse, P.R. 1971: A text book of soil chemical analysis. John Murray, (Pub) Ltd. London. 377.

Hooker, J.K. 1872-1897: *The Flora of British India.* Vol. I – VII, L Reeve and Co., London.

Jackson, M.L. 1967: Soil chemical analysis. Prentice Hall of India (Pvt.) New Delhi.

Joshi, S.J. 2000: Medicinal Plants. *Oxford and IBH Publishing Co., Pvt. Ltd.*, New Delhi.

MacLean, D.A. and Wein, R.W. 1978: Weight loss and nutrient changes in decomposing litter and forest floor material in New runswick forest stands. *Can. J. Bot.*, 56: 2730-2749.

Mishra, B.P. 2010: A study on the micro-environment, litter accumulation on forest floor and available nutrients in the soils of broad-leaved, mixed pine and pine forests at two distinct altitude in Meghalaya, northeast India. *Curr. Sci.*, 90(12): 1829-33.

Perala, D.A. and Alban, D.H. 1982: Biomass, nutrient distribution and litterfall in Populus, *Pinus* and *Picea* stands on different soils in Minnesola. *Plant Soil*, 64: 177-192.

Piper, C. S. 1965a: Soil and Plant Analysis (Indian Reprint). Hans Publishers, Bombay.

Piper, C.S. 1966: Soil and Plant Analysis (Indian Reprint). Hans Publishers, Bombay.

Rosch, C., Mergel, A. and Bothe, H. 2002: Biodiversity of denitrifying and dinitrogen-fixing bacteria in an acid forest soil. *Applied Environ. Microbiol.*, 68: 3818-3829.

Singh J.N., and Mudgal, V. 2000: Assessment of mineral content of tree leaf litter on Nokrek Biosphere Reserve and its impact on soil properties. *Trop. Ecol.*, 41: 225-232.

Singh, K., Borana, J. and Srivastava, S. 1999: Effect of thiram on root growth, root nodules and nitrogen fixation in Glycine max (L.) merril by *Bradyrhizobium japonicum. Journal of Soil Biology and Ecology*, 19: 11-14.

Singh, N.P., L. Narasimhan, S. Karthikeyen and P.V. Prasanna, 2001. *Flora of Maharashtra. BSI Publication, Kolkata.*

Stafford, W.H.L., Baker, G.C., Brown, S.A., Burton, S.G. and Cowan, D.A. 2005. Bacterial diversity in the rhizosphere of poteaceae species. *Environ. Microbiol.*, 7: 1755-1768.

Subbaih, B.V. and Asija, G.L. 1956: A rapid procedure for estimation of available nitrogen in soil. *Curr. Sci.*, 25 – 259-260.

Tondon, H.L.S. Ed. 1993: Methods of Analysis of Soil, Plants, Water and Fertilizers, FDCO, New Dehli, India.

Ugemuge, N.R. 1986: *Flora of Nagpur District. Shree Prakashan, Nagpur,* India.

2013, Climate Change Effects on Agriculture and Economy *Pages* ***103–111***
Editors: **D.R. Khanna, A.K. Chopra, Vikas Singh, Rakesh Bhutiani & Gagan Matta**
Published by: **BIOTECH BOOKS, NEW DELHI**

Chapter 13

Principle of Prevention and Management of Pollution Related Health Hazards with Special Reference to Pollution of Metal Industry

Deepa Sethi Singh
Department of Zoology, Hindu College, Moradabad, M.J.P. Rohilkhand University, Bareilly, U.P.

There are variety of cell systems with specific characteristics, structure and functions. The metallic composition of various types of cell systems is different *i.e.* relative concentration level of a metal in different types of cell system is also different. Various types of cell systems mutually communicate through 'single ionic' channel of different types of metals present in them. When normal concentration level of a metal in a cell system is reduced to deficient state its communication with adjoining cell system become abnormal. It is either partially or totally block.

A cellular metal bears different levels of its concentration in different types of cell system that actively participate in normal functioning of specific types of mechanism that operate for particular purpose in the body *e.g.* thermo- regulatory, excretory, sensory cell renewal, healing *etc.* With the significant change in the level of a cellular metal in the body, the operation level of related mechanism is adversely affected *i.e.* at different levels of deficiency of a cellular metal; the nature of adversity affected mechanism will be different. The abnormal functioning of a mechanism tends to increase the rate of production of free redials in the body. With the change in normal level of body fluid pH, the rate of occurrence of certain biochemical reactions changes that modifies the profile of physiological environment and adversely affects biochemical activities of specific types of enzymes/mettallo enzymes. For restoring normalcy in level of body fluid pH, certain amount of specific type of cellular metal depletes.

Keywords: *Pollution, Metals, Health hazards, Managements.*

Introduction

The proper operation of a cell depends upon specific range of temperature, pH and its communication level with adjoining cell systems that maintain its normal structure and functions. A significant change in the working limits of them tends to develop abnormality in structure and functions of the cell. A cell has several components each of these components has a specific role in creating life through biochemical synthesis, using particular type of cellular metal/metals in specific level of its concentration.

An infected cell tends to change its loyalty from host to parasite. The level of change depends upon the nature of localized infection. This change is indicated in the form of abnormality developed in the functional character of specific type of regulatory mechanism such as thermo- regulatory, cell-renewal, pigmentory *etc. i.e.* temperature of an infected cell system is either raised or lowered, rate of cell renewal becomes very slow or stops, pigments formation/distribution decreases. These facts are observed in different types of diseases *e.g.* leprosy is associated with lowering of temperature as well as degraded level of cell renewal mechanism of infected cells while leucoderma is connected with low level of formation/distribution of pigment and cancer is responsible for uncontrolled growth of tissue cells. The abnormality developed in a cell, adversely affect synthesis, character of the cell *i.e.* biochemicals synthesized by an abnormal cell systems are either poor in quality or quantity *e.g.* insulin secretion from pancreas in diabetic persons.

There are variety of cell systems with specific characteristics, structure and functions. The metallic composition of various types of cell systems is different *i.e.* relative concentration level of a metal in different types of cell system is also different.

Various types of cell systems mutually communicate through 'single ionic' channel of different types of metals present in them. When normal concentration level of a metal in a cell system is reduced to deficient state, its communication with adjoining cell system become abnormal. It is either partially or totally blocked.

Due to different types of metallic composition in various types of cell system, their pH level are likely to be different *i.e.* normal requisite level of pH in different types of cell system is also different and with the change in normal requisite concentration level of any of the metals present in a cell, its pH level is likely to be changed. The altered pH level in cells not only adversely affect normal structure and functions but also provide favorable ground for the invasion of infection of specific nature.

Cellular metals play a pivotal key role in the development of personality (sum of physical, physiological, emotional, spiritual aspects) of animal/human. The personality enhances confidence decision taking ability. Power to face adverse situation. The personality affects longevity of life and is regulated by one's life style *e.g.* heart disease and cancer both are related to unhealthy life style.

Cellular metals undergo metabolism either through liberation or consumption of specific amount of energy and are accordingly referred as hyper caloric or hypo caloric metals. These metals are present in specific ratio of their concentration levels and regulate energy status of the body. The energy produced by hyper caloric metals during their metabolism meet various requirements of the body. When relative concentration levels of hypo and hyper caloric metals are changed, it tends to change (a) energy status of body (b) level of metabolism of hypo caloric metals (c) level of energy stores in the body. This cause degradation in normal level of nutrition, metabolism, enzyme functions, immune response in specific type of cell systems and renders them prone to the attack of infections and develops abnormality of specific nature in their structure and functions related to the nature of metals (hypo or hyper caloric).

The deficiency state of a cellular metal is associated with the (a) development of abnormality in functional level of specific type of mechanism such as respiratory, sleep, digestive, excretory *etc.* (b) change in normal level of temperature, pressure, pH, electro- kinetic-potential *etc.* of cellular fluid (c) invasion of disease causing infection that are related to the nature and level of deficiency of cellular metal (d) change in level of various physiological functions of the cellular metal in animal/human body.

The environmental infection consistently attack but fail to exercise their injurious effects due to high level of nutritional status of cell system *i.e.* an infection invades nutritionally deficient cell system only during the course of onset of a health problem.

The localized infection in a cell tends to break immune system of the cell besides simultaneously organizing an immune system for itself to evade the curative effects of drugs to be used for its management. It also tends to (a) Change loyalty of cell from host to parasite by developing partial or total obstruction in specific types of 'single-ionic' channels.

A healthy cell is immune to the invasion of environmental infections, environmental infection invade only nutritionally deficient cell. The deficient state of a metal in a cell renders it prone to the attack of disease causing infections. The nature of infections is related to the nature and level of deficiency of the cellular metal *i.e.* with different nature of the metals the nature of invading infection changes and with different levels of deficiency of a metal the nature of invading infection is also changes during the course of onset of different types of health problems or different levels of deficiencies of a cellular metal are related to different types of diseases.

The invading infection tends to exercise following effects:

1. Brings structural and functional changes related to its nature.
2. Breaks immune system of the cell and organize immune system for itself.

3. Interrupts communication with adjoining cell systems through development of obstructions (partial/total) in specific type of 'single-ionic' channels.
4. Changes loyalty of cell from host to parasite *i.e.* develops isolation.
5. Further decreases the concentration level of deficient cellular metal.
6. Reduce the level of immune response of the cell.
7. As set of various types of cell systems participate in the operation of different types of mechanisms, the infected cell adversely affects operation level of related mechanisms that alarms signals of specific sings and symptoms of particular diseases.
8. Changes pH level of cellular fluid to certain extent so that infection may survive.

The developed hyper acidity leads to the onset of health problem in animal/human body that is related to the nature of cell system involved. The nature of cell system involved depends upon the chemical nature of acid (in hyper state) that is responsible for developed hyper acidity. Variety of acids (organic and inorganic) is present in body fluid in specific concentration levels. The hyper state of any of them can develop hyper acidity. Thus chemical nature of acid, which develops hyper acidity, bears a close correlation with the nature of cell system affected and nature of cellular metal depleted. The abnormal pH level of body fluid, abnormal concentration level of a cellular metal, in a cell, renders it nutritionally deficient and prone to the attack of infections. The nutritional level, not only changes with the changed in nature of cellular metal in deficient state but also changes with different levels of deficiency of a metal. The abnormality developed in a cell system due to hyper acidity may be due to (a) deficiency of different types of cellular metals or (b) different levels of deficiency of a cellular metal in the body *i.e.* deficiency of different types of cellular metals affects different types of cell system and different levels of deficiency of a cellular metal also affects different types of cell systems.

The deficiency of a macro cellular metal tends to reduce the absorption rate of micro/trace cellular metals and enhances the accumulation rate of heavy metals of specific nature besides increasing the absorption rate of its antagonistic metals.

The therapeutic effects of cellular metals are governed by certain notions:

1. Deficiency of a macro cellular metal not only degrades normal level of nutrition, metabolism, enzyme functions, immune response related to itself but also fail to counter balance adverse effects of its antagonistic cellular metals and reduce absorption rate of specific type of micro/trace metals. This reduces immune response of body.
2. For restoration of normalcy in disease related deficient trace metal, it is essential that concentration levels of macro cellular metals are within normal limits in the body *i.e.* concentration of trace metals cannot be raised unless macro metals are within normal limits.
3. Deficiency of a macro metal leads to the development of acidity while deficiency of a trace metal adds strength to acidity and cause hyper acidity.

The potential effects of hyper acidity include (a) hair fall (b) loss of appetite (c) disturbed sleep (d) nausea. These effects occur in human body at different levels of hyper acidity. The effects occur in different types of cell systems due to hyper level of acids of different chemical nature that deplete cellular metal/metals of specific nature in them and render them prone to the attack of particular type of infection.

Methodology

The patients suffering from a diagnosed health problem in the initial stage belonging to different age groups, sex and food habits were enrolled in the study.

Biopsy samples (hair/blood) among enrolled persons were collected on random basis and was marked with specific sign of identification and was sent for elemental analysis for assessing the status of different types of cellular metals. The elemental analysis was conducted by using atomic-absorption spectra-photo-meter technique. The analytical data so obtained was compared with the standard values for a healthy person of the same age group, sex and food habits. The nature of imbalanced (deficient) cellular metal thus identified.

The suitable salt of (therapeutic importance) of disease related cellular metal/ metals were prepared by vapour phase reaction between oxide of cellular metal and ammonium-chloride in a sealed container. The product obtained, after vitro-vivo study was used in the clinical management of health problems related to its deficiency.

The status of imbalanced (deficient) cellular metal was determined in biopsy material on random basis, before instituting, during the course of and on termination of the treatment. The progressive improvement towards normalcy in the status of the cellular metal at conclusion of the treatment revealed that signs and symptoms of the disease are gradually disappeared.

The blood samples collected on random basis before instituting, during the progress and on conclusions of the treatment were sent for virological examination in a pathology lab to identify the nature of disease causing infection and their intensity/ load.

The workers working in micro environment, in different sections of metal industry and suffering from a particular disease were enrolled in the study. The blood samples collected on random basis were subjected for determination of pH level using pH meter. The values obtained were tabulated against the nature of disease along with other particulars such as age, sex, duration of service, section of work.

The identification of exact chemical nature of the acid in hyper state in body fluid of a worker, related to a health problem is difficult because no suitable technique is developed for it so far. It is assessed in terms of, the nature of cellular metal and its level of deficiency developed under the influence of industrial effluents.

Conclusion

With the change in environmental profile the normal level of body fluid pH undergoes a significant change due to change in normal concentration level of specific types of bio-metal/metals and renders the body susceptible to disease causing infections of particular character.

A close co-relation exists between the level of environmental profile, body fluid pH, nature of disease causing infections and nature of health problem onset.

At particular level of body fluid pH, different types of diseases occur due to (a) different nature of acid in hyper state (b) invasion of disease causing infections of different nature, which are related to the deficiency of specific types of cellular metals at particular level of its deficiencies in animal/human body.

Discussion

Human/animal health depends upon the supply rate of nutrients particularly those nutrients that play a key role in the process of growth development and maintenance of body. Any discrepancy in requisite ratio of concentration of metals in an antagonistic pair leads to the onset of a health problem due to development of deficiency of a metal.

The developed deficiency of a metal in human/animal body, alters normal level of body fluid pH by developing hyper acidity that adversely affect structure and functional of particular cell system.

Animal/human body has strong tendency to refuse foreign material including infection and synthetic medicines. A foreign substance when enters the body, it does so with degradation in normal levels of specific characteristics of body that are essential for health keeping process. The cellular metals in their requisite adequate concentration levels protect body from the attack of environmental infections.

A significant change in normal level of cellular fluid pH in a cell system is an essential condition for onset of a health problem. The persistent change in cellular fluid pH is associated with the alteration in normal concentration level of cellular metal of specific nature in particular type of antagonistic pair that adversely affects characteristics of the cell. The abnormal pH level related change in concentration of cellular metal/metals in cellular fluid of a cell system may lead to the onset of different types of diseases depending upon the nature of cellular metal and its various levels of deficiencies.

The level of cellular fluid pH changes due to change in concentration levels of acids of different chemical nature present in the cellular fluid. The hyper level of any of them can develop hyper acidity in particular type of cell system related to its chemical nature.

The acids of different chemical nature in the hyper level deplete different types of cellular metals of specific character from the cellular fluid. The amount of depleted cellular metal depends upon its hyper level *i.e.* with different hyper levels of an acid; the depleted amounts of a cellular metal are different.

A cellular metal bears different levels of its concentration in different types of cell system that actively participate in normal functioning of specific types of mechanism that operate for particular purpose in the body *e.g.* thermo- regulatory, excretory, sensory cell renewal, healing *etc.* With the significant change in the level of a cellular metal in the body, the operation level of related mechanism is adversely affected *i.e.* at different levels of deficiency of a cellular metal; the nature of adversity affected mechanism will be different. The abnormal functioning of a mechanism tends to increase the rate of production of free redials in the body. With the change in normal level of body fluid pH, the rate of occurrence of certain biochemical reactions changes that modifies the profile of physiological environment and adversely affects biochemical activities of specific types of enzymes/mettallo enzymes. For restoring normalcy in level of body fluid pH, certain amount of specific type of cellular metal depletes.

The clinical potential of a cellular metal depends upon chemical structure of its salt *i.e.* metallic salt with different chemical constitutions have clinical potential of different orders. The clinical efficacy of a salt depends upon its level of bio-availability factor that varies considerably with the change in its chemical structure *i.e.* salts of different chemical nature; consequently differ in the clinical effects due to different levels of their bio-availability.

The metallic salts that contain nitrogen have relatively higher level of clinical potential and has large number of diseases in its clinical spectrum than other salts of metals without nitrogen. As health problems onset in the body under the influence of various types of factors and are associated with the change in normal level of (a) concentration of cellular metal/metals (b) body fluid/cellular fluid pH (c) 'Nitrogen balance', that are inter-related. Nitrogen salts of a metal on the one hand improve concentration level of the cellular metal and on the other hand tend to restore normalcy in the disturbed level of 'Nitrogen balance'.

The detailed study on bio- chemical aspects of cellular metals revealed that metal are safe and effective in the control of growth, development and maintenance of human/animal health and can be used in the clinical management of diseases that are associated with their deficiencies in the body.

The environmental infection becomes active only when conditions are favorable to it with particular reference to temperature, humidity *i.e.* for onset of a health problem, infection requires particular range of temperature. The temperature range for activation of infection of different nature is different. This variation in temperature for activation of different infections play a key role in the onset of seasonal health problems *i.e.* specific type of infection became active in particular season so long as environmental temperature is above or below their activation range for an infection, the infection remain passive and do not cause infectious disease.

The temperature in microenvironment of an industry is generally higher due to the presence of effluents of various chemical natures and favors the activation of specific type of infection for onset of particular type of health hazards.

During the course of onset of health problem, infection invades nutritionally deficient cell system only in the body. The nature and level of nutritional deficiency in the cell bears a close co-relation with the nature of infection *i.e.* different levels of nutritional deficiency in a cell are sensitive to infection of different nature and nutritional deficiency of different types in a cell, are also sensitive to different types of infection during onset of different types of health problems. To minimize the effect of pollution on workers, the knowledge of chemical nature of effluents and nature of imbalanced (deficient) cellular metal in animal/human body is essential. The harmful effects of pollution can be prevented either by dietary modification or through regular intake of suitable salt of the pollution affected cellular metal. The dietary modification aims at restore normalcy in the concentration level of hypo caloric and hyper caloric cellular metals by increasing the amount of dietary items that contain relatively higher concentration of pollution related cellular metal/metals in highly bio-available form and decreasing the amount of dietary items that contain rich amount of its antagonistic cellular metal in daily diet.

The dietary modification or regular intake of salt of pollution related cellular metals/metals would keep the profile of physiological environment (through maintaining level of body fluid pH) in the target cell system at such a level that localized infection goes to passive state or is eliminated from the cell and body will remain within healthy working limits. Biopsy samples were sent for virological examination in a pathology lab to identify the nature of disease causing infection and their intensity/load.

The identification of exact chemical nature of the acid in hyper state in body fluid of a worker related to a health problem is difficult, because no suitable technique is developed for so far. It is assessed in terms of the nature of cellular metal and its level of deficiency developed under the influence of industrial effluents.

References

Bourai, A.A.: 2011: 'Health and environmental effect of radium' abstract National conference on emerging trends in chemistry- biology interface, Nainital, November 03-05, 2011 page 52.

Gupta, S. 2011: '*Invitro* studies of effects of various heavy metals on haematological parameters' abstract National conference on emerging trends in chemistry-Biology interface, Nainital, November 03-05, page 73.

Sethi, D. 2010: 'Occupational health problem causing effects of industrial effluents on body fluid pH' abstract National seminar on management of environmental pollution for sustainable energy production, Bareilly, 92.

Sharma, D.K. 2011: 'Analysis of trace elements in spinach (*Spinacea oleracea*) and its biological and environmental importance' abstract National conference on emerging trends in chemistry- biology interface, Nainital, November 03-05, 2011 page 80.

Singh D.S. 2011: 'Role of cellular matter in prevention and control of animal health problems with special reference to liver disorder (jaundice) abstract National conference on emerging trends in chemistry- biology interface, Nainital, November 03-05, page 31.

Singh, D.S- 'Clinical potential of mollusca shell with special reference to the management of pollution related health hazards' abstract National seminar on recent advancement in life sciences: potential benefits and Ricks, Meerut 05th - 06th 2011.

Singh, D.S. 2011: 'Antibiotic' behaviour of cellular metals during management of infectious 'diseases in human' abstract International conference on new horizons in microbial bio technology and pharmaceutical sciences Pounta Sahib (India) April 24th -25th, page 80.

Singh, D.S. 2011: 'Health hazards related to industrial waste treatment (E-waste) for extraction of noble metals', abstract national seminar on New innovation in parasitological research, Dehradun, 21-30th March, Abs.no.63.

Singh, D.S. 2011: 'Prevention of adverse effect of industrial pollution on human health through maintenance of body fluid pH within normal limits' abstract national conference on biodiversity *vis-à-vis* environmental degradation in hilly terrains, Dehradun, 15-16th May, page 66.

Singh, D.S. 2011: 'Role of heavy metals in biological system be redefined with particular reference to their clinical potential'; abstract National conference on emerging trends in chemistry- biology interface, Nainital, November 03-05, page 32.

Vidushi, 2011: 'Investigations of chromium, lead, nickel, zinc, cobalt, copper and cadmium in hair samples of auto drivers and bricks furnace workers by atomic absorption spectra photo meter technique' abstract National conference on emerging trends in chemistry- biology interface, Nainital, November 03-05, 2011 page 47.

2013, Climate Change Effects on Agriculture and Economy *Pages* **113–134**
Editors: **D.R. Khanna, A.K. Chopra, Vikas Singh, Rakesh Bhutiani & Gagan Matta**
Published by: **BIOTECH BOOKS, NEW DELHI**

Chapter 14

Status of Drinking Water Quality in Raniganj Coalfield Area of West Bengal in India

Alokkumar De[1], Chinmoy Chatterjee[1], Krishna Bardhan Ghosh[1], Saumendra Sankar Desarkar[1], Narayan Ray[1], Bidhan Chandra Mahatha[1], Manali Sengupta[1], Ansuman Ray[1], Suvra Dey[1], Avay Kumar Singh[2], Srijit Bhattacharya[3], Sudhee Ranjan Banerjee[4], Supriya Mukhopadhyay[4], Surajit Pal[4] and Deepak Pandit[4]

[1]*Raniganj Girls' College, Raniganj, Burdwan, West Bengal*
[2]*Central Institute of Mining and Fuel Research, Dhanbad, Jharkhand*
[3]*Barasat Govt. College, Barasat, N-24 Pgs, Kolkata, West Bengal*
[4]*Variable Energy Cyclotron Centre,*
1/AF Bidhan Nagar, Kolkata, West Bengal

Frequent occurrence of water borne diseases severely affected the people in Raniganj Coalfield area of West Bengal in the recent past which point towards remarkable deterioration of drinking water quality in the area. Discharge of mining effluents, domestic sewage and industrial wastes into the water bodies and their seepage into the subsurface sources of water seem to be the prime causes of this deterioration. With an aim to assess the present status of drinking water quality of Raniganj Coalfield area, physico-chemical as well as bacteriological parameters were measured and examined rather extensively, involving about 60 drinking water sources in the area. The sources include exposed dug wells, bore wells and municipality supplies. Attention was also paid on the drinking water from the storage-tanks of the consumers like institutions, health centers, hotels and sweet shops where mass people use to consume water almost every day. Parameters measured were pH, dissolved oxygen (DO), free carbon dioxide (Free CO_2), total hardness (TH), Chloride (Cl), total dissolved solids (TDS), sulphate (SO_4^{3-}), nitrate (NO_3^-), electrical conductivity (EC), total coliform bacteria (MPN) and *E. coli* bacteria. Correlation coefficients and water quality indices were computed for the sources concerned. Presence of strikingly high amount of TDS (~2100 mg/l) and MPN (~2400 dl^{-1}) was observed for a large number of sources studied. This may be attributed to the cause of sufferings of the people of the area from water borne diseases. Preventive and cautionary measures have been and are being adopted involving several administrative and other bodies concerned. Measurements have also been carried out and results discussed in detail considering about 20 drinking water sources from non-industrial zones to extract the sole characteristics of the industrial one.

Keywords: *Drinking water quality, Physico-chemical, Bacteriological parameters, Correlation coefficients, Water quality indices, Water borne diseases.*

Introduction

The nature, its essence and beauty is completely linked with its proper realization by the best creature of the Universe – the human being. Survival of mankind in nature strictly depends on the consumption of safe drinking water. The very proverb: "The Water is Life", speaks itself the very importance of maintaining and monitoring adequate standards in drinking water quality.

Among several factors, lack of proper waste management as well as lack of awareness may lead to severe sufferings of the mankind from water borne diseases like diarrhoea, dysentery, cholera, jaundice and typhoid. These water borne diseases claim nearly 15 lakh lives annually in India (Rahman, 2003). Though 70 percent of the ailments emanates from water, 1/3rd of the total beds in hospitals are occupied by the people suffering from water related diseases (Ghazali, 1992). According to Richariya and Mishra (1998), in India it is estimated that two million deaths occur every year due to water borne diseases and more than fifty million people are partially incapacitated annually due to these ailments. Frequent occurrence of water borne diseases severely affected the people in Raniganj Coalfield area of West Bengal in the recent past which point towards remarkable deterioration of drinking water quality in the area. The present paper aims at assessing the present status of drinking water quality of this area.

Review of Literature

Investigations on water pollution involving coalfield areas in different parts of our country have been and are being carried out since long ago on different water quality parameters: physico-chemical as well as microbial ones. Richariya and Mishra (1998) carried out bacteriological studies of groundwater in Rewa district of MP and observed bacteriological contamination of groundwater caused due to seepage of polluted surface water from industries, mining and domestic wastes. Selenium contamination in groundwater as well as in mine discharge much above ISI limits has been recorded by Singh (2000) in Rawanwara colliery area. High concentration of selenium has been reported to exert adverse impacts on CNS, liver, heart and lungs of the consumers. Singh *et al.* (2007) studied aquatic geochemistry of industrial and coal mining area of Dhanbad district of Jharkhand. The studies indicate that mine waters and waters of mining areas have high concentration of SO_4^{3-}, chloride and TDS. The values of TDS and EC in some surface and subsurface water samples exceeded the WHO limits for drinking water. So far as the Raniganj (23° 36′ N and 87° 07′ E) coalfield area is concerned, several studies have been conducted involving the drinking water sources there. Studies conducted by Ghosh and Kundu (1999) in some selected sampling sites of eastern coalfield area in Burdwan district reveals arsenic contamination in some groundwater sources. Basu and Mitra (2002) reported the presence of arsenic, lead and cadmium in Damodar River receiving effluents from collieries and thermal power station in Asansol-Raniganj area. Studies by Khosal (2007) pointed towards overall degradation of water quality due to the extensive mining operation in the area. A study on drinking water quality in Raniganj town conducted by Chatterjee and De (2008) reveals that quality of groundwater in some pockets of this town is not at all safe for drinking. Very recently, *Singh et al.* (2010) analyzed 77 mine water samples from the Raniganj coalfield to assess water quality and suitability for domestic, industrial and irrigation uses. So far as the domestic use is concerned, their observation reveals that much of the mine water, especially of the Barakar formation area contains high TDS, total hardness and SO_4^{3-} concentration. Incidences of severe sufferings from diarrhoea in different parts of the Raniganj coalfield area affecting about 3000 people and thereby causing deaths of about 11 people (including child) among them have been reported in the recent past (Anandabazar Patrika, 2008, 2007). A close scrutiny of the existing literature strongly demands a thorough investigation on water quality involving a wide range of physico-chemical as well as bacteriological parameters in surface water, groundwater and mine water encompassing the vast rural and urban areas of the Raniganj coal belt. Special attention on the ultimate storage of drinking water is also called for to pin point the origin of the contamination involved.

Materials and Method

The present work consists of four different sets of drinking water sources involving about 80 sources as a whole. Water samples consisting set-1 were collected from ten dug wells at and around Baktarnagar in the Raniganj coalfield area during pre-monsoon period. Physico-chemical and bacteriological parameters of the sources measured are pH, free carbon dioxide (Free CO_2), dissolved oxygen (DO), total hardness (TH), Chloride (Cl^-), total dissolved solids (TDS), sulphate (SO_4^{3-}), nitrate (NO_3^-), electrical conductivity (EC), total coliform bacteria (MPN) and *E. coli* bacteria. Set-2 corresponds to studies on TDS only involving 10 different drinking water sources at and around Searsole of the Raniganj coalfield area. Keeping into background that microbial contamination alone may lead to a lot of deleterious effects on human health, like fever, typhoid, dysentery, diarrhoea and cholera, microbial examination was carried out involving about 40 drinking water sources (set-3, say) in the Raniganj Municipality. Sites like Institutes, Health Centers, Hotels and Sweet Shops were chosen where mass people use to consume drinking water almost every day. Emphasis was given on the contributions of consumers' storage tanks with respect to those of the immediate supplies from the Municipality. With an aim to investigate on whether the results obtained with set-3 are the sole characteristics of the (Raniganj) coalfield area, we measured the microbial parameters in drinking water collected from Burdwan town, a non-coalfield area, which is about 80 km away from Raniganj. The sources (set-4, say) there are again Institutes, Health Center, Hotels and Sweet Shops as were also studied in Raniganj coalfield area. The experiments were carried out at the Environmental Research Laboratory, Raniganj Girls' College by following the standard methods prescribed by APHA (1980). The particular methods followed for the detection of water quality parameters are presented in Table 14.1.

Table 14.1: Methods Followed/Equipment Used for Measuring the Water Quality Parameters

Measured Parameter	*Method Followed/Equipment Used*
pH	Measured at the sampling places using digital p^H meter.
Free CO_2	Titrimetric method using 0.05N NaOH solution and phenolphthalein indicator.
DO	Azide modification of Winkler's method.
TH	EDTA titrimetric method.
Cl^-	Titrimetric method using 0.02 (N) $AgNo_3$ solution and 5 per cent K_2CrO_4 indicator.
TDS	Evaporating the filtered sample and taking the weight of residue after evaporation.
SO_4^{3-}	Turbidimetric method using spectrophotometer.
NO_3^-	Phenoldisulphonic acid method.
MPN	Multiple tube fermentation technique using McConkey's Broth and incubating at 37.0°C for 48 hours.
E. coli	Streaking EMB agar plates from positive tubes (resulting from MPN test) with a sterile inoculating needle and incubating the inoculated plates at 35°C for 24 hours.
EC	Conductivity meter.

Statistical Analysis

Pearson's coefficients of correlation were computed between possible pairs of water quality parameters by using the formula:

$$r_{xy} = \frac{\sum(x-\bar{x})(y-\bar{y})}{\sqrt{\sum(x-\bar{x})^2 \sum(y-\bar{y})^2}} \qquad (1)$$

where,

x and y represent two variants and $\bar{x}$ and $\bar{y}$, the corresponding means.

Overall quality of water for each of the sources represented in Table-II was determined by computing Water Quality Index (WQI) following the weighted arithmetic index method (Brown *et al.*, 1972). The formula used is:

$$WQI = \frac{\sum_{n=1}^{N} q_n W_n}{\sum_{n=1}^{N} W_n} \qquad (2)$$

where,

W_n and q_n are the unit weight and sub index, respectively of the nth parameter and N, the total no of parameters.

Results and Discussion

The experimentally measured parameters of the drinking water sources specified by set-1 are displayed in Table 14.2, along with the corresponding standard permissible values (WHO, 1993; Kavitha, 2010). The study reveals the following:

1. Except TDS and MPN, the values of all other measured parameters are either well within or nearly equal to their corresponding standard permissible values.
2. Most of the sources are of slightly alkaline character as are evident from their measured pH values.

Table 14.2: Observed Values of Water Quality Parameters of different Drinking Water Sources (Set-1) at and around Baktarnagar in Raniganj Coalfield Area of West Bengal in India

Source	Parameters										
	pH	*FCO_2*	*DO*	*TH*	*Cl*	*TDS*	*SO_4^{3-}*	*NO_3^-*	*MPN*	*EC*	*E. coli*
S01	7.3	2.3	3.6	152	262.7	340	23.32	7.98	1600	1058.4	A
S02	7.0	2.1	2.8	236	266.2	560	37.55	20.37	920	1156.4	P
S03	7.0	3.1	3.4	244	131.3	1115	55.78	18.77	240	1205.4	A
S04	7.2	2.2	2.2	122	106.5	560	34.89	13.44	79	1146.6	A
S05	7.1	3.4	3.1	348	145.0	950	56.45	2.15	1600	1362.2	A
S06	7.1	2.6	3.4	325	149.1	1060	40.61	20.89	33	1264.2	P
S07	7.1	1.8	4.3	294	142.0	845	43.80	17.12	350	1323.0	A
S08	7.2	2.7	3.6	280	142.0	2215	43.80	18.51	e"2400	1352.4	A
S09	7.2	3.2	2.5	224	117.1	1050	32.90	17.99	1600	1156.4	A
S10	7.1	4.2	4.2	182	92.3	230	21.05	17.41	e"2400	1058.4	A
Standard permissible value	8.5	–	5.0	300	250	500	250	20.0	00	1400	0

FCO_2 to NO_3 in the Table are expressed in mg/l, MPN, in dl^{-1} and EC, in ìmho/cm. P: Present; A: Absent

3. Chloride (Cl^-) values for two sources (S01 and S02) are found to be significantly high. This point towards contamination by sewage and man

and animal excreta (WHO, 1978). Higher concentration of Cl may cause unpleasant test to water and the people who are not accustomed to high chloride can be subjected to laxative effects. Cardiovascular problems may also arise due to high Cl content (Guru Prasad, 2003)

4. NO_3^- for most of the sources is found to be high. This indicates probable involvement of nitrogenous fertilizers, organic manures, human and animal wastes as well as industrial effluents through the biochemical activities of microorganisms. Drinking water with high NO_3^- may lead to 'Infant Cyanosis' or Methemoglobinemia (Blue Baby Syndrome) if it is ingested by infants less than six months old. Elevated NO_3^- in groundwater may also be a human teratogen leading to prenatal death due to congenial malformations (Scragg *et al.*, 1982).
5. Measured TDS values are excessively high for most of the sources studied; some of them exceeding double or even four times the standard permissible values. Measurements on TDS in another set of ten drinking water sources (set-2) at Searsole in the Raniganj coalfield area show similar results (Table 14.3). TDS values for all the twenty sources are plotted in Figure 14.1. As could be seen, the TDS values for almost all the sources exceed very largely the corresponding highest desired limit set by the WHO. According to Chandra *et al.* (2000) pollution has direct relation with the dissolved solids. Higher TDS values could be due to low water level within the aquifer and sediment effect. Dissolution of soil particles containing minerals under slightly alkaline condition favours the TDS concentration in groundwater. Water containing high TDS concentration may cause laxative or constipation effects (Kumaraswami, 1991). High TDS could also lead to formation of kidney stone and heart diseases.
6. All the sources studied are found to be contaminated with total coliform bacteria. Excessively high values of MPN are observed for most of the sources studied, including the presence of *E. coli* in two among them. Higher values of MPN indicate higher degrees of defecation, leakage from septic tanks and contamination by animal wastes. Observed very high values of MPN may be attributed to frequent sufferings of the people of the area from diarrhea and such other diseases.

Table 14.3: Observed Values of TDS of different Drinking Water Sources (Set-2) at and around Searsole in Raniganj Coalfield Area

Source	*S11*	*S12*	*S13*	*S14*	*S15*	*S16*	*S17*	*S18*	*S19*	*S20*
TDS (mg/l)	2238	1570	522	1814	1722	2004	1638	1830	636	2937

Correlation coefficients, computed by using formula (*1*), are displayed in Table 14.4. As could be discerned from the Table, as well as from a glance on the Figures 14.2(a,b,c and d), TDS (Figure 14.2a), TH (Figure 14.2b), SO_4 (Figure 14.2c) and EC (Figure 14.2d) are highly correlated with EC; EC and SO_4^{3-} EC and TH and TH, SO_4^{3-} and TDS, respectively. These seem to be reasonable because an increase or decrease

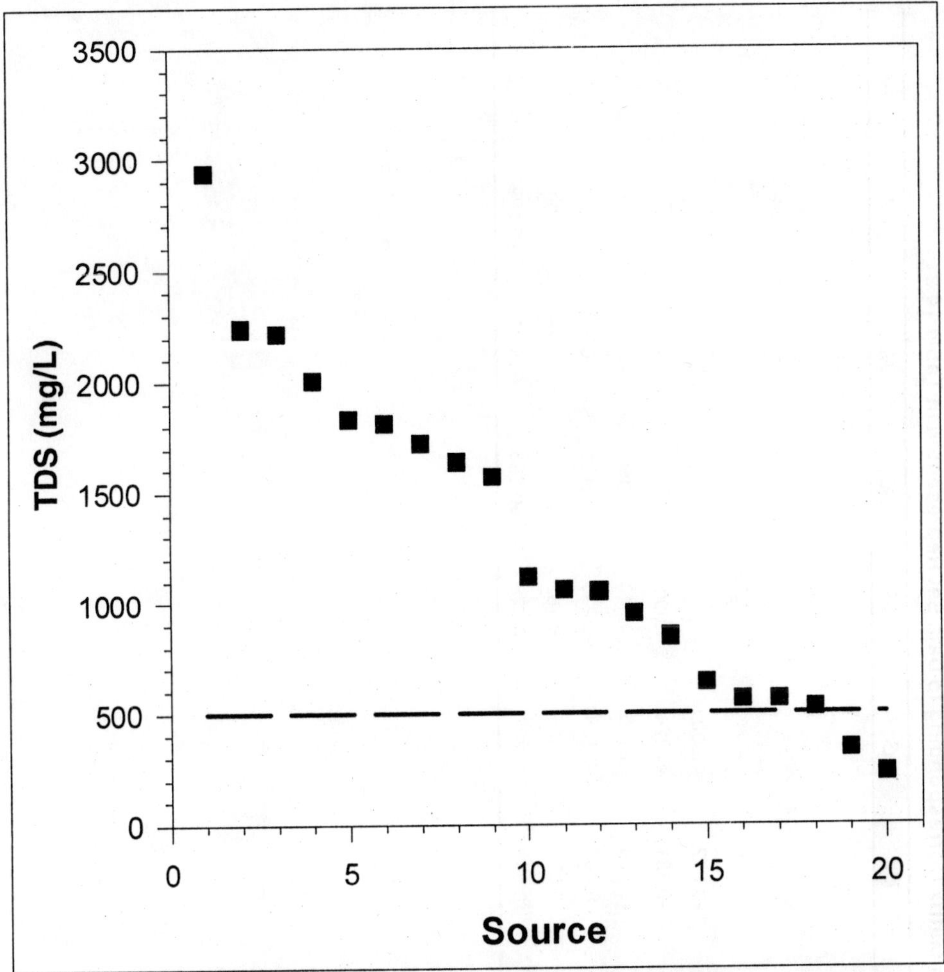

Figure 14.1: Measured TDS for Twenty Drinking Water Sources Studied in Raniganj Coalfield Area of W.B. in India. The sources have been plotted as per their TDS values in decreasing order. The dashed line indicates the highest desired limit for the parameter as per WHO.

in conductivity of particular water source usually results in a similar increase or decrease in other water quality parameters such as TDS, SO_4^{3-} and metals. Higher EC corresponds to presence of higher amount of ionic substances, indicating polluted quality of the water. Strong positive correlation of TH with SO_4 indicates that hardness of water is mainly due to sulphates. Frequent occurrence of diarrhea among the consumers might be due to the presence of this SO_4^{3-}.

WQI calculated for each of the sources corresponding to set-1, by using formula (2) are displayed in Table 14.5. Observed WQI values, based on pH, DO, TH, Cl, TDS, SO_4^{3-}, NO_3 and EC ranged from 69.27 to 84.47. As per the status (Table 14.6) of water

Table 14.4: Correlation Matrix Corresponding to Data Set Represented in Table 14.2

	pH	*FCO_2*	*DO*	*TH*	*Cl^-*	*TDS*	*SO_4^{3-}*	*NO_3^-*	*TC*	*EC*
pH	1.000									
FCO_2	–0.126	1.000								
DO	–0.109	+0.150	1.000							
TH	–0.433	+0.045	+0.265	1.000						
Cl^-	+0.053	–0.517	–0.038	–0.072	1.000					
TDS	+0.032	–0.047	+0.006	+0.531	–0.224	1.000				
SO_4^{3-}	–0.516	–0.082	–0.085	+0.708	–0.156	+0.546	1.000			
NO_3^-	–0.383	–0.123	+0.092	–0.007	–0.128	+0.235	–0.121	1.000		
TC	+0.350	+0.566	+0.262	–0.057	–0.005	+0.162	–0.352	–0.228	1.000	
EC	–0.222	–0.178	+0.127	+0.849	–0.207	+0.728	+0.803	–0.091	–0.109	1.000

Figures 14.2(a), (b),(c) and (d): Correlation Coefficients of TDS, TH, SO_4^{3-} and EC with Other Parameters Involving Ten Drinking Water Sources at and Around Baktarnagr in Raniganj Coalfield Area

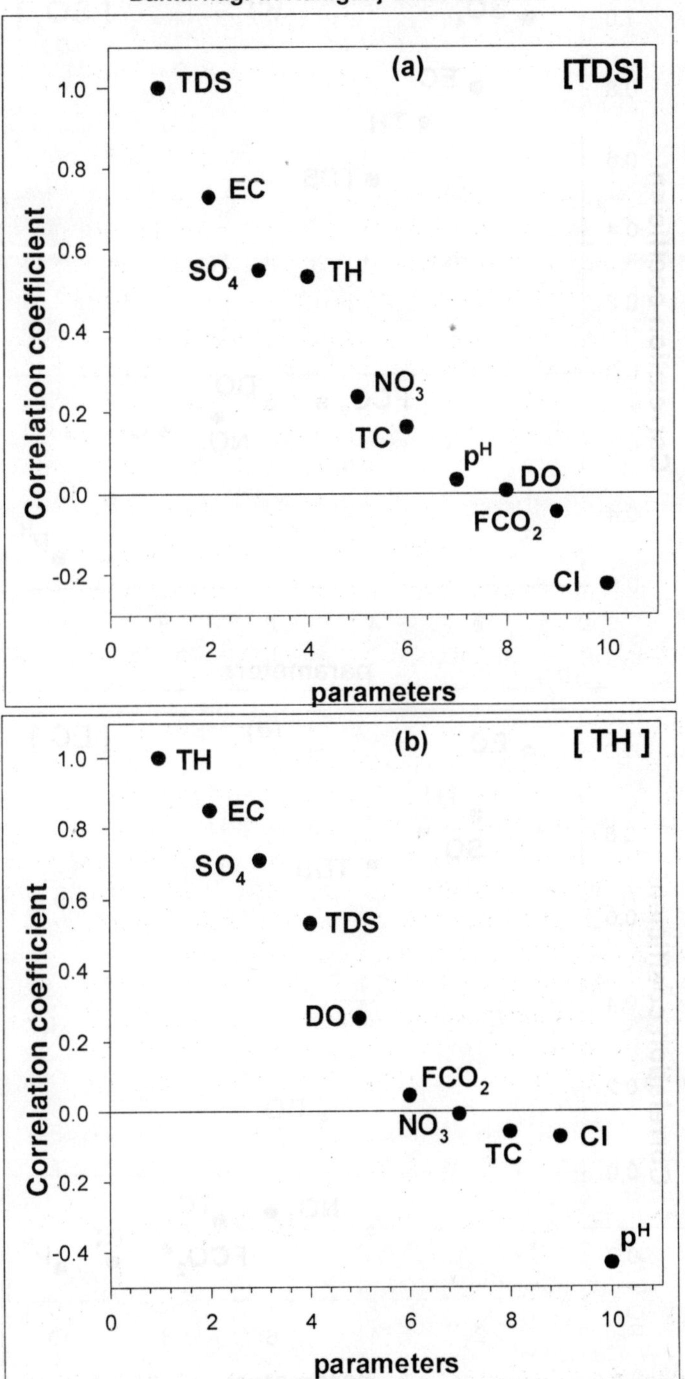

Contd...

Figures 14.2–*Contd...*

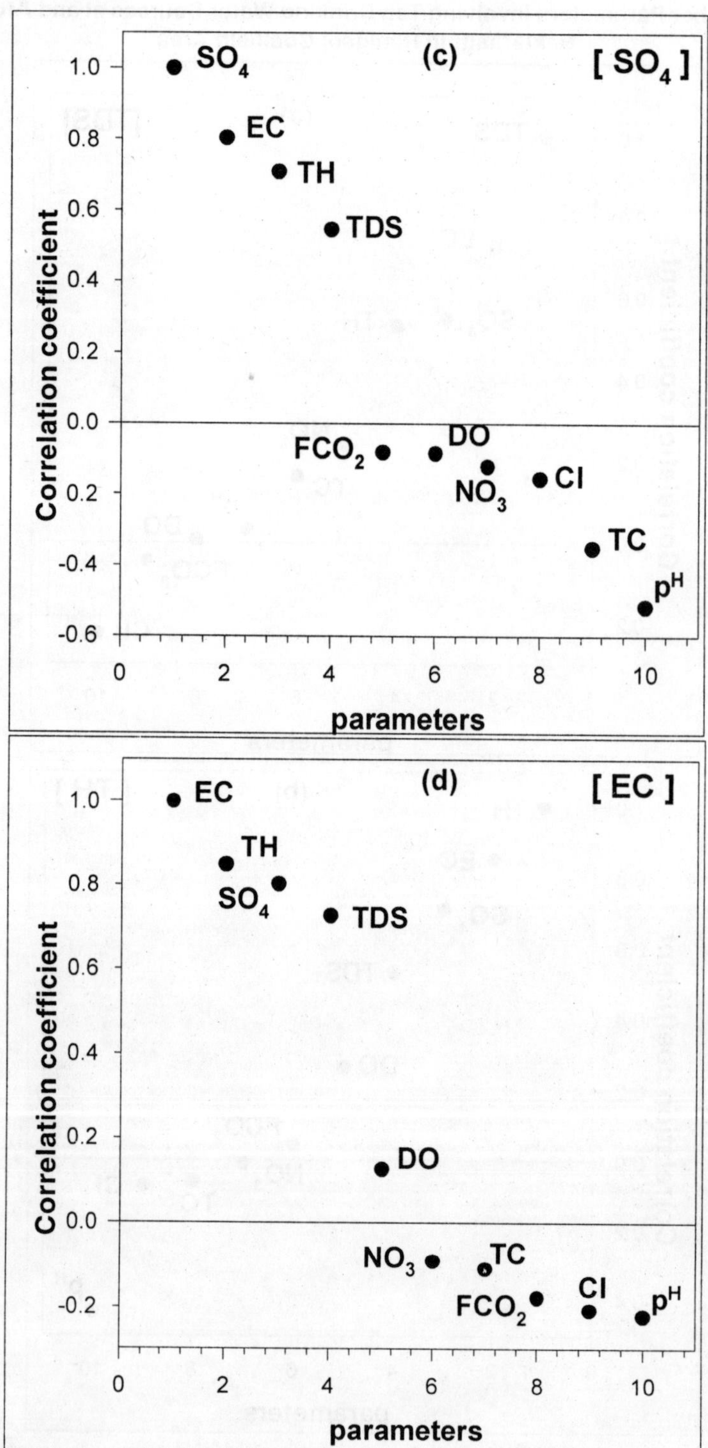

quality based on WQI (Mishra and Patel, 2001) the sources studied fall in the poor to very poor categories (Table 14.5)

Table 14.5: Observed WQI Values and the Status of Water Quality Corresponding to Data Set-1

Source	*WQI*	*Status*
S01	73.57	Poor
S02	80.45	Very poor
S03	76.25	Very poor
S04	82.28	Very poor
S05	69.27	Poor
S06	79.89	Very poor
S07	72.18	Poor
S08	80.37	Very poor
S09	84.47	Very poor
S10	71.61	Poor

Free CO_2, MPN and *E. coli* were excluded in the calculation.

Table 14.6: Status of Water Quality Based on WQI (Mishra and Patel, 2001)

WQI	*Status*
0-25	Excellent
26-50	Good
51-75	Poor
76-100	Very poor
Above 100	Unsuitable for drinking

Results of investigation on microbial contamination involving about 39 sources (set-3) which include different Institutes, Health Centers, Hotels and Sweet Shops are presented in Tables 14.7(a), (b) and (c) and Figures 14.3(a), (b), (c), (d) and 14.4. As is evident from Table 14.7(a), all types of sources studied are highly contaminated with total coliform bacteria. Twelve among the 30 sources studied show MPN e" 2400 per dL. To differentiate contamination arising from the consumer's storage tank itself from that of the Municipality supply water at the immediate vicinity, *i.e.*, just before entering into tank, we present six cases of such studies in Table 14.7(b) as well as in Figure 14.4. As the result shows, the storage tanks of the Institues: IN2, IN5 and IN7 seem to be themselves sources of higher contamination. Similar is the case for the Sweet Shop SW1. However, SW3 is found to contain lower concentration of MPN than that for the direct Municipality supply. This may be due to possible cleaning and cautionary measures (applying bleaching etc) taken in the tank beforehand, by the consumer concerned. For the health Center HC2, MPN $<$ 2 per DL at the storage tank corresponds to treated water there within the centre. Whereas MPN $\geq$ 2400 per dL for the Municipality supply direct, involving the same institute, may be due to contamination somewhere on the way from treatment plant to the health centre. This is because for the water of the treatment plant itself, the MPN is found to be equal to 22 per dL (Table 14.7b). Similar are also the cases involving Table 14.7(a) where MPN values at the storage tanks are found to be less than 22. That the treated water is really free from total coliform was tested and verified. This is represented in Table 14.7(c).

Results of investigation based on MPN involving 22 drinking water sources (set-4) at Burdwan town (a non-coalfield area) are displayed in Table 14.8. A comparison of these data with those in Table 14.7(a) shows that microbial contamination in the (Raniganj) coalfield area is significantly higher than that of a non-coalfield area. A comparison of the average values (Table 14.9) depicts the same. Discharge of domestic sewage and industrial wastes directly into the water bodies

Figure 14.3: A Comparison of the Observed Total Coliform (MPN) [dL^{-1}] in Municipality-Supply Drinking Water Collected from Tanks of Different (a) Institutes (b) Health Centers, (c) Sweet Shops and (d) Hotels at Raniganj

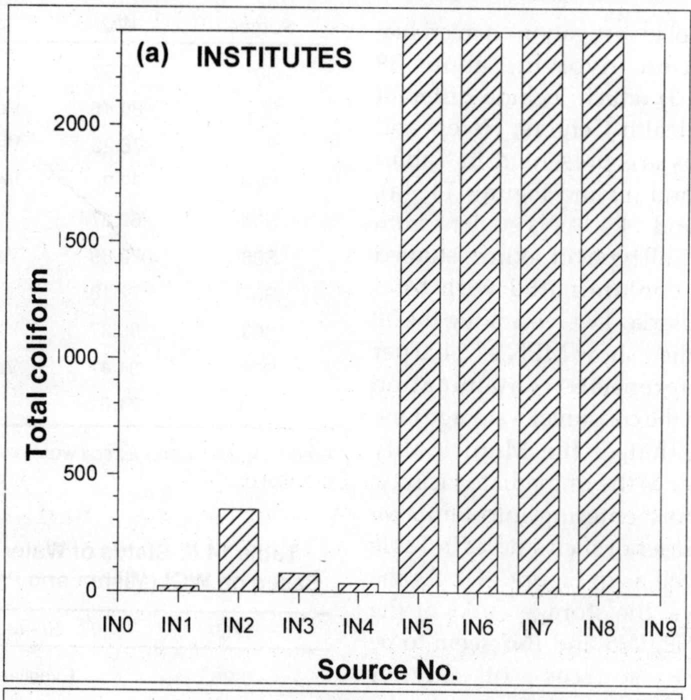

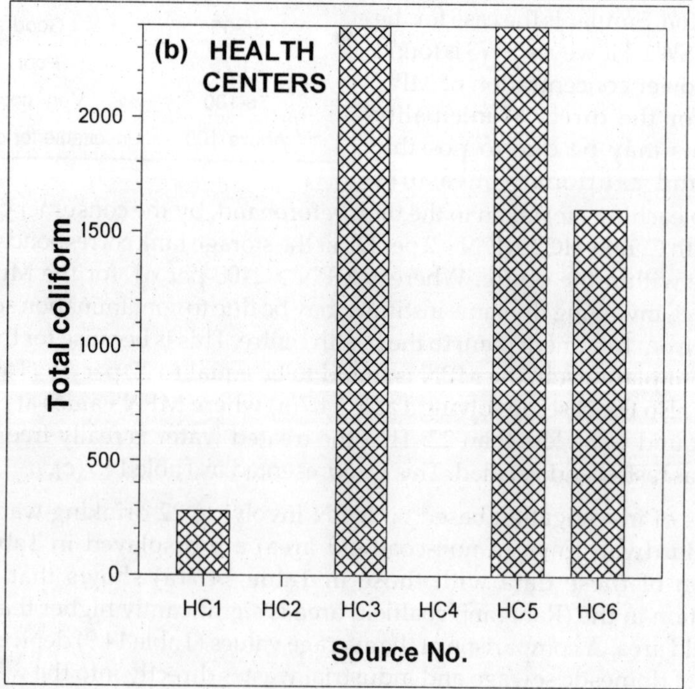

Contd...

Figure 14.3–*Contd...*

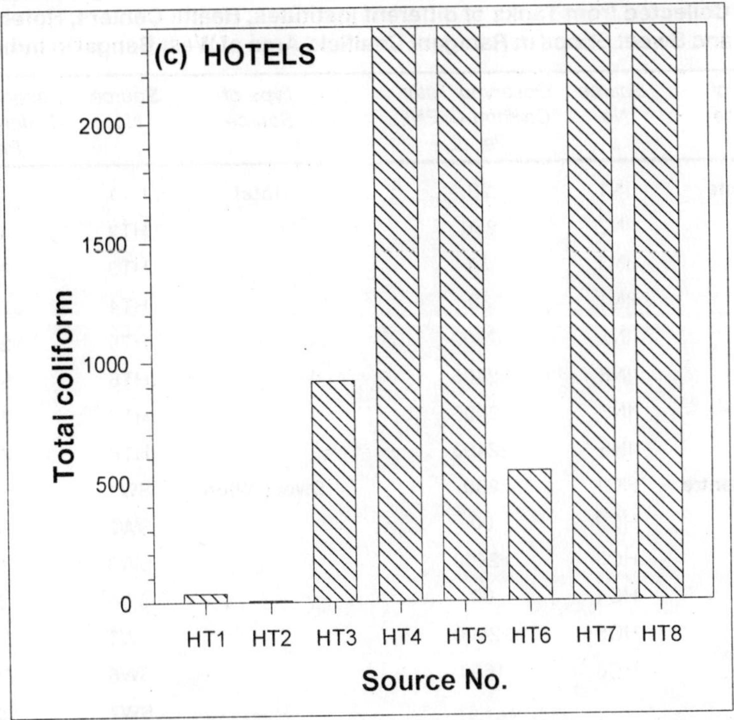

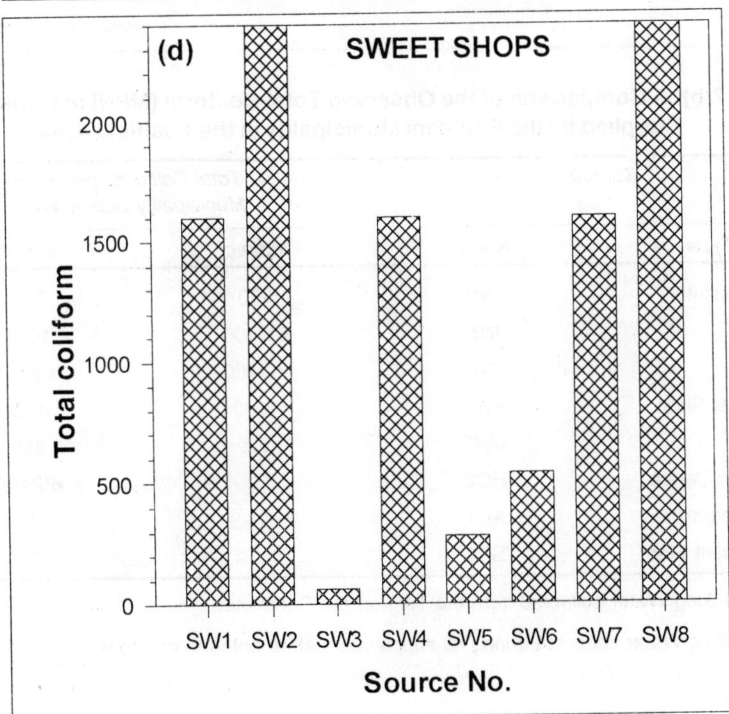

Table 14.7(a): A Comparison of the Observed Total Coliform (MPN) in Drinking Water Collected from Tanks of different Institutes, Health Centers, Hotels and Sweet Shops in Raniganj Coalfield Area of West Bengal in India

Type of Source	*Source No.*	*Observed Total Coliform (MPN) Per dL*	*Type of Source*	*Source No.*	*Observed Total Coliform (MPN) Per dL*
Institute	IN1	23	**Hotel**	HT1	34
	IN2	350		HT2	04
	IN3	79		HT3	920
	IN4	33		HT4	≥2400
	IN5	≥2400		HT5	≥2400
	IN6	≥2400		HT6	540
	IN7	≥2400		HT7	≥2400
	IN8	≥2400		HT8	≥2400
Health Centre	HC1	280	**Sweet Shop**	SW1	1600
	HC2	00		SW2	≥2400
	HC3	≥2400		SW3	63
	HC4	00		SW4	1600
	HC5	≥2400		SW5	280
	HC6	1600		SW6	540
				SW7	1600
				SW8	≥2400

Table 14.7(b): A Comparison of the Observed Total Coliform (MPN) in Drinking Water Supplied by the Raniganj Municipality in the Coalfield Area

Source		*Total Coliform (per dL) in Municipality Supply Water*	
Type	*Name*	*Indirect*	*Direct*
Institute	IN2	350	110
	IN5	e”2400	1600
	IN7	e”2400	21
Sweet Shop	SW1	1600	130
	SW3	63	350
Health Centre	HC2	<2	e”2400
Arbitrary	AR1		22
Treatment Plant	SP1		22

Indirect: Drinking Water collected from the Tank of the Consumer.

Direct: Drinking Water collected during its supply and before entering into tank.

Table 14.7(c): A Comparison of the Observed Water Quality Before and after treatment

Consumer	*Source*				
	Before Treatment				*After Treatment*
	Municipality Supply		*Dug Well*		*Aqua Guard (originally, from Tank filled with Dug Well water)*
	Direct into Tank	*After Storing*	*Direct into Tank*	*After Storing*	
Institute: IN7	21	≥2400	≥2400	≥2400	<2

Table 14.8: A Comparison of the Observed Total Coliform (MPN) and *E. coli* in Drinking Water Collected from different Institutes, Health Centers, Hotels and Sweet Shops in Burdwan Town which is about 80 km Away from Raniganj 'A' Signifies 'Absent'

Type of Source	*Source No.*	*Observed Total Coliform (MPN) per dL*	*E. coli*
Institute	IN1	<02	–
	IN2	1400	A
	IN3	33	A
	IN4	<02	–
	IN5	<02	–
Health Centre	HC1	<02	–
Hotel	HT1	<04	A
	HT2	49	A
	HT3	14	A
	HT4	<02	–
	HT5	05	A
	HT6	≥2400	A
	HT7	≥2400	A
	HT8	≥2400	A
	HT9	350	A
Sweet Shop	SW1	350	A
	SW2	<02	–
	SW3	≥2400	A
	SW4	920	A
	SW5	≥2400	A
	SW6	34	A
	SW7	350	A

'A' signifies 'Absent'.

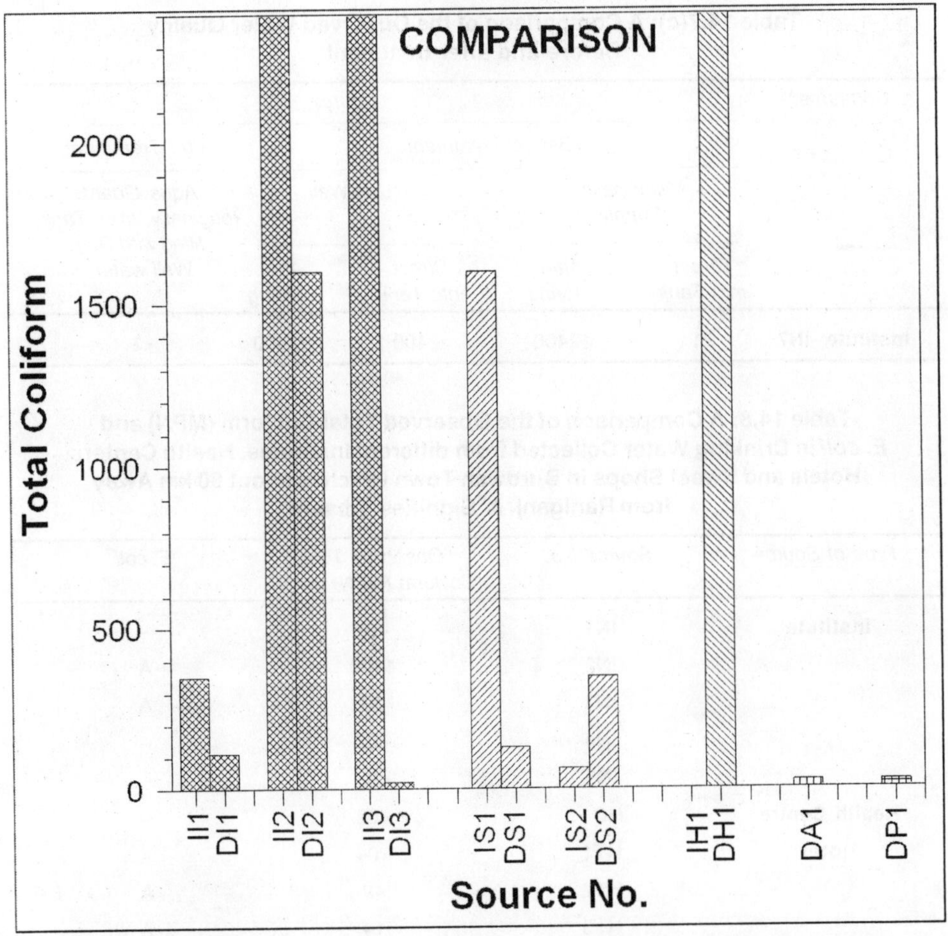

Figure 14.4: A Comparison of the Observed Total Coliform (MPN) [dL^{-1}] in Municipality-Supply Drinking Water.

'I'ndirect: Drinking Water collected from the Tank of the Consumer; 'D'irect: Drinking Water collected during its supply and before entering into tank. I1, I2 and I3 correspond to Institutes; S1 and S2 correspond to Sweet Shops; H1, A1 and P1 correspond to Health Center, Municipality supply at arbitrary site and Municipality supply treatment plant, respectively.

and their seepage into the subsurface sources of water in the coalfield area seem to be prime causes behind such findings.

Lack of awareness as well as scarcity of drinking water in some particular seasons seems to be some import causes of frequent sufferings of the people of the Raniganj coalfield area. This inference stems from the following points which were yielded during our survey with the people of different parts of the coalfield area. The survey reveals the following facts:

1. Safe drinking water, especially in summer season, is not sufficient to meet the demand of the people.

Figures 14.5(a) and (b): Pit Water which is Reported to be Used for Drinking Purpose during Ceremonial Gatherings, without any Prior Treatment of the Same. Bathing, washing and such other domestic activities go on regularly with the water.

Figure 14.6: A Dug Well which is Reported to be Used, though Very Rarely, by the Poor People of the Area to Battle against Severe Scarcity of Drinking Water. It is very hard to believe that the well is properly cleaned before using its water for drinking purpose.

2. People are forced to drink groundwater (Dug well, Tube well) which is unpalatable for many a cases.
3. People use pit water during ceremonial gatherings (marriage ceremony *etc*) without prior treatment of the same. Bathing, washing and such other domestic activities go on regularly with the water (Figures 14.5a and b).
4. To battle against severe scarcity of drinking water, people use dug well which is in a very bad situation (Figure 14.6) due to its remaining unused for a long time. It is very hard to believe that the well is properly cleaned by the poor people of the area before using its water for drinking purpose.
5. Some people incline to use water from exposed dug well, irrespective of knowing whether it is polluted or not, only because that the water is colder than the available municipality supply water (This is just lack of awareness).
6. Cow milk and its very business involves the use of a dug well which was found to lie in a very dirty situation (Figures 14.7a and b).

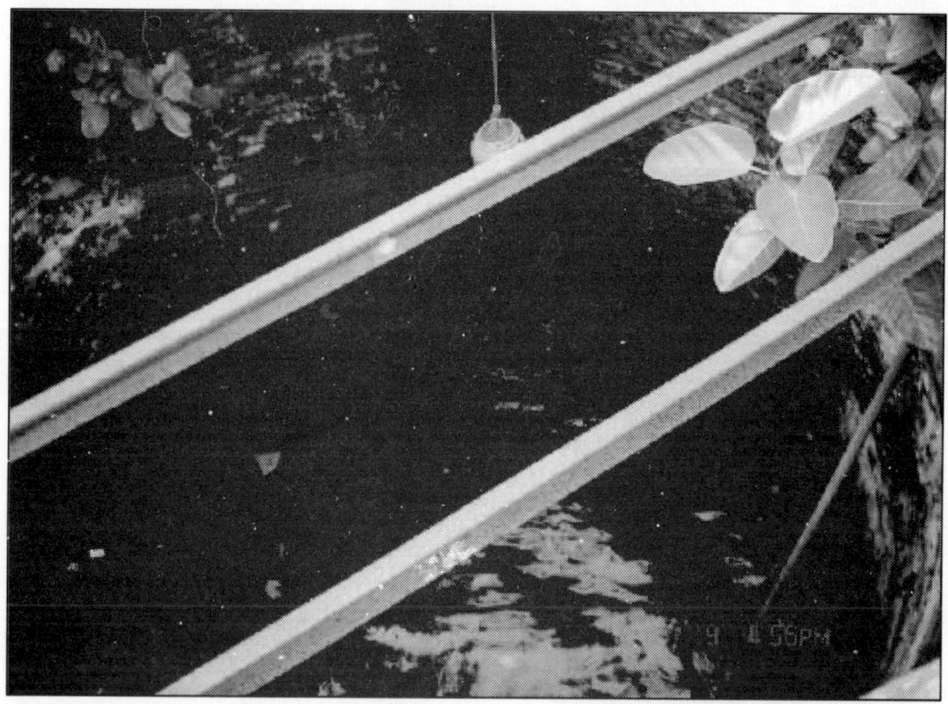

Figures 14.7(a) and (b): Cow Milk and its Business Involve the Use of water from such a Filthy Dug Well

Table 14.9: A Comparison of Average Values of Observed Total Coliform (MPN) of different Types of Drinking Water Sources in Raniganj Coalfield Area with those Observed in Burdwan Town (A non-coalfield area) which is about 80 km Away from Raniganj

Type of Source(s)	*Region*			
	Raniganj Coalfield Area		*Burdwan Town*	
	Total No. of Source(s) Studied	*Average Value of Observed Total Coliform (MPN) per dL*	*Total No. of Source(s) Studied*	*Average Value of Observed Total Coliform (MPN) per dL*
Institutes	08	≥1261	05	≈288
Health Centers	06	≥1447	01	≤02
Hotels	08	≥1375	09	≥847
Sweet Shops	08	≥1310	07	≈922

Preventive/Cautionary Measures

Some preventive/cautionary measures have been taken with reference to awareness program, by

1. Arranging seminars at the college sharing shoulders with members of the NSS units.
2. Discussing with the people of the locality and owners of the shops, authorities of the institutes etc
3. Making aware of the results of investigation with municipality officials.

Stresses have been given on using bleaching powder in a regular manner and use of proper water treatment and tank cleaning devices whenever and wherever affordable.

Conclusion

We have measured physico-chemical as well as microbial water quality parameters of drinking water sources of Raniganj coalfield area in an aim to monitor as well as to investigate on the cause of frequent sufferings of the people of the area. Among the measured parameters TDS and MPN values are seen to be excessively high. Very high values of NO_3^- and Cl^- also add to them. A comparison of the coalfield area (Raniganj) with a non-coalfield one (Burdwan town) on the basis of MPN pollution shows coalfield area to be comparatively more polluted. The frequent sufferings of the people of the Raniganj coalfield area may be attributed to the highly polluted character of the concerned drinking water sources. Scarcity of sufficient amount of potable drinking water in some areas and poverty, in conjunction with lack of awareness, also seem to be contributing factors towards the suffering.

Acknowledgement

Financial support in the form of research project [No.: 892(Sanc)/ST/P/S and T/13G-7/2008] from the Department of Science and Technology, West Bengal, India, is gratefully acknowledged.

References

Anandabazaar Patrika, Kolkata. 13th October, 2008: P-09.; 17th July, 2008: P-09; 15th July, 2008: P-09; 5th May, 2008: P-09; 21st July, 2007: P-11.

APHA, 1980: Standard methods for examination of water and wastewater. American Public Health Association, 15th Edn., Washington. D.C.

Basu,, M. and Mitra, A.K. 2002: Studies on pollution of river Damodar at Burdwan with special reference to metals, COD and pH. *Nat. Env. Poll. Tech.* 01(04): 397-399.

Brown, R.M., McCleiland, N.J., Deininger, R.A. and ÒConner, M.F. 1972: A Water Quality Index crossing the Psychological Barrier (Jenkin, S. H. ed) Proc. Int. Conf. on water Poll. Res., Jerusalem, 6: 787-797.

Chandra, R., Bahadur, Y. and Sharma, B.K. 2000: Monitoring of water quality of River Ramganga at Bareilly. p: 260-267. In: (Trivedi, R. K. ed.) Pollution and Biomonitoring of Indian Rivers. ABD Pub. Jaipur.

Chatterjee, C. and De, A. 2008: Impact of Anthropogenic Activities on Drinking Water Quality: A Case Study in Raniganj Coalfield Area (W. B.). *J. Nat. Env. Poll. Tech.* 7(3):451-459.

Ghazali, F.A. 1992: Poisoned waters, mindless industrialization polluting rivers. Nation and the World. 15th Feb., pp. 28-29.

Ghosh, M.K. and Kundu, N.K. 1999: Status of arsenic contamination of water in eastern coalfield areas and its removal process. J. IWWA, vol. XXXI(1):63-64.

Guru Prasad, B. 2003: Physico-chemical and bacteriological quality of groundwater at Tadepalli Madal of Guntur district. A. P. *J. Nat. Env. And Poll. Tech.* 2(2): 175-178.

Kavitha, R. and Elangovan, K. 2010: Groundwater quality characteristics at Erode district, Tamilnadu, India. *International Journal of Environmental Sciences,* 1(2):145-150.

Koshal, A.K. 2007: Environmental problem analysis of coalmining in Raniganj and Asansol blocks (West Bengal) using remote sensing and GIS. Map World Forum, Hydrabad, India (Internet source).

Kumarswamy, N. 1991: An approach towards assessment of dug well water quality by physico-chemical characteristics-A case study. *Poll. Res.* 10 (1): 13-20.

Rahaman, A. 2003: Assessing water quality from Jal Nigam hand pumps in Aligargh city, India. J. Nat. Env. *Poll. Tech.* 2(2): 241-244.

Richariya, L.K. and Mishra, R. 1998: MPN and *E. coli* in groundwater samples of Rewa area (M.P.) *India. J. Env. and Poll.* 5(1):73-77.

Scragg, R.K.R., Dorsche, M.M., McMichael, A.J. and Baghurst, P.A. 1982: Birth defects and household water supply. *The Medical J. Australia,* 2: 577-579.

Singh, A.K., Mahato, M.K., Neogi, B. And Singh, K.K., 2010: Quality Assessment of Mine Water in the Raniganj Coalfield Area. Earth and Env. Sc. Mine Water and the Environment, 29 (04): 248-262.

Singh, Amarjeet K. 2000: Dissolved Selenium in surface and subsurface waters of the Rawanara colliery area of Penchvalley coalfield and its impacts on human health. *J. Env. and Poll.* 7(4): 333-336.

Singh, A.K., Mondal, G.C., Singh, S., Singh, P.K., Singh, T.B., Tewary, B.K. and Sinha A. 2007: *J. Geological Society of India.* Vol. 69: 1088-1102.

WHO. 1993: Guidelines for Drinking Water Quality, Recommendations, World Health Organisation, Geneva.

WHO. 1978: Environmental Health Criteria, World Health Organisation, Geneva.

2013, Climate Change Effects on Agriculture and Economy *Pages* ***135–143***
Editors: **D.R. Khanna, A.K. Chopra, Vikas Singh, Rakesh Bhutiani & Gagan Matta**
Published by: **BIOTECH BOOKS, NEW DELHI**

Chapter 15

Comparative Analysis of Physio-chemical Properties of Effluents Discharged from Different Industries of Faridabad, Haryana

Jyoti Luhach, Manju Sharma, Himani, Smita Chaudhry
Institute of Environmental Studies, Kurukshetra University, Kurukshetra, Haryana

Growing industrial establishments without proper attention on pollution control measures have resulted in adverse impacts on surrounding environment. Industrial activities are mainly responsible for discharge of waste containing toxic and hazardous substances, which are detrimental to human health and aquatic systems. Evaluation of pollution load in industrial effluent is necessary so that required modifications can be incorporated in designing of effluent treatment plants. In present study, effluent from three different industries of Faridabad, Haryana *i.e.* Textile dying industry, Paint industry and Plastic industry was analyzed for physicochemical properties. The analysis was done before and after treatment in effluent treatment plants of respective industry to determine the extent of industrial pollution and to check the treatment level of effluent. Before treatment, the effluent from plastic industry showed higher alkalinity, chloride, total dissolved and suspended solids, total calcium and magnesium content followed by textile dye industry and paint industry. The heavy metals concentration and sodium - potassium level was exceptionally high in textile dying industries as compared to others. The pH, temperature, BOD and COD level was also high in textile dyeing industry indicating high organic and inorganic pollution load. After treatment of effluents, some parameters lower down to permissible level by CPCB for effluent to be discharged in inland water bodies. But the ranges of some parameters like

BOD, COD and temperature was still high. There was no treatment method for heavy metals removal from effluent. These industries need to comply with government norms for permissible limits of these parameters before discharging effluents in water bodies to avoid adverse impact on the health of residing inhabitants.

Keywords: *Industrial effluent, Faridabad, Pollution, Physico-chemical properties.*

Introduction

In the beginning of 21st century environmental pollution has emerged as a measure concern for survival of mankind throughout the world. Increased urbanization, industrial revolution, technological development and human population have given birth to undesirable changes in the physical, chemical and biological characteristics of air, land and water with in an ecosystem. These changes are so acute that they are causing large scale ecological imbalance in ecosystem (Singh *et al.*, 2006). The problem of effluent disposal of a wide variety of wastewater has been attracting the global attention. The industries continues to be one of the most significant cause of the pollution of aquatic system due to diverse kind of waste produced (Trivedi and Goel, 1984). In India 85 per cent of water pollution is from the industrial waste/effluent. Most of the world rivers receive millions of liters of the sewage, oil, grease, domestic waste, industrial and agricultural effluents containing substances varying from simple nutrients to highly toxic substances like N, P, K, Na, Cd, Pb, Hg, Cu, Fe, Zn, Ni, weedicides, pesticides, chlorinated hydrocarbons *etc.* (Aggarwal, 1989). Even in advanced countries like U.S.A., the important rivers and lakes have become severely contaminated.

In India, fourteen major rivers including Ganges, Gomati, Cooum, Cauvery, Damodar and Minimahi have become polluted. Aggarwal (1989) reported that water of these rivers is heavily loaded with a variety of pollutants from industrial and petrochemical waste of cities located along the banks, viz. Haridwar, Delhi, Allahabad, Baroda, Kanpur, Varanasi, Patna, Madras, and Kolkata. This has adversely affected the growth of economically beneficial aquatic flora and fauna. All along nearly 1550 miles journey to the sea, the holy Ganga river receiving 902 million liters of wastewater from 27 major cities with 312 industrial units daily alone affects the health of 250 million people in the northern India. Water quality studies of rivers Ganga (Shrivastava *et al.*, 1991), Betwa (Polestya *et al.*, 2007), Patra, Subernarakha (Singh *et al.*, 2004), Brahamni Kali (Bharti and Krishna Murti, 1990), St. Lawrence river and Nile river have indicated that their water is highly toxic and unsuitable for human consumption and domestic purposes because of industrial effluents (CPCB, 1986). These chemicals are not only poisonous to humans but also found toxic to aquatic life (WHO, 2002) and they may result in food contamination (Novick, 1999). It is concluded that, at present, out of more than 5000 major and medium industries, only 1000 have some kind of wastewater treatment facility. The rest allow their completely untreated waste to mix with open water bodies.

There are 2338 industrial units in the state of Haryana comprising large, medium and small scale units causing the pollution. Haryana State Pollution Control Board has categorized these units in to highly polluting industries. These units are required to acquire a no-objection-certificate before operation. As of now 221 highly polluting, 1698 polluting and 419 large and medium industries are operating in Haryana (CPCB, 2002).

According to Water Pollution Act (1974), most of the industries are now required to treat their waste to a degree as fixed by set conditions of CPCB (Central Pollution Control Board) before discharging them in to any water body or disposing them on land. So, there is an immense need to identify the various industrial hazards and their adverse effects by analyzing water quality parameters. Considering this fact, for the present study three industries (Textile dying, Paint and Plastic) of Faridabad district, Haryana were selected. In the view of above, the aim of the study was to examine and compare various physio-chemical parameters in effluent from paint, plastic and textile dye industries before entering in effluent treatment plant and after treatment by ETP. So, it can be recommended that the effluent is suitable for irrigation or should be disposed in water bodies.

Materials and Method

Study Area

Faridabad district of Haryana located on south eastern part of Haryana state lies between 270 39′, 280 31′ north latitude and 760 40′ and 77′32′ east longitudes. In the north it is bordered by the Union Territory of Delhi in the east by Uttar Pradesh, in the North-West by Mewat Gurgaon districts of Haryana. Total geographical area of the district is 2151 sq. km. Faridabad is the 9th biggest industrial town in India. During the last few years, about 70 units of textiles, dyeing and printing have come up in these industrial complexes.

Sampling

Samples of effluent were collected in triplicate from three selected industries of Faridabad *i.e.* Textile dying, paint and Plastic industry. Samples were collected from effluent treatment plant before and after treatment of effluent from textile dying, paint and plastic industries. All samples were placed in thoroughly cleaned and tightly closed polyethylene bottles. The samples were placed in a cooler box and then taken to the laboratory for analysis.

Methodology for Physio-chemical analysis

Various parameters like pH, temperature, total dissolved solids, total suspended solids and dissolved oxygen of effluent were analyzed by multi-parameter analyzer (Multi 340i by Merck) for field analysis in the effluent channel of each industry. Other parameters like alkalinity, chlorides, total calcium and total magnesium were analyzed by complexometric titration method, sodium and potassium by flame photometery (Model 381, ESICO), BOD by conventional incubation method and COD by reflux method (APHA, 1996). Heavy metals (Pb, Cr, Ni, Co and Cd) were determined by atomic absorption spectrophotometer (ASC – 7000, Shimadzu Corporation) (Table 15.1).

Table 15.1: Methodology Used for Physico-chemical Analysis of Effluent

pH,Temp.,Total Suspended solids (TSS) Total Dissolved solids (TDS)	Multiparameter Water analyser
Alkalinity and Chlorides	Titration Method
Total Calcium and Total Magnesium	Complexometric Titration
Sodium and Potassium	Flame Photometer
BOD at 20°C (5 days)	Incubation Method
Chemical Oxygen Demand	Reflux Method
Lead, Chromium, Nickel, Cobalt, Cadmium	Atomic Absorption Spectrophotometery

Results and Discussion

The results of physio–chemical analysis of effluent from different industries showed variation in parameters from industry to industry. The maximum permissible limits by CPCB for industrial effluent discharges in to different sinks (inland surface water, in public severs and on land for irrigation) were given in Table 15.2 and the results of physiochemical analysis of effluent before and after treatment in effluent treatment plants were given in Table 15.3.

Table 15.2: Maximum permissible limits by CPCB for industrial effluent discharges

Parameters	*Inland Surface Water*	*Into Public Sewers*	*On land for Irrigation*
pH	5.50–9.00	5.50–9.00	5.50–9.00
Temp. (°C)	40	45	–
Total Suspended Solids (mg/l)	100	600	200
Total Dissolved Solids (mg/l)	2100	2100	2100
Chlorides (mg/l)	1000	1000	600
Dissolved Oxygen (mg/l)	4-6	–	–
Chemical Oxygen Demand (mg/l)	250	–	–
BOD at 20.0 °C (5 Days) (mg/l)	30	350	100
Lead (mg/l)	0.10	1.00	–
Chromium (mg/l)	0.10	2.00	–
Nickel (mg/l)	3.00	3.00	–
Copper (mg/l)	3.00	3.00	–
Cadmium (mg/l)	2.00	1.00	–

Source: CPCB, 2006.

The values of pH represent acidic or basic nature of effluent. The hydrogen ion concentration is influenced by biological activities. Besides in addition of chemical substance, presence of organic matter in large quantity may lower the pH value due to release of CO_2 and SO_2. In this study, the effluent of paint and plastic industry before treatment in ETP was slightly acidic (5.7 and 6.2) and of textile dying was slightly basic (7.9) in nature (Table-3). After treatments in ETP plant the values for paint, plastic and textile dying industries were 6.3, 6.8 and 7.1, respectively. The values of pH were in the range of permissible limits by CPCB before and after treatment (Table-3). The wide variation in the pH value of effluent can affect the rate of biological reaction and survival of various microorganisms. Thorat and Wagh (1999) reported the pH of the sludge sample as 8.4 whereas Rao *et al.* (1993) have reported the pH of the textile industry effluent varying from 8.0 to 11.0. Temperature is also an important factor in effluent from any industry because high temperature causes thermal pollution in water bodies in which effluent is to be discharged. The temperature of effluent from three industries ranged from 51.0 °C to 58.0 °C. After treatment in ETP plant the temperature range declines (43.0 °C to 51.0 °C) but this range was also above the permissible limit by CPCB (40.0 °C).

Table 15.3: Physico-chemical Parameters of Effluent Before and After Treatment in ETP from different Industries

Parameters	*Max. Permissible Limit for discharge in Inland Water Bodies (CPCB)**	*Paint Industry*		*Textile Dying Industry*		*Plastic Industry*	
		Before	*After*	*Before*	*After*	*Before*	*After*
pH	5.5 – 9.0	5.7	6.3	7.9	7.1	6.2	6.8
Temp. (°C)	40	53	43	58	51	51	43
Total Suspended solids (TSS) (mg/l)	100	152	92	310	135	386	184
Total Dissolved solids (TDS) (mg/l)	2100	2425	1164	3210	2100	4533	2280
Alkalinity (as $CaCO_3$) (mg/l)	–	145	102	168	120	180	110
Chlorides (mg/l)	1000	1025	832	1565	1135	1870	954
Total Calcium (mg/l)	–	142	120	210	193	235	210
Total Magnesium (mg/l)	–	96	72	125	115	153	128
Sodium (mg/l)	–	390	210	935	615	430	220
Potassium (mg/l)	–	138	110	472	225	264	180
Dissolved Oxygen (mg/l)	4 to 6	3.2	4.6	3.8	4.8	3.4	4.1
BOD at 20.0 °C(5 days) (mg/l)	30	286	115	465	165	340	160
COD (mg/l)	250	685	460	1050	655	865	580

Source: CPCB, 2006. Max. Permissible Limits for effluent discharge.

TSS and TDS represent total solids in effluent, which further determine the turbidity status of wastewater. Total solids were highest in plastic industry (TSS-386 mg/land TDS-4533 mg/l) followed by textile dying (TSS-310 mg/land TDS-210 mg/l) and paint industry (TSS-152 mg/land TDS-2425 mg/l). After treatment effluent of paint industry minimize total solids below permissible limits by CPCB but in case of other two total solids were higher than permissible level (Table 14.3).

The BOD and COD level indicate severity of organic and inorganic water pollutants in effluent. The presence of dissolved oxygen is essential to maintain a variety of forms of biological life in water and the effects of the water discharged in water body are largely determined by oxygen balance of the system. Non-polluted surface water remains normally saturated with the dissolved oxygen. Oxygen can be rapidly removed from the water by addition of oxygen demanding waste. According to present study, the value of both BOD and COD were higher in textile dying industry (465 mg/l and 1050 mg/l respectively) as compared to plastic industry (340 mg/l and 865 mg/l respectively) and paint industry (286 mg/land 685 mg/l respectively). This indicates that textile dying industry had higher organic and inorganic pollutants followed by plastic and paint industry. After processing in effluent treatment plant, the BOD and COD level declined but the values were still high as prescribed by CPCB for disposal in inland water bodies. Trivedi *et al.* (1986) have reported the BOD values of mixed effluent of textile industry ranging between 320 mg/l to 720 mg/land of final effluent ranging from 80 mg/l to 640 mg/l.

Table 14.4: Concentration of Heavy Metals of Effluents from different Industries (mg/l)

Parameters	*CPCB (Permissible Limit)*	*Paint Industry*	*Textile Dying Industry*	*Plastic Industry*
Lead as Pb (mg/l)	0.10	0.52	0.72	0.24
Chromium as Cr (mg/l)	0.10	0.63	0.46	0.34
Nickel as Ni (mg/l)	3.00	3.03	5.05	1.01
Copper (mg/l)	3.00	1.01	2.82	1.64
Cadmium (mg/l)	2.00	2.85	4.05	2.70

Source: CPCB, 2006. Max. Permissible limits for effluent discharge.

The observed values of Sodium and Potassium were also higher in textile dying industry (935 mg/l and 472 mg/l) followed by plastic (430 mg/l and 264 mg/l) and paint industry (390 mg/land 138 mg/l). This is due to use of sodium and potassium salts during bleaching and fibre dying in textile dying industry. The values of total Calcium and Magnesium represent hardness properties of effluent. In this study, higher Ca and Mg content was observed in plastic industry (235 mg/l and 153 mg/l) respectively, followed by textile dying (210 mg/l and 125 mg/l) respectively and paint industry (142 mg/l and 92 mg/l) respectively.

Heavy metals contamination is an environmental risk (Wang *et al.*, 2007). The concentrations of lead was higher in textile dying industry (0.72 mg/l) as compared to paint and plastic industry, but the concentration of chromium was higher in paint industry (0.63 mg/l). The concentration of nickel, copper and cadmium was also higher in textile dyeing industry (5.05 mg/l, 2.82 mg/land 4.05 mg/l) respectively followed by paint and plastic industry for nickel and cadmium. These concentrations were higher than the permissible limits prescribed by CPCB for the effluent before discharging in water bodies and the concentrations of Nickel and Cadmium were even higher then the permissible limit for disposal in public severs. In all the three

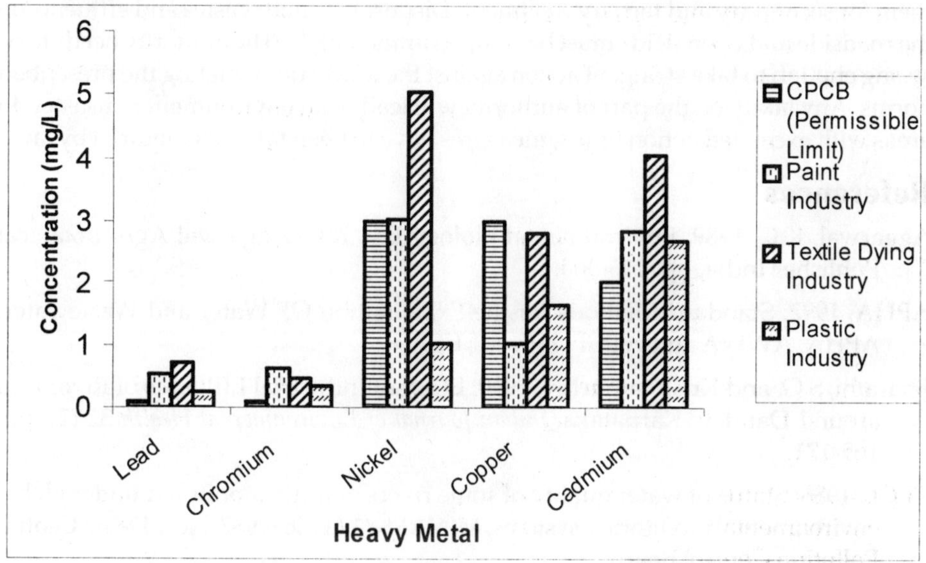

Figure 15.1: Concentration of Heavy Metals in Effluent from different Industries

industries, there were no strategies in the effluent treatment facility for removal of heavy metal from effluent and presently wastewater is being dumped as such in nearby water bodies. Heavy metals are highly toxic for aquatic ecosystems because high concentration severally affects aquatic microbiota.

Conclusion

According to present study, heavy metals and organic and inorganic pollutants were higher in textile dying industry as compared to other two industries. The chemicals used in dying contain heavy metals like lead, cadmium and chromium. The dyeing substances that are used in textile industry give colour to the water. Coloured textile wastewater accumulates in aquatic environment and deteriorates the aesthetic appearance of the water and decreases the light permeability. Decrease in the light permeability and the amount of decomposed oxygen causes the extinction of living beings and restricts the reuse of the water. For these reasons treatment of the textile industry wastewater is very important.

The concentrations of many physico–chemical parameters even after treatment in effluent treatment plants from paint and plastic industries were much above the permissible limits by CPCB for the effluent to be discharged in water bodies. Hence, necessary modifications are required in designing of treatment plants so as to reduce the concentrations of toxic substances to permissible levels.

Recommendations

Major effluent generating industries that have primary effluent treatment facilities available must maintain a pH level of about 7.5 in their effluents before transporting them for secondary and tertiary treatment. Disposal of solid wastes and effluents on the roadside and open fields must be stopped immediately. The authority needs to be strong enough to take stringent action against the industries violating the prescribed norms. Any laxity on the part of authority will lead to an environmental disaster. In areas with exceeded action limits, measures have to been taken as required by law.

References

Aggarwal, K.C. 1989: Environmental Biology Ed., K.C. Aggarwal Agro Botanical Publisher India, pp. 246-305.

APHA, 1992: Standard Methods for the Examination OF Water and Wastewater. APHA, AWWA and WPCF. 16th ed., USA.

Bharathi, S.G. and Krishnamurthy, 1990: Effect of Industrial Effluents of River Kali around Dandeli, Karnataka. *Indian Journal of Environmental Health* 32 (2): pp. 165-171.

CPCB, 1986: Status of water quality of some rivers of India monitored under global environmental monitoring systems, MINARS/1/1986–1987. New Delhi: Central Pollution Control Board.

CPCB, 2002: Status of water quality of some rivers of India monitored under global environmental monitoring systems, New Delhi: Central Pollution Control Board.

CPCB, 2006: Pollution Control Acts, Rules, and Notification issued hereunder. Volume, pp.311-312, New Delhi: Central Pollution Control Board, Ministry of Environment and Forests, 501.

Novick, R. 1999: Overview and the Health in Europe in the 1990s, World Health Organization, Europe Regional Office, Copenhagen, EUR/ICP/EH/CO 02 02 05/6, pp 20.

Polestya, A., Sharma, S. and Reddy, A.S. 2007: Physico-chemical characteristics of the east bein river water. *Eco Res J Biosci*, pp. 18-22.

Rao, A.V., Jain, B.L. and Gupta, I.C. 1993: Impact of textile Industrial effluents on agricultural land – A case study. *Indian J. Environ Health.* Vol. 35 (2), pp. 13-138.

Singh, K.P., Malik, A., Mohan, D. and Sinha, S. 2004: Multivariate statistical techniques for the evaluation of spatial and temporal variations in water quality of Gomti River (India)–a case studies. *Water Research, 38*, pp. 3980–3992.

Singh, K.P., Malik, A., Mohan, D., Singh, V.K. and Sinha, S. 2006: Evaluation of groundwater quality in northern Indo-Gangetic alluvium region. *Environ. Monit. Assess.*, pp. 211-230.

Srivastava, M., Srivastav, A.K., and Rawat, M.S.M. 1991: Physico-chemical and phycological assessment of water of Ganga at Rishikesh and Hardwar. Advances in Himalayan Ecology, pp.255–265.

Thorat S.P. and Wagh S.B. 1999: Physico chemical analysis of tannery water. *Jr. Industrial Poll. Cont.*6 (1), pp, 107-109.

Trivedi, R.K. and Goel, P.K. 1984: Chemical and biological methods for water pollution studies Karad Environmental Publication, pp. 1-251.

Trivedi, R.K., Khatavkar, S.B., Goel, P.K. 1986: Characterization, treatment and disposal of wastewater in a textile industry. *Ind. Poll. Cont.* 2 (1). pp, 1- 12.

Wang, Y., Shi, J., Wang, H., Lin, Q., Chen, Y. 2007: The influence of soil heavy metals pollution on soil biomass, enzyme activity, and community composition near a copper smelter. *Ecotox. Environ.* Sat 67: 75-81.

WHO, 2002: Water Pollutants: Biological Agents Dissolved Chemicals, Non-dissolved Chemicals, Sediments, Heat, WHO CEHA, Amman, Jordan.

2013, Climate Change Effects on Agriculture and Economy *Pages* ***145–151***
Editors: **D.R. Khanna, A.K. Chopra, Vikas Singh, Rakesh Bhutiani & Gagan Matta**
Published by: **BIOTECH BOOKS, NEW DELHI**

Chapter 16

Hydrological Conditions of Upper Stretches of River Beas in Himachal Pradesh, India

Ashok Kumar[1] and V.K. Saxena[2]
[1]*A.D.Hydro Power Limited, Prini, Manali, Kullu, Himachal Pradesh*
[2]*Department of Zoology, Narain College Shikohabad, Uttar Pradesh*

The present communication highlights the hydrological regime of upper stretches of river Beas in Kullu district, Himachal Pradesh (India) which was studied at four sites during December, 2002 to November, 2004. Flow of current ranged between 378 - 432 cubic feet/sec. Depth of river ranged between 1–4 mts. Width of the river ranged between 40-50 mts. Water temperature ranged between 5.2 °C -17.7 °C. pH showed only little fluctuation (7.0-7.7). Transparency of the water was high throughout the year except rainy season. Dissolved oxygen was high and it showed an inverse relationship with water temperature. Total alkalinity was also high.

Keywords: *Hydrological, River Beas, Physico-chemical factors, Himachal Pradesh.*

Introduction

Himachal Pradesh is situated between 32° 22' 40" to 33° 12' 20" north latitude and 75° 47' 55" to 79° 04' 20" east longitude. The altitudes in Himachal Pradesh, Himalayas range from 350 to 6975 meters above mean sea level. The five main rivers *viz.*, Beas, Satluj, Ravi, Chenab and Yamuna have their catchments in Himachal Pradesh. These rivers rise in the Himalayas and have perennial source of water being fed by snow during summer. Except, for the Yamuna, the other rivers meet the Indus, falling in the Arabian Sea. Beas River is one of the main river of the Himachal Pradesh.

The river Beas originates from southern slope of Rohtang pass (Beas Kund) at an elevation of 4062 m (msl). The river receives number of tributaries both on right and left banks during its downward, drift of over 470 km. Its principal tributaries are Solang, Manalsu, Sujjain, Fojal and Sarvari on the right bank and Allain, Chhaki, Parbati, Tirthan and Sainj on the left bank. Out of its total length, a stretch of 60 km between Manali and Bhuntar is the trout stream which is divided into four beats *i.e.* Bhuntar, Kullu, Patlikuhl and Manali. Till date, very limited work has been done on the ecology of river Beas. Shah (1975) reported the food and feeding habits of brown trout in river Beas. Dhanze and Dhanze (1998) reported impact of habitat shrinkage of fish fauna of river Beas due to drainage system. Observations were therefore made to study the hydrological conditions of upper stretches of the river Beas. The disposition of the river in Himachal Pradesh is shown in Figure 16.1.

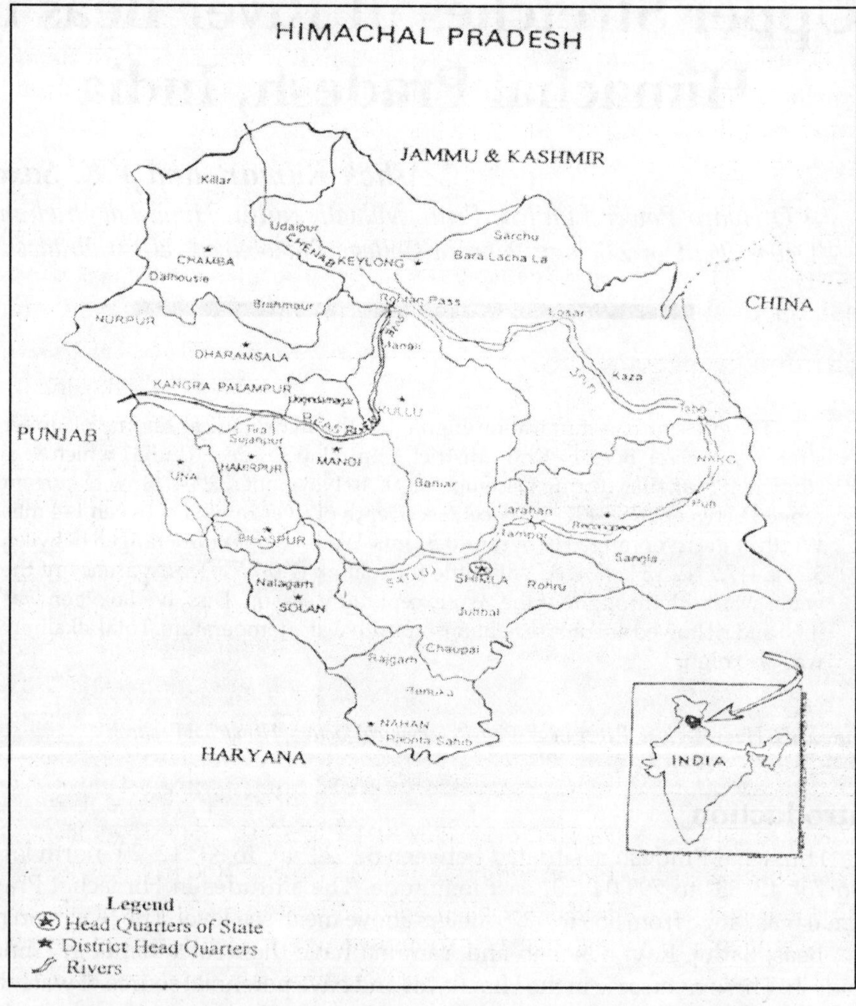

Figure 16.1: Disposition of the River Beas in Himachal Pradesh

Materials and Method

In the river Beas, four sampling sites were chosen namely, Bhuntar, Kullu, Patlikuhl and Manali. Collections were made from these sites at fixed time and almost remained uniform throughout as for as possible. The hydrological conditions of the river Beas *viz.*, water current, depth, width, longitude, latitude and altitude. Water current (cubic feet/sec) was calculated by cork floating method. The flow of current was measured by the following formula: R = WDaL/T. The river depth was calculated with the help of a marked bamboo stick. It was recorded from atleast 5-6 points having different depths and the average of all these points was considered as the mean depth. The river width was measured with the help of measuring tape. Longitude, latitude and altitude (msl) were determined with the help of Magellan GPS.

The physico-chemical parameters of water of the river Beas *viz.* water temperature, pH, transparency, dissolved oxygen, total alkalinity were analyzed fortnightly following APHA (1998). Due to shallow depth, stony bottom and fast current, the transparency was measured by bright pin head method (Saha *et al.*, 1971). Hydrogen ion concentration of water was determined by digital pH meter. For dissolved oxygen, unmodified Winkler's method was adopted (Welch, 1948), while carbonate and bicarbonate (total alkalinity) were determined by titration method with N/50 Sulphuric acid using phenolphthalein and methyl orange as indicators.

Results and Discussion

Hydrological Factors

Flow of Current

The flow of water current in river Beas was not same throughout the year. In rainy season, the rate of flow was recorded 432 cubic feet/sec and 378 cubic feet/sec was recorded during winter season.

Depth

The average depth of the river Beas was 1.0–1.5 metre in winter season and 3.0–4.0 meters in summer and monsoon season.

Width

The average width of the river Beas was 50 meters.

Longitude, Latitude and Altitude (msl)

Longitude, latitude and altitude (msl) were determined with the help of Magellan GPS.

Physico-chemical Parameters

The physico-chemical factors of river water was studied from December, 2002 to November, 2004 and its seasonal variations have been depicted in Tables 16.1 and 16.2.

Water Temperature

Temperature is one of the most important factor, which affects the biota as well as other factors. Low temperature reduces metabolic activity. Feeding, breeding, respiration and all other physiological activities of fish are influenced by temperature. Results of the present study showed that the fluctuations of water temperature ranged from 5.2 °C-17.7 °C in the first year while between 6.8 °C-17.2 °C during the second year. It was high during monsoon and low during winter. The value of water temperature showed regular decrease from September to January and then showed increasing trend during both years of the investigation.

Transparency

River water was clear and transparent during winter and low during monsoon (4.5 cm) which was due to fast current and greater inflow of muddy rain water. Results of the present study showed that the transparency of the river water ranged between 6.1 cm - 53.2 cm during first year. The highest value was recorded during November and the lowest in month July. In second year, transparency varied from 4.5 cm - 62.3 cm being maximum in December where as minimum was recorded in August.

Hydrogen Ion Concentration (pH)

pH of water is an important factor that influences metabolism of organisms inhabiting it. Therefore, study of this factor is highly important. The pH of the river water was always found to be in the alkaline range (7.0-7.7). During the first year the range of pH was recorded between 7.0 - 7.7 with a mean of 7.35 while during the second year from 7.0 - 7.6 with a mean of 7.29. The minimum pH was observed in the month of September and October and maximum in August during first year. During second year, minimum and maximum pH was observed in the month of October and January respectively. The mean of pH for all months indicated alkaline water, which was quite suitable for fish.

Dissolved Oxygen

Dissolved oxygen is essential for the metabolism of all aerobic aquatic organisms and oxygen distribution is important for the direct need of many organisms. Hence, determination of this factor is essential. Dissolved oxygen was quite high and showed wide fluctuation. During the first year of the study, range of dissolved oxygen was recorded from 9.2 ppm - 12.6 ppm while during second year it ranged from 9.7 ppm - 12.8 ppm. The minimum and maximum DO was recorded in August and December respectively during first year. During second year, minimum and maximum DO was observed in July and December respectively.

Total Alkalinity

The alkalinity is a term that refers to the buffering capacity of the carbonate system in water. Alkalinity or acid combining capacity, which is generally caused by carbonates and bicarbonates of calcium and magnesium, is an important parameter for assessing the productivity of freshwater. In the first year total alkalinity ranged from 53.7-85.0 ppm while it varied from 60.0-85.6 ppm during second year of the study. Maximum alkalinity was observed in February in both the years where as the minimum alkalinity was observed in August in first year and May, June and August in second year.

Table 16.1: Monthly Fluctuations in Physico-chemical Factors in River Beas during December, 2002–November, 2003

Months	*Water Temp. (°C)*	*Transparency (cm)*	*pH*	*Dissolved Oxygen (ppm)*	*Alkalinity (ppm)*
December, 02	6.2	51.5	7.5	12.6	75.6
January, 03	5.2	47.8	7.6	12.0	70.6
February, 03	7.1	43.3	7.4	11.2	85.0
March, 03	9.2	46.1	7.4	11.2	83.1
April, 03	11.3	51.5	7.3	10.0	80.1
May, 03	10.0	29.3	7.1	10.8	61.2
June, 03	12.8	10.2	7.3	10.2	57.5
July, 03	16.0	6.1	7.5	9.7	59.3
August, 03	17.7	7.7	7.7	9.2	53.7
September, 03	17.1	9.9	7.0	9.9	56.2
October. 03	12.9	29.5	7.0	10.9	67.5
November, 03	8.1	53.2	7.4	11.4	75.6
Mean	11.13	32.17	7.35	10.76	68.78

Table 16.2: Monthly Fluctuations in Physico-chemical Factors in River Beas during December, 2003–November, 2004

Months	*Water Temp. (°C)*	*Transparency (cm)*	*pH*	*Dissolved Oxygen (ppm)*	*Alkalinity (ppm)*
December, 02	8.2	62.3	7.5	12.8	75.6
January, 03	6.8	51.1	7.6	12.5	84.3
February, 03	8.5	38.1	7.4	10.8	85.6
March, 03	9.9	33.0	7.3	10.6	82.5
April, 03	12.5	31.5	7.1	10.3	80.6
May, 03	13.8	28.2	7.3	9.9	60.0
June, 03	15.8	15.2	7.1	10.1	60.0
July, 03	17.0	5.5	7.4	9.7	61.2
August, 03	17.2	4.5	7.3	9.8	60.0
September, 03	15.5	9.5	7.1	10.8	63.7
October. 03	14.0	25.3	7.0	11.0	68.7
November, 03	8.7	43.7	7.4	11.2	82.5
Mean	12.32	29.00	7.29	10.79	72.05

Most of the physical and chemical factors varied at different periods of the years. The value of water temperature ranged between 5.2-17.7°C showed regular decrease from September to January during investigation period and then showed increasing trend. Maximum turbidity was observed in rainy season which results in low transparency (4.5 cm). The higher value of dissolved oxygen ranged between 9.2-12.6 ppm, which is pretty good for aquatic life. The water is highly alkaline in nature ranging between 53.7-85.6 ppm. High value of dissolved oxygen, total alkalinity and alkaline pH along with nutrients and salts being received from the inflowing water made the river highly productive which provided a suitable habitat for fishes especially *Schizothorax richardsonii* comprising the major of catch of the river.

The variation in physico-chemical parameters of water body has been generally attributed to the rainwater. Carter (1960) pointed out that amount and type of rainfall play a significant role in regulating the seasonal, biological rhythm in tropical water. The pattern of fluctuation in temperature was more or less similar at all the sites. This wide variation is mainly due to the change in altitude. Kumar and Bhagat (1978) also reported similar trend of water temperature fluctuation in Sind and Lidder (1.5 °C-17.3 °C), streams of Kashmir. Gurumayum *et al.* (2002) have reported the same seasonal variations while studying the cold water streams of East Khasi hills of Meghalaya.

Vass *et al.* (1977) reported that the pH value of river Jhelum was alkaline throughout the year at all the sites (ranged between 7.0-7.6). Kumar and Bhagat (1978) also observed neutral to slightly alkaline pH (7.1-7.8) round the year at all the sites of Sind and Lidder streams of Kashmir. Sunder (1988) also observed alkaline nature of river Jhelum. Dhanze (2002) reported the range of transparency from 13.0-85.0 cm in different tributaries of the river Beas. Kumar and Bhagat (1978) observed turbidity in Sind and Lidder streams of Kashmir from nil to 27.0 cm.

In the present study, high values of dissolved oxygen was recorded in winter season when the water temperature was low. It was comparatively lower in summer season and lowest during the monsoon season. The same has been observed by Gurumayum *et al.* (2002), while studying the physico-chemical quality of cold water streams of Meghalaya. Total alkalinity was observed to be high during winter months. The same has trend of alkalinity has been observed by Kumar (1987) in two trout streams of Kashmir, Gurumayum *et al.* (2002) in rivers of Meghalaya and Agarwal *et al.* (2004) in the river Bhilangana of Garhwal region. On the basis of above said findings it can be concluded that the water of river Beas is moderately alkaline.

Conclusion

The above mentioned hydrobiological conditions of river Beas are found to be suitable for fish fauna and insect fauna. The following fish fauna reported are *Schizothorax richardsonii* represented the most dominant species, followed by *Oncorhynchus myskiss, Salmo trutta fario, Nemacheilus rupecola* and *Garra gotyla gotyla, Glyptothorax indicus*. On the other hand various hydro power projects running on the river Beas and its tributaries causing great damage to fish breading ground. Human intervention in river, road construction on the sides of the rivers, construction of hotels at tourism places, clearing of forest area for hydro power projects, stone crushers, removal of stones from the river bed and grazing of cattle. All these activities have depleted the fish habitat. The scale of human intervention in disturbing the river ecology is on the increase every year and is directly proportional with the development activities.

Acknowledgement

We are highly indebted to Dr. K. Kumar, former Director cum Warden of Fisheries, Himachal Pradesh for his keen interest, encouragement and critically going through the manuscript and offering valuable suggestions to undertake the study. We are also thankful to Mr. Khem Singh, Fisheries Officer (Kullu), Mr. Rajan Sud Assistant Director Fisheries for their kind cooperation.

References

Agarwal, N.K., Rawat, U.S., Thapliyal, B.L., Raguvanshi, S.K. and Saini, Vandana. 2004: Proceeding national symposium: Status of cold water fisheries with reference to fragile Himalayan aquatic ecosystem. 98-111 pp.

APHA, 1998: Standard methods for the examination of water, sewage and industrial wastes 20th Edition. American Public Health Association.

Carter, G.S. 1960: Tropical climates and biology. *Nature,* London: 187-843.

Dhanze, J.R. 2002: Water quality analysis of Hill streams in Himachal Pradesh with focus on mahsheer fishery.

Gurumayum, S., Daimari, D.P., Goswami, B. S., Sarkar, A. and Choudhury, M. 2002: Physico-chemical quality of water and Plankton of selected rivers in Meghalaya. *J. Inland Fish. Soc.* India, 34(2): 36-42.

Johal, M.S. 2001: Final Report U. S. Fish and wild life service,1-149 p.

Kumar, K. and. Bhagat, M.J. 1978: Observation on the ecology of the two trout streams in Kashmir and its possible effects on the brown trout (Salmo trutta fario Lin.) catches. *J. Inland Fish. Soc.* India, Vol. 10, 1978.

Kumar, K. 1987: Observation on seasonal variations of benthic organisms in two trout streams of Kashmir. Proc. Indian Nati. Sci. Acad. B 53 No. 3 pp. 227-234.

Saha, G.N., Sehgal, K.L., Mitra, Eva and Nandy, A.C. 1971: Studies on the seasonal and diurnal variations in physico- chemical or biological conditions of a perennial freshwater pond. *J. Inland fish.Soc.* India, 3: 79-102.

Sundar, S. 1988: Monitoring of water quality of a stretch of river Jhelum, Kashmir. pp 131-161. In Ecology and Pollution of Indian Rivers (R.K. Trivedy ed) Ashish Publishing House, New Delhi.

Vass, K.K., Raina, H.S., Zutshi, D.P. and Khan, M.A. 1977: Hydrobiological studies of river Jhelum. *Geobios,* (4) 238-242.

Welch, P. S., 1948: Limnology Method;The Blakiston company, Philadelphia Toranto. 199-229.

2013, Climate Change Effects on Agriculture and Economy *Pages* ***153–158***
Editors: **D.R. Khanna, A.K. Chopra, Vikas Singh, Rakesh Bhutiani & Gagan Matta**
Published by: **BIOTECH BOOKS, NEW DELHI**

Chapter 17

Water Quality Assessment of Tadoba Lake, Tadoba Andhari Tiger Reserve Chandrapur, Maharashtra

***P.M. Telkhade*[1], *N.R. Dahegaonkar*[1], *P.J. Khinchi*[2] and *S.B. Zade*[3]**

[1]*Department of Zoology, Arts, Commerce and Science College, Tukum, Chandrapur, Maharashtra*

[2]*P.G. Department of Zoology, Janta Mahavidyalaya, Chandrapur, Maharashtra*

[3]*P.G. Department of Zoology, R.T.M. Nagpur University, Nagpur, Maharashtra*

Today with rapid increase in population, over exploitation for different purposes the water quality has been deteriorating at an alarming rate, which ultimately results in depletion of aquatic biota. The present paper deals with the monthly variation of physico-chemical characteristics of Tadoba Lake. Tadoba Lake is free from human activities and is surrounded by a dense forest. The lake water is utilized for only wild life drinking. The study was carried out for twelve months in between June, 2005 to May, 2006. Sampling sites were chosen by keeping in view the accessible area of the lake. Four different sites were selected for periodical sampling of lake water and analysis of physico-chemical parameters. The main aim of the present study was to find out the water quality status of the lake. The parameters conducted were Temperature, pH, Conductivity, Transparency, Dissolved Oxygen, Carbon dioxide, TSS, TS, TDS, Hardness, Ca-hardness, Mg-hardness, COD, BOD, Phosphate, Nitrate and Sulphate. The present study concluded that all the four sites were free from pollution. The physico-chemical parameters were within the permissible limit.

Keywords: *Tadoba lake, Tadoba Andhari Tiger Reserve, Pollution, Water quality.*

Introduction

Throughout history, water has been considered as a natural resource critical to human survival. Being a natural resource it has been a primary concern of man throughout his existence. Human history can infact be written in terms of interactions and interrelations between human and water (Biswas, 1964). In aquatic ecosystem, physico-chemical environment exert profound influence on its biotic components. Today, with the rapid increase in population and over exploitation for different purposes, the quality of water has been deteriorating at an alarming rate, which ultimately results in depletion of aquatic biota. In many developing countries, governments are having difficulty, coping with fecal contamination and drinking water, the degradation of freshwater resources and hazardous waste pollution. Several investigators from abroad and India have contributed their efforts in studies of various aspects of physico-chemical conditions in freshwater, Gause (1932), Lindeman (1942), Ayyappan and Gupta (1985), Khatavkar *et al.* (1989), Dhamaji and Jain (1994), Raghunathan *et al.* (2000), Khanna and Bhutiani (2003) and Kumar (2007).

Materials and Method

The study was carried out for 12 months from June, 2005 to May, 2006, during different seasons at four different sites S_1, S_2, S_3 and S_4. The main aim of present study is to investigate the physico-chemical characteristics of water. Sampling sites were chosen from the lake keeping in view the accessible area of the lake. Water samples for physico-chemical analysis were collected in five liter plastic cans during 8:30 AM to 10:30 AM. Sample for dissolved oxygen was collected in 300 ml capacity BOD bottle and fixed by Winkler A and Winkler B solution at the site. The parameters like Temperature, pH and Conductivity were analyzed with the help of Thermometer and water analysis kit developed by EIPRODUCTS, (MODEL -161 – E) respectively Transparency was measured by Secchi disc. Physico-chemical parameters were analyzed with the help of the procedures given in APHA (1998) and Bhalerao (1998).

Results and Discussion

Water Temperature

The water temperature is one of the important parameter in aquatic ecosystem. In the present study of Tadoba Lake, difference in the fluctuation of water temperature was maximum 33.60 °C to minimum 23.01 °C. The observation is in conformity with earlier reports of Vyas and Kumar (1968) who studied Indrasagar Lake, Udaipur.

pH

Hutchinson (1957) opined that nearly neutral pH of water is regulated by carbon dioxide and bicarbonates. In the present study, throughout the year the pH was alkaline at all the four sites. The pH value was recorded in the range of 7.04 to 8.60. The maximum pH was recorded in summer season. Maya *et al.* (2007) observed the pH range between 7.7 to 8.6 at Yellamallappa Chetty Lake, Bangalore.

Conductivity

Conductivity of water was fluctuated within the range of maximum 0.417 mmhos/cm in summer to minimum 0.203 mmhos/cm in winter season. The values are attributed to restricted outflow that leads to stagnation and sedimentation of dissolved salts.

Transparency

Transparency is light penetration capacity of the water. The colour of water is due to concentration of suspended organic and inorganic particles. Transparency varied from maximum 91.5 cm to minimum 39.5 cm. The maximum transparency was recorded in winter months while minimum in monsoon and moderate in summer. Shashikant and Raina (1990) and Kaushik and Saksena (1999) have reported low transparency in summer and rainy season in the different water bodies in Jammu.

Total Alkalinity

Total alkalinity of water is mainly due to cations of Calcium, Magnesium, Sodium and Potassium. It is also due to combined carbonate or bicarbonate or occasionally hydroxides. The total alkalinity of water is a measure of weak acid present in it. The total alkalinity ranges between maximum 334.0 mg/l to minimum 115.0 mg/l in monsoon and summer season, respectively.

Total Hardness, Ca Hardness and Mg Hardness

Mairs (1996) has found that the hardness and alkalinity are closely related with each other. Like alkalinity the hardness is important in decreasing the toxic effect of poisonous elements, it has also been suggested that whenever hardness is lower than alkalinity, it is due to carbonate. During the present study, total hardness ranges from 68.0 mg/l to 132.0 mg/l during summer season and winter season respectively. Lower value of hardness could be due to massive growth of phytoplankton. In earlier studies, the total hardness has been reported by Devi (1996) as 156.0 mg/l. Ca hardness of lake water was maximum in winter season. The maximum value of Mg-hardness was 59.0 mg/ltr in the summer season and minimum 24.0 mg/l in the winter season. The magnesium hardness was always observed lower than Ca-hardness.

Total Solids, T.D.S and T.S.S.

Total solids cause ecological imbalance in the aquatic ecosystem by mechanical abrasive action. In the present study, the total solids range between 512 mg/l to 1010.0 mg/l. The total dissolved solids include salt and variety of organic substances, which readily dissolve in water and often impart a degree of hardness. The value of total dissolved solids ranged between minimum 99 mg/l to maximum 267 mg/l. The maximum seasonal value is observed in monsoon and minimum value in winter season. Moreover the low value of total dissolved solids in summer and high in monsoon was observed by Trivedy *et al.* (1988). Total suspended solids in water depend upon the rate of settlement of suspended particles in the water. T.S.S. was recorded maximum during monsoon season due to high turbidity, leaching of rocks and surface flow.

Dissolved Oxygen

Dissolved oxygen is one of the most important parameter of water quality directly affecting survival and distribution of flora and fauna. The range of dissolved oxygen was from 3.90 mg/l to 9.90 mg/l. It was maximum during winter and minimum during summer season. Chavan *et al.* (2004) have recorded the range of dissolved oxygen 5.1 mg/l to 9.9 mg/l in Manjara project reservoir, Maharashtra.

Free CO_2

The free CO_2 is released during the decomposition of certain substances and metabolic activities of the living organism. The range of free CO_2 was 2.10 mg/l to 5.76 mg/l. Free CO_2 was recorded maximum during summer season and minimum in winter. In winter season the free CO_2 is low because the temperature of water decreases and high planktonic growth take place.

Chloride

The maximum value of chloride was recorded 89.41 mg/l in summer season and minimum 46.95 mg/l during winter season. Chaurasia and Adoni (1985) stated that the high concentration of chloride in summer might be due to high evaporation rate because of high ambient temperature.

B.O.D and C.O.D

Biological Oxygen Demand is a direct measure of O_2 requirement and indirect measure of biodegradable organic matter. The maximum B.O.D. was 15.20 mg/l in the month of May and minimum 3.10 mg/l in the month of January. Seenaya and Zafar (1979) similarly observed B.O.D. value minimum 2.32 mg/l during January and maximum 20 mg/l in May. Chemical Oxygen Demand indicates the extent of chemical pollution mainly from industrial effluents. The C.O.D. value ranged between 16.02 mg/l to 51.13 mg/l. The maximum value was observed in winter season and minimum in monsoon season. Singh and Roy (1995) observed minimum and maximum range of C.O.D. during the month of September to February. The higher level of C.O.D. was mainly due to high concentration of pollutants and hence the biological resistant organic substance and level of toxicity was greater in most of the months.

Sulphate

Sulphate is produced by biological oxidation of sulphur content of organic matter. The sulphate values ranged between 21.35 mg/l in month of May to minimum 9.89

mg/l during month of January. Angadi *et al.* (2005), minimum concentration of sulphate was in the month of November and maximum in March.

Phosphate

The phosphate value ranged between 0.840 mg/l to 0.137 mg/l. The phosphate content was observed to be maximum in summer and minimum during winter while moderate phosphate in monsoon. These findings were supported by the observations of previous workers (Cynthia, 1980).

Nitrate

In the present study, nitrate values ranged between 0.190 mg/l to 1.450 mg/l. Maximum of nitrate values were reported during monsoon and summer and minimum during winter. The high values of nitrate were observed in S_1 and S_4 stations during monsoon and S_2 and S_3 during summer months. Gohram (1961) observed maximum nitrate in monsoon, which is linked to heavy run-off of the organic matter from the catchments.

From the present study of physico-chemical parameters at four sampling stations it may be concluded that the lake water is not polluted and is free of contamination.

References

Angadi, S.B., Shiddamallayya, N. and Patil, P.C. 2005: Limnological studies of Papanash pond, Bidar (Karnataka)., *J. Env. Biol.* 26 (2): pp.213 -216.

APHA (1998): Standard Method for Analysis of Water and Wastewater. American Public Health Association, 17th edition. Washington DC. pp:- 1-1193.

Ayyappan, S. and Gupta T.R.C. (1985): Limnology of Ramsamudra tank primary production. Proc. Nat. Symp. Pure and Appl. Limnology (Ed) Adoni A.D. Bull. Bot. Soc. 32: pp.-16-88.

Bhalerao B.B. (1998): Manual of standard methods, Reva Environ Systems Pvt. Ltd. Nagpur.

Biswas, S. (1964): Species of Daphidae (Crustacea : Cladocera) from Shimla Hill in India, with a new record of Alona costata (Sars) from Nepha. *J. Zool. Soc. India.* 16: pp.- 92.

Chavan, R.J. Sawant, R.J., Hiware C.T. and Tat M.B. (2004): Studies on water quality of Manjara Project reservoir, Beed district, Maharashtra. *J. of Aqua. Biol.* Vol 19 (2).pp.78-84

Chourasia, S.K. and Adoni A.D. (1985): Zooplankton dynamics in shallow eutrophic lake. In: proc. Nat. symp. Pure and Appl. Limnology (Ed. A.D. Adoni), Bull Bot. Soc. Sagar, 32: pp-30-39.

Cynthia, M (1980): Ecological investigations of two lakes in Hyderabad development area ph.D Thesis, Osmania University, Hyderabad. Andhara Pradesh, India.

Devi, P.J.S.T (1996): Studies on phyco toxins of objectionable algae of ponds and reservoirs of Hyderabad, Ph.D. Thesis, Osmania University, Andhra Pradesh, India.

Dhamaji, S.K. and Yatish Jain (1994): Variations in the physico-chemical characteristic of lentic water bodies, Jabalpur. (M.P.), Env. And pollution 1 (3 and 4): pp.125 – 128.

Gause, G.F. (1932): Ecology of population. *Quartz Rev. Biol* 7: pp. 27-46.

Gohram (1961): The chemical composition of some waters from Dune slacks at Sandscale, North Lancshire. *J. Ecology*, 49 (1): pp.79-82.

Hutchinson, G.E. 1957: A treatise on limnology vol. II John Wiley and Sons. New York, 1015.

Kaushik, S. and Saksena D.N. 1999: Physico-chemical limnology of certain water bodies of central India. In: freshwater ecosystems of India (Ed. K. Vijay Kumar) Daya Publ. House. Delhi. pp.-1-58.

Khatavkar, S.D., Kulkarni A.Y. and Goel P.K. 1989: Limnological studies on two lentic freshwater bodies at Kolhapur with special reference to pollution. *Ind. J. Env.Prot.* Vol. 9 No. 3. pp.- 198-203.

Khanna, D.R. and Bhutiani, R. 2003: Ecological status of Sitapur pond at Hardwar (Uttaranchal), India. Indian J. Environ and Ecoplan.72 (1): pp.175-178.

Kumar, P. 2007: Studies of some aspect of ecology of certain fish ponds of Muzaffarpur with relation to the seasonal variation in physico-chemical characteristic of water and biotic community. *Ind.J. Env. and Eco.* (14): pp-195-199.

Liendman, R.L. 1942: The trophica- dynamic aspects of ecology. Ecol. 23:pp. - 399-418.

Mairs, D.F. 1996: A total alkalinity atlas for marine lake waters. Limnol. Ocenogr 11: pp.- 68-72.

Maya, C., Raviprasad, G., Krishnaram, H. and Lakshman, M. 2007: Limnological studies on Yellamallappa chetty lake, Banglore, Karnataka: Ind.J. Env. and Eco. (14): pp-115-118.

Raghunathan, M.G., Mahalingam, S. and Vanitadevi, K. 2000: A study on physico-chemical characteristics of Otteri Lake and Palar river water in Vellore Town, Tamil Nadu, India. *J. Aqua. Biol.* 15: (1 and 2): pp.-56–58.

Seenaya, G. and Zafar, A.R. 1979: An ecological study of Mir Alam Lake, Hyderabad, India. *Indian. J.Bot.* 2002), pp. 204 – 220.

Kant, Shashi and Raina, A.K. (1990): Limnological studies of two ponds in Jammu. *J. Env. Biol.* 11(2): pp-137-144.

Singh, J.P. and Roy, S.P. (1995): Limnobiotic investigation of Karwar Lake, Begusarai, Bihar. *Env. Eco.* 13: pp.-330 – 335.

Trivedy, R.K., Goel, P.K. Shrotri, A.C., Ghadage, M.R. and Khatavkar, S.D. 1988: Prospective in Aquaculture. *Biol.*, pp: 15-18.

Vyas, L.N. and Kumar, H.D. 1968: Studies on the phytoplankton and other algae of India Sagar tank, Udaypur, India. *Hydrobiologia.* 31: pp.421 – 434.

2013, Climate Change Effects on Agriculture and Economy *Pages* ***159–165***
Editors: **D.R. Khanna, A.K. Chopra, Vikas Singh, Rakesh Bhutiani & Gagan Matta**
Published by: **BIOTECH BOOKS, NEW DELHI**

Chapter 18

Analysis of Physico-chemical Characteristics of Junona Lake, Dist. Chandrapur, (M.S.) India

P.J. Khinchi[1], P. M. Telkhade[2], N.R. Dahegaonkar[2] and S.B. Zade[3]

[1]P.G. Department of Zoology, Janta Mahavidyalaya, Chandrapur, Maharashtra
[2]Department of Zoology, Arts, Commerce and Science College, Tukum, Chandrapur, Maharashtra
[3]P.G. Department of Zoology, RTM Nagpur University, Nagpur, Maharashtra

Reservoirs and lakes are becoming very important resources. Lakes are important resources for human life and human civilization grow along the lakes. At present man is exploiting water resources for various purposes such as agriculture, hydropower, industries, municipal supplies fishes and recreational uses. The present paper deals with monthly variations of physico-chemical characteristics of Junona Lake at Dist. Chandrapur. The lake has historical background and it was renovated by Water Resources Department of Maharashtra for irrigation purpose. Presently villagers of Junona utilize this lake water for domestic use and wildlife also use lake for drinking purpose. The present study was carried out during Feb., 2006 to Jan., 2007. Four different sites were selected for periodical sampling of physico-chemical parameters. The main aim of this study is to find out the water quality status of the lake. The parameters conducted were Temp., pH, Conductivity, Transparency, Dissolved Oxygen, Carbon dioxide,

TSS, TS, TDS, Hardness, Ca- hardness, Mg- hardness, COD, BOD, Phosphate, Nitrate and Sulphate were conducted for the period of the one year. Monthly range of these parameters is within the permissible limit at all sites except little variation were noted at site S_2.

Keywords: *Junona lake, Physico-chemical, Conductivity, COD, BOD.*

Introduction

The earth is unique in the solar system for the enormous quantities of water. About three fourth of the earth's surface (71.0 per cent) is covered with hydrosphere. It has been estimated that the hydrosphere contains about 1.46 × 109 cubic kms of water. Of this (93.0 per cent) is in the oceans 2.0 per cent in the glaciers and polar ice and 0.052 per cent is in the freshwater lakes, rivers and atmospheric moisture. Pollution from human activities is the major issue in conservation of aquatic ecosystems. The pollution invariably alters water quality, in turn, influencing diversity, biomass and survival of biotic communities. Any change in the physico-chemical environment has got its effect on biotic communities due to the fact that different species of flora and fauna exhibit variation in their responses to the altered water quality. A historical resume of abroad and India, Yoshimura (1932), Pennak and Gerpen (1947), Berg (1938), Tebo and Hassler (1961), Roy (1954), Chopra and Rehman (1995), Prakasham and Joseph (2000), Telkhade (2007), Keshre *et al.* (2008)

Materials and Method

The present work was carried out for one year *i.e.* from February, 2006 to January, 2007 on Junona Lake. Four different sites were selected for periodic sampling of water for analyzing physico-chemical parameters during 8.30 to 10.30 AM. The water samples were collected in 5 liter plastic can, brought to the laboratory and analyzed. The parameters like Temperature, pH, Conductivity and Transparency were measured on the spot. The results were calculated as per standard formulas and methods APHA (1985), NEERI (1986), Khanna (1993) and Bhalerao (1998).

Results and Discussion

Physical Parameters

Water Temperature is a measure of intensity of heat, stored in a volume of water. The temperature of water bodies varies with climatic conditions, light intensity and depth. In the present investigation, the minimum water temperature is 23.40 °C which was recorded at site S_1 and S_4 in the month of December and maximum 37.90 °C at site S_2 and S_4 in the month of May. Yeole and Patil (2005) recorded minimum temperature in the month of August and maximum in the month of May in Yedshi Lake (M.S).

pH is the value expressed as negative logarithm of the hydrogen ion concentration. Any change in pH of water is accompanied by changes in the other physico-chemical aspects of the medium. Minimum pH 7.01 was recorded at site S_1 in the month of November and maximum 8.31 at site S_4 in the month of June. Rana (1991) recorded the alkaline pH 7.0 to 8.6 in the Lake Udaisager, Udaipur. In the present investigation, the maximum pH recorded at site S_1 and S_3 may be due to decreased volume of water by evaporation. The low pH in winter may be due to short day length and decrease in photosynthetic activity.

Transparency of water body is affected by the factors like planktonic growth, rainfall, sun's position in the sky, angle of incidence of rays, cloudiness, visibility and turbidity due to suspended inert particulate matter. In the present investigation, the minimum transparency 27.50 cm was recorded at site S_4 in the month of August and maximum 86.50 cm at site S_3 in the month of March. Pulle (2000) observed the transparency values in the range of 28.4 cm to 87.8 cm in Isapur dam. In the present investigation, the low values of transparency were recorded, which might be due to influx of rain water from catchment area, cloudiness, less penetration of light and high turbidity due to suspended inert particulate matter. However, high values of transparency in summer may be due to clear atmosphere and high light penetration.

Conductivity and Dissolved Solids are directly proportional to each other and these are mainly due to ionic composition of water. In the present investigation, the minimum conductivity 0.119 mmhos/cm was recorded at site S_3 in the month of August 2006 and maximum 0.385 mmhos/cm at site S_4 in the month of May 2006. Lubana *et al.* (2007) reported the conductivity of average fifteen months, 0.363 mmhos/cm in Beechwal reservoir, Bikaner. The seasonal variation in conductivity may be due to increase in concentration of salts entering the lake with sewage.

Chemical Parameters

Dissolved Oxygen is one of the most important parameter of water quality directly affecting survival and distribution of flora and fauna in an ecosystem. In the present investigation, the minimum dissolved oxygen 3.12 mg/l was recorded at site S_4 during the month of August, 2006 and maximum 10.30 mg/l at site S_1 during the month of January 2007. Jain *et al.* (1996) recorded the range of dissolved oxygen from 5.80 mg/l to 10.00 mg/l in Haldi reservoir of District Vidisha (M.P.). The high dissolved oxygen value during winter period could be due to increased rate of photosynthesis by increased density of phytoplankton in relatively stable environmental condition.

Free CO_2 plays a vital role in aquatic ecosystem as it influences the metabolic and developmental rates and distribution of some aquatic invertebrates. Respiratory activity of aquatic organism and process of decomposition are important sources of CO_2 in water bodies. In the present investigation, the minimum CO_2 was recorded 2.23 mg/l at site S_1 in the month of December and maximum 5.70 mg/l at site S_2 in the month of March. The maximum values recorded during summer season may be due to increase in atmospheric temperature and biological oxidation of organic matter. However, low values of free CO_2 during winter may be due to maximum utilization of CO_2 by more algal blooms in winter, decreased water temperature and low diffusion from air. Bandela *et al.* (1998) recorded free CO_2 range between 3.08 to 7.48 mg/l in Barul dam water (M.S.).

Biological Oxygen Demand represents the amount of O_2 required for the microbial decomposition of organic matter in the water. There is a close relationship between dissolved oxygen and organic matter. In the present investigation, the minimum value of B.O.D. 4.01mg/l was recorded in the month of December at site S_2 and maximum 15.26 mg/l in the month of June at site S_4. The maximum B.O.D was recorded in summer season may be due to highest biological activities and minimum during winter due to lower temperature. Solanki (2006) reported B.O.D. values from 24.2 mg/l to 89.0 mg/l in Pandu Lake of Bodhan (A.P.).

Chemical Oxygen Demand is related to Biological Oxygen Demand. Degree of organic pollution can be assessed by Chemical Oxygen Demand. It is the measurement of oxygen required in oxidizing organic compound by involving a chemical oxidant as potassium chromate or potassium permanganate. In the present study, minimum COD 12.20 mg/l was found in the month of January and December and maximum value was observed 44.30 mg/l respectively. The maximum values in summer may be due to less quantity of water, death and decay of aquatic flora and fauna and minimum in winter was due to settlement of organic matter and dilution effect. Similarly, Bhatnagar *et al.* (2007) studied Jhamri dam water and recorded the value of C.O.D between 1.2 mg/l to 3.8 mg/l.

Chloride occurs in all natural waters in widely varying concentration. Chloride content normally increases as the mineral content increases. The ecological significance of chloride lies in its potential to regulate the salinity of water and exerts consequent osmotic stress on biotic communities. In the present investigation, the minimum value of chloride 38.20 mg/l was found in month of November, 2006 and maximum 105 mg/l during May, 2006. From the above observation, it can be concluded that there was no definite pattern of chloride fluctuation, lower value during winter could be attributed to dilution effect and renewal of water mass, alter summer stagnation and also may be due to high sedimentation rate on relatively stable environmental condition. Subbamma and Ramasarma (1992) noticed the range of variation in the chloride content from 43.88 to 180.00 mg/l in Temple pond near, Machilipatname.

In the present investigation, the minimum total hardness 93.00 mg/l was recorded at site S_3 during the month of December and maximum 191.00 mg/l at site S_2 during the month of June. Pawar and Pulle (2005) recorded total hardness range from 128.00 to 168.00 mg/l in Pethwadaj dam, Nanded (M.S.). During the present investigation, the Ca-hardness values varied between 57.00 mg/l to 115.00 mg/l. Ca-hardness values were observed minimum at site S_1 while maximum at Site S_4. Anitha (2002) observed the range of Ca-hardness between 15.00 to 98.00 mg/l in Mir Alam Lake, Hyderabad. In the present investigation, the minimum Mg-hardness 31.00 mg/l was recorded during December and maximum 81.00 mg/l in month of June. Rajashekhar *et al.* (2007) reported the Mg-hardness values between 18.00 to 34.00 mg/l in minor reservoir Nandergul, Ranga Reddy District (A.P).

Total Solids, turbidity and dissolved matters are found closely interrelated with one another and cause common effect upon the pond and its aquatic life (Khanna, 1993). The total dissolved solids include salts and variety of organic substances which readily dissolve in water and often impart a degree of hardness. The excessive total dissolved solids generally affect palatability. In the present investigation the minimum total dissolved solids 112.00 mg/l was found in the month of January at site S_4 and maximum 340.00 mg/l in July month at site S_2. The maximum TDS was may be due to the release of decaying matter from aquatic vegetation and surface run off from catchment area. In the present investigation, T.S. and Total Suspended Solids were recorded minimum 410.00 mg/l and 250.00 mg/l in the month of December at site S_1 and maximum 1020.00 mg/l and 702.00 mg/l in the month of July at site S_2. Salve and Hiware (2006) recorded the range of total solids from 200.00 to 570.00 mg/l in Wanaprakalpa reservoir, Nagpur, near Parli Vaijanath Dist. Beed, Maharashtra.

Sulphate usually occurs in natural water and is the second most common anion being derived from most sedimentary rocks. In lakes, sulphur is cyclic and involve organically reduced from gas as well as the common free sulphate ion. In the present investigation, minimum sulphate was recorded 10.80 mg/l at site S_1 in the month of December and maximum 27.35 mg/l at site S_4 in the month of May. The maximum sulphate concentration was noted during summer may be due to biodegradation and low water level. However, the low sulphate concentration during winter may be due to the dilution and utilization of Sulphate by aquatic plants. In the present investigation, the minimum values of phosphate was observed 0.295 mg/l during December, 2006 at site S_2 and maximum 1.300 mg/l during May 2006 at site S_3. Kumavat and Jawalekar (2003) studied fish pond at Anjale, District Jalgaon and recorded the values of Phosphate between 0.440 mg/l to 0.315 mg/l.

The formation of nitrates from ammonium ions in water is mediated by aerobic bacteria, chiefly *nitrobacter*, which utilizes the fall in the free energy accompanying the reaction to reduce carbon dioxide to organic carbon. In the present investigation, the minimum nitrate concentration 0.210 mg/l was recorded at site S_3 during the month of December, 2007 and maximum 1.500 mg/l at site S_4 during the month of June, 2006. Solanki (2006) recorded nitrate values between 0.62 ppm to 1.60 ppm in Bellal Lake, Bodhan (A.P).

From the above study, it may conclude that all the physico-chemical parameters were in permissible limit at all the sites, except little variation at site S_2. It is suggested that proper measures are necessary to avoid contamination in the lake.

References

Anitha, G. 2002: Hydrography in relation to benthic macro-invertebrates in Mir-Alam Lake Hyderabad, Andhra Pradesh, India. Ph.D. Thesis submitted to Osmnaia University, Hyderabad. A. P. India.

APHA, 1985: Standard Methods for the Examination of Water and Wastewater, 16th Edn. American Public Health Association, Washington D.C.

Bandela, N.N., Vaidya, D.P. and Lomte, V.S. 1998: Seasonal temperature changes and their influence on the level of Carbon dioxide and pH in Barul dam water. *J. Eco. Biol.*13 (1): 43-46.

Berg, K. 1938: Studies on the bottom animals of Esrom Lake - K. Dan Vid. Selsk. Skr. Nat. Mat. Aid. 9 (8): 1-255.

Bhalerao, B.B. 1998: Manual of standard methods, Reva Environ. Systems Pvt. Ltd. Nagpur.

Bhatnagar, Chhaya, Sharma, Vinita, Jani, Karnika and Gill, Nidhi, 2007: Plankton and ichthyo-fauna of Jhamri Dam, Udaipur, Rajasthan. NSL: 236 - 238.

Chopra, A.K. and Rehman, A. 1995: A study on self-purification of Physico-chemical properties of Ganga canal water at Jwalapur, Hardwar. *Him. J. Env. Zool.* (9): 11 - 13.

Jain, S.M., Sharma, M. and Thakur, R. 1996: Seasonal variations in physico-chemical parameters of Halali reservoir of Vidisha District, *India. J. Ecobiol.*, 8 (3): 81-188.

Keshre, V., Mudgal, L.K., Khanna, D.R., Matta, G., Kumar, D. and Singh, V. 2008: A study of Plankton population in the Moghat reservoir at Khandwa Dist. (M.P.). *Environment Conservation Journal*, 9 (1 and 2): 73-79.

Khanna, D.R. 1993: Ecology and Pollution of Ganga River. *Ashish Publishing House*, Delhi. 1-241.

Kumavat, D.A. and Jawale, A.K. 2003: Phytoplankton Ecology of a fish pond at Anjale Distt. Jalgoan (M.S.). *J. Aqua. Biol.* 18 (1): 9-13.

Lubana, Harbhan Kaur, Rathore, N.S., Singh, Y., Khatri, A.K., Sharan, L. and Saxena, M.M. 2007: Transformation in planktonic component and nutrient regime in a reservoir in the Rajasthan desert over a decade. NSL: 336-339.

NEERI., 1986: Manual on water and waste watger Analysis. National Environmental Engineering Research Institutte. Nehru Marg, Nagpur, India.

Pawar, S.K. and Pulle, J.S. 2005: Qualitative and quantitative study of zooplankton in Pethwadaj Dam, Nanded, District. (Maharashtra), India. *J. Aqua. Biol.* 20(2): 53-57.

Pennak, R.W. and Gerpen E.D.V. 1947: Bottom fauna production and physical nature of the substrate in Northern Colorado trout stream. *Ecology* (28): 42–48.

Prakasham, V.R. and Josepeth, M. 2000: Water quality of Sasthmacotta Lake, Kerala in relation to primary productivity and pollution from anthropogenic sources. *J. Envn. Biol.* 21 (4): 305–307.

Pulle, J.S. 2000: Biomonitoring of Isapur Dam water. Ph.D. Thesis, Swami Ramanand Teerth Marathwada University, Nanded.

Rajashekhar, A.V., Lingaiah, M.S., Rao, Satyanarayana., Piska, Ravi Shankar. 2007: The studies on water quality prameters of a minor reservoir, Nadergul, Rangareddy district, Andhra Pradesh. *J. Aqua. Biol. Vol.* 22 (1): 118-122.

Rana, B.C. (1991): Limnology of a Eutrophic Lake udaisagar. Indian J. Ecol. 18 (2): 161- 165.

Roy, J.C. (1954): Periodicity of the planktonic diatoms of the Chilka Lake for the years 1950- and 1951. *J. Bomb. Nat. Hist. Soc.* (32): 112-122.

Salve, B.S. and Hiware, C.J. 2006: Studies on water quality of Wanparakalpa reservoir, Nagapur, near Parli Vaijnath, Dist. Beed. Marathwada region. J.Aqua. Biol. 22 (2): 113-117.

Solanki, V.R. 2006: Ecological studies on Bellal and Pandu Lakes of Bodhan, A.P. India. Ph.D. Thesis submitted, Osmania University, Hyderabad.

Subbamma, D.V. and Ramasarma D.V. 1992: Studies on the water quality characteristics of a temple pond near Machilipatnam, Andhra Pradesh, *J. Aqua. Biol.* (7): 22 - 27.

Tebo, L.B. and Hassler W.W. 1961: Seasonal abundance of aquatic insect in western North Carolina trout streams. *J. Elisha Mitchell Scient. Soc.* (77): 249–259.

Telkhade, P.M. 2007: Hydrobiology and Biodiversity of Tadoba Lake, Chandrapur. Ph.D. Thesis submitted to R.T.M. Nagpur University, Nagpur.

Yeole, S.M. and Patil, G.P. 2005: Physico-chemical status of Yedshi Lake in relation to water pollution. *J. Aqua. Biol.* 20 (1): 41- 44.

Yoshimura, S. 1932: Seasonal Variation in the content of nitrogenous compounds and phosphate in the water of Takasuka pond. Saitma. *Japan. Arch. Hydrobiol.* 24 (1): 155-176.

2013, Climate Change Effects on Agriculture and Economy *Pages* **167–172**
Editors: **D.R. Khanna, A.K. Chopra, Vikas Singh, Rakesh Bhutiani & Gagan Matta**
Published by: **BIOTECH BOOKS, NEW DELHI**

Chapter 19

Physico-chemical Aspects of River Erai Near Chandrapur Maharashtra

N.R. Dahegaonkar[1], *P.M. Telkhade*[1] *and W.R. Bhandarkar*[2]
[1]*Department of Zoology, Arts, Commerce and Science College, Tukum, Chandrapur, Maharashtra*
[2]*Department of Zoology, N.H. College, Brahmapuri, Dist. Chandrapur, Maharashtra*

Growing population densities, progressive industrialization and intensification of agriculture are contributing to increase surface water pollution. An evaluation of physico-chemical conditions prevailing in any lotic ecosystem indicates various aspects of water quality and its primary requirements to put the water to specific use and also express the nature of biota therein.

The present investigation attempts to study the water quality of river Erai near Chandrapur. River Erai, the main tributary of Wardha River is the major source of drinking water which receives sewage, municipal and industrial effluents from neighboring industries. Although number of rivers have been extensively studied for water quality in India, except few there is no much work on this aspect on river Erai. Therefore, in the present study the physico-chemical characteristic of water quality was investigated.

The water quality and pollution load was monitored at monthly intervals for a period of two years (June, 2005 to May, 2007). The data is compiled statistically to study the seasonal response of the river for domestic pollution. The quality of water was assessed by analyzing various physico-chemical parameters, such as Water Temperature, pH, Conductivity, Total alkalinity, Total Dissolved Solids, Dissolved Oxygen, Free CO_2 etc. The high value of BOD and COD is an indication of high degree of organic contamination.

Keywords: *Water quality, Erai River, Chandrapur, Physico-chemical.*

Introduction

Manmade, natural, chemical, biological and physical processes together with hydrology influence water quality. Growing population densities, progressive industrialization and intensification of agriculture are contributing to increased pollution of surface water around the globe (Soman *et al.*, 1997). There is a need for regular monitoring of river systems to assess river water quality. Assessing seasonal variation in the river water quality is important in drinking water supply as well as pollution load of the receiving water.

In India, numbers of studies on physico-chemical characteristics of river water have been extensively carried out (Olaniya *et al.*, 1976; Mitra, 1982; Shyam Sunder, 1988; Shivanikar, 1999; Mane and Pawar, 2007). River Erai, the main tributary of river Wardha is the major source of drinking water, which receives sewage, municipal and industrial effluents from neighboring industries and baring few reports (Sawane 2002, 2004 and 2007), no attempts were made to analyze the water quality of the river.

Materials and Method

The water samples were collected in polythene bottles (APHA, 1975) once in a month for two years *i.e.* from June, 2005 to May, 2007, from the selected sampling site for analyzing the water quality parameters. The recorded data was segregated in three seasons. The concentration of D.O. was measured by modified Winkler's method and expressed in mg/l and BOD was measured by keeping sample in controlled environment for 5 days at 20.0 °C.

Results and Discussion

The physical factors influencing water temperature include basin morphometry, altitude, topography and vegetation (Unni, 1990), current, velocity and discharge (Smith, 1972). During the present investigation, the maximum value of water temperature was observed in summer months and minimum in winter months. The low water temperature in winter can be explained on the basis of high water level and lower solar radiation whereas higher in summer was on account of low water level, greater solar radiation and clear atmosphere (Vyas, 1968).

During the study, a very good pattern of pH change was noted. Its low values were recorded in winter and high in rainy season. High conductivity values are not suitable for agricultural or for drinking purposes (Atthappan *et al.*, 1992). In the present investigation, high value of pH was recorded during summer. High concentration of municipal wastes and domestic activities and less flow of river during summer is responsible for increase of ionic contents in water with consequent increase in conductivity. Bobori (1996) has reported that high EC values indicate water is rich in electrolytes.

Natural bodies of water in tropics usually show a wide range of fluctuations in total alkalinity values depending upon the location, season, nature of bottom deposits, waters of hill stream, sandy rocky or clayey areas. During study, high values of alkalinity were recorded during summer and low during winter. Adebisi (1981) in seasonal upper Ogun River, Ibadafan, while reporting the seasonal variation in alkalinity reported minimum total alkalinity during the winter and maximum during summer.

In the present investigation, higher values of TDS were recorded during summer and are due to receiving large quantity of domestic and industrial effluents. The total suspended solids are very important to know water quality of the river. In the present study minimum values of TSS was during winter and maximum during monsoon. It is maximum during monsoon might be due to rainfall and influx of dissolved contents from the catchment. The high content elevates the density of water, influences osmoregulation of freshwater organisms and reduces solubility of gases like oxygen and utility of water for drinking, irrigation and industrial purpose (Harishkumar, 1998).

Higher value of dissolved oxygen was recorded during winter, which might be due to fact that the solubility of D.O. increases with decrease in water temperature. Campbell (1970), Agarwal *et al.* (1976) also observed similar seasonal fluctuation. Pophali *et al.* (1990) reported gradual decrease of dissolved oxygen due to higher consumption for oxidation of organic matter by microorganisms at polluted stations.

Table 19.1: Seasonal Mean Values of Physico-chemical Parameters of River Erai during 2005–2006

Sl.No.	*Parameters*	*Monsoon*	*Winter*	*Summer*	*Average*
1.	Temperature	30.45±1.38	26.53±1.48	31.20±±2.79	29.39±2.05
2.	pH	7.85±0.038	7.62±0.04	7.80±0.05	7.75±0.10
3.	Conductivity	0.277±0.02	0.300±0.01	0.369±0.02	0.315±0.03
4.	Transparency	58.87±4.43	79.50±10.34	101.25±6.75	79.87±17.30
5.	Alkalinity	176.50±28.73	159.75±12.34	232.00±45.04	189.42±30.87
6.	Hardness	155.50±41.3	80.75±10.28	141.50±34.30	125.92±33.13
7.	Ca-Hardness	105.75±22.3	50.50±4.72	88.25±14.65	81.50±23.05
8.	Mg-Hardness	49.75±16.93	30.25±6.30	53.25±19.83	44.42±10.11
9.	T.S.	1142.0±163.59	620.3±117.1	783.8±184.4	848.7±217.9
11.	T.D.S.	233.0±56.86	181.3±22.52	302.5±61.63	238.9±63.34
10.	T.S.S.	909.0±142.4	439.0±134.1	481.3±125.4	609.8±160.8
12.	D.O.	5.03±0.62	7.12±0.56	5.43±1.02	5.86±0.90
13.	CO2	3.84±0.93	3.20±0.79	5.59±0.31	4.21±1.01
14.	Chloride	170.50±40.73	43.35±8.12	56.78±5.10	90.21±57.03
15.	B.O.D.	7.39±1.8	5.03±1.55	9.18±1.28	7.20±1.69
16.	C.O.D.	21.78±3.56	45.35±9.13	43.93±4.51	37.02±10.79
17.	Sulphate	34.86±9.75	17.28±3.82	22.23±9.34	24.79±7.40
18.	Phosphate	0.732±0.18	0.380±0.15	0.765±0.23	0.626±0.17
19.	Nitrates	0.970±0.23	0.350±0.08	0.650±0.08	0.657±0.25

Table 19.2: Seasonal Mean Values of Physico-chemical Parameters of River Erai during 2006–2007

Sl.No.	Parameters	Monsoon	Winter	Summer	Average
1.	Temperature	30.50±1.54	26.28±1.19	31.03±2.92	29.27±2.12
2.	pH	7.97±0.17	7.66±0.14	7.89±0.05	7.84±0.13
3.	Conductivity	0.261±0.05	0.293±0.00	0.327±0.03	0.294±0.02
4.	Transparency	61.75±9.90	82.50±12.80	95.00±10.68	79.75±16.08
5.	Alkalinity	164.25±19.74	125.75±14.79	222.50±47.19	170.83±39.77
6.	Hardness	174.00±50.45	106.25±11.99	157.50±27.54	145.92±29.66
7.	Ca-Hardness	110.00±25.52	71.75±10.35	103.50±17.74	95.08±16.71
8.	Mg-Hardness	64.00±26.79	34.50±5.94	54.00±2.69	50.83±12.11
9.	T.S.	1163.8±97.0	681.5±153.1	849.3±78.4	898.2±199.9
11.	T.D.S.	241.0±32.13	219.8±16.02	277.0±51.60	245.9±35.94
10.	T.S.S.	922.8±92.9	461.8±137.2	572.3±32.2	652.3±173.6
12.	D.O.	5.02±0.60	7.29±0.51	5.63±1.26	5.98±0.95
13.	CO2	4.42±0.76	3.19±0.25	5.61±0.47	4.41±0.98
14.	Chloride	44.50±14.89	27.40±9.08	32.93±5.40	34.94±7.12
15.	B.O.D.	8.80±1.15	4.75±1.18	6.48±0.46	6.68±1.65
16.	C.O.D.	30.28±4.74	42.70±2.48	39.48±9.52	37.49±5.26
17.	Sulphate	39.88±3.79	25.83±5.98	40.25±4.85	35.32±6.71
18.	Phosphate	0.712±0.22	0.249±0.04	0.798±0.26	0.586±0.20
19.	Nitrates	0.880±0.16	0.380±0.13	0.560±0.17	0.607±0.20

Shyam Sunder (1988) reported peak of CO_2 in summer months which might be due to accelerated rate of decomposition of organic matter by microbes in the bottom resulting in the rapid production of free CO_2. During the study minimum value of free CO_2 was recorded during winter and maximum during summer. The minimum value of free CO_2 during winter months might be due to it's utilization through photosynthetic activity by the aquatic macrophytes and phytoplankters. However, the higher values of CO_2 can be attributed to the higher rate of decomposition with consequent increased release of free CO_2, decrease in utilization in photosynthetic activity and high respiratory activity of benthos and microbes.

Palharya and Malviya (1988) observed lowest B.O.D. during winter and reported it to be due to lower temperature and in summer highest biological activities contributed to higher BOD values. In present investigation also, low BOD value was recorded during winter and high during summer.

The C.O.D. values always recorded higher than B.O.D. values clearly indicate that the non-biodegradable oxygen demanding pollutants are present in the water, which comes into the system as effluents. During present study maximum COD values were recorded in winter and minimum in monsoon. Shah (1988) reported maximum values during December and stated that high values in winter were probably, as a result of death and decay of plants during winter and the consequent increase in organic matter.

In the present investigation, maximum sulphate values were recorded during the monsoon season and minimum during winter season. Singh and Singh (1990) reported maximum value during monsoon month due to municipal sewage in Subarnarekha River at Ranchi (Bihar).

Phosphate in large quantities in freshwaters indicates pollution through sewage and industrial wastes. It promotes growth of nuisance causing microorganisms. During study maximum phosphate was recorded during summer and minimum during winter. Ansari (1993) reported high values of soluble phosphates in summer due to enhanced rate of decomposition in river Godavari at Nanded.

Significant sources of nitrates in water are from chemical fertilizers, domestic effluents and industrial discharges and due to changes in land used patterns. The maximum values of nitrates recorded during rainy season due to influx of nitrogen rich floodwater that brings large amount of contaminated sewage water.

It is inferred that, river is sustaining the present pollution load. However, logarithmic growth of human population and habitat expansion in near future is likely to influence the water quality of the river system.

References

Adebisi, A.A. 1981: The physicochemical hydrology of tropical seasonal river Upper Ogun., Hydrobiologia. 79: 157-165.

Agarwal, D.K., Gaur, S.D., Tiwari, I.C., Narayanswamy, N. and Marwah, S.M. 1976: Physico-chemical characteristics of the Ganges at Varanashi. *Indian J. Environ. Hlth.* 18: 201-206.

Ansari, M.A. 1993: Hydrobiological studies of Godavari river water at Nanded., Ph. D. Thesis submitted, Marathwada University, Aurangabad.

APHA, 1975: Standard methods for the examination of water and wastewater, 14th Ed. APHA-AWWA- WPCF. Washington DC. 20036.

Atthapan, P.R., Sethuraman, K. and Kannan, N.(1992: A study on the pollution of river Vaigai at Madurai. *Indian J. Env. Proct.* 12 (11): 818 - 823.

Bobori, D. 1996: Bioaccumulation of heavy metals in the ecosystem of lake Koronia (Macedonia, Greece). Bios. (Macedonia, Greece). 4: 171-174.

Campbell, I.C. 1970: A biological investigation of an organically polluted Surbon stream of Victoria. Aust., March freshwater Res. 29: 275-291.

Harishkumar, 1998: Modern methods of pollution analysis vol. 2., 1st edition, 1998, Sarup and Sons publication, New Delhi, ISBN.

Mane, A.M. and Pawar, S.K. 2007: Some physicochemical properties of Manar river in Nanded Dist. Maharashtra. *J. Aqua. Biol.* 22(2) 88- 90.

Mitra, A.K. 1982: Chemical characteristics of surface water at selected gauging stations in the river Godavari, Krishna and Tungabhadra., *Ind. J. Environ. Hlth.* 24 (33): 165-179.

Olaniya, M.S., Saxena, K.L. and Sharma, H.S. 1976: Pollution studies of Chambal river and its tributaries at Kota. *Indian J. Environ. Hlth.* 18: 219-266.

Palharya, J.P. and Malviya, S. 1988: Pollution of Narmada river at Hoshangabad in Madhya Pradesh and suggested measures for control., Ecology and Pollution of Indian rivers (Ed. Trivedy R. K.), Ashish Publishing House, New Delhi. 55- 86.

Pophali, S., Siddiqui, S. and Khan, L.H. 1990: Study of some physico-chemical parameters to asses pollution load in Patra river. *Indian J. Env. Proct.* 10(3): 203-207.

Sawane, A.P. 2002: Impact of industrial pollution on water quality parameters form Erai river, Chandrapur (M.S.), Ph. D. Thesis, submitted North Maharashtra University.(M.S.)

Sawane, A.P., Puranik, P.G. and Bhate, A.M. 2004: Assessment of water quality of river Irai (Dist. Chandrapur) on the basis of seasonal fluctuations in DO and BOD., *J. Eco.. Occ. Hlth.* 4: 117-122.

Sawane, A.P., Puranik, P.G. and Bobdey, A.D. 2007: Seasonal distribution of trace elements in water of Irai river, Dist. Chandrapur, Maharashtra. NSL - 2007. 66: 376-379.

Shah, A.R. 1988: Physicochemical aspects of pollution in river Jhelum Kashmir during 1981-83., Ecology and Pollution of Indian rivers, (Ed. Trivedy R. K.), A. P. H. New Delhi.132-163.

Shivanikar, S.V., Patil, P.M., Vaidya, D.P. and Bandela, N.N. 1999: Environmental temp. fluctua-tion determine dissolved oxygen level in Godavari river water. *Poll. Res.* 18 (4): 415-418.

Shyam, Sunder. 1988: Monitoring the water quality in a stretch of river Jhelum (Kashmir). Ecology and Pollution of Indian rivers, (Ed. Trivedy R. K.) A. P. H. New Delhi.114 -131.

Singh, D.K. and Singh, C.P. 1990: Pollution studies on river Subarnarekha around industrial belt of Ranchi (Bihar). *Indian J. Environ. Hlth.* 32 (1): 26-33.

Smith, K. 1972: River water temperatures: An environmental review. *Scot. Geogr. Mag.* 211-220.

Soman, K., Mahamaya, C. and Ouseph, P.P. 1997: Status of riverine pollution in S.Kerala and its relation to physiography and land use. Proc.IX Kerala Sci. Cong. Thiruvananthapuram 93-95.

Unni, K.S. 1990: Ecology of river Narmada., A. P. H. Publishing House, New Delhi.

Vyas, L.N. 1968: Studies on phytoplankton ecology of Picchola lake Udaipur. Proc. Symp. Recent Adv. Trop. Ecol. 334-337.

2013, Climate Change Effects on Agriculture and Economy *Pages 173–186*
Editors: **D.R. Khanna, A.K. Chopra, Vikas Singh, Rakesh Bhutiani & Gagan Matta**
Published by: **BIOTECH BOOKS, NEW DELHI**

Chapter 20

Copper Induced Phytotoxicity and Differential Tissue Specific Responses of Antioxidant System as Plant Defense in *Vigna radiata* L. Wilkzek

Shaifali Mathur and G.S. Shekhawat
Department of Bioscience and Biotechnology,
Banasthali University, Banasthali, Rajasthan

Ecological contamination by heavy metals is one of the major contributors of plant abiotic stress. Toxic effect of heavy metal pollution on plant growth and antioxidant system were evaluated in hydroponically grown seedlings of *Vigna radiata* under Cu stress after 72 and 120 hrs. Reduction in seedling growth in the presence of Cu was observed as root length and dry weight at 200 µM decreased 3 folds and 1.6 folds in comparison to control. Increased level of MDA content in treated seedlings reveal the oxidative stress conditions caused by Cu. Chlorophyll content decreased throughout at all treatments however, protein content in shoot showed a slight increase at lower Cu concentrations. At low Cu treatments proline and ascorbic acid content increased significantly. In roots antioxidant enzymes (SOD, APX) exhibit significant induction at low Cu treatments however, CAT activity decreased after 5µM. In shoots activities of these enzymes increased even at higher Cu concentrations (50-100 µM) suggesting significant tissue differences in response to Cu stress in *Vigna radiata* is closely related to differences in activities of antioxidant enzymes and compounds. Conclusively Cu stress has a pronounced effect on growth of *Vigna radiata* L. Wilkzek and antioxidants as well as enzymes play a major role in plant defense system when exposed to Cu pollution.

Keywords: *Vigna radiata L. Wilkzek, Copper toxicity, Oxidative stress, Antioxidant enzymes.*

Introduction

Heavy metal contamination has become one of the main abiotic stress agents that severely inhibit plant growth and productivity because of their increasing use in industrial development and high bioaccumulation and toxicity. Crop improvement under metal stress conditions requires an understanding of the mechanisms that enable metal tolerance. Copper (Cu) is an essential micronutrient for normal growth and metabolism in plants. Nevertheless excess Cu is a potential metal pollutant in soils, that impairs important cellular processes, causing chlorosis, necrosis, stunting, leaf discoloration, die-back, reduced photosynthetic activity, inability to produce seeds and inhibition of root growth in agriculturally important crops (Shekhawat *et al.*, 2009). Cu being a redox active metal catalyzes the production of ROS *via* Fentons-Harbour-Weiss reaction and hence create oxidative stress conditions. Overproduction of these ROS indiscriminately attack virtually all macromolecules, leading to serious damage of lipid membranes (peroxidation), protein denaturation, and often leads to impermeable metabolic dysfunctions (Li *et al.*, 2005). Plants get hold of efficient metal tolerance machinery including enzymatic and non-enzymatic components, enabling them to cope with metal toxicity in tissue and ROS formation induced by metal presence, for adequate functioning of metabolic pathways (Cho *et al.*, 2003). Among them, synthesis of antioxidant compounds such as proline and ascorbic acid and enhancement of antioxidant enzyme activities are most important. In addition tissue damage occurs when the capacity of antioxidant systems become lower than the amount of ROS generated. Therefore regulated production of ROS and maintenance of redox homeostasis are essential for the physiological health of organisms (Verma *et al.*, 2008). The deleterious effects of ROS may be alleviated by several antioxidative enzymes. Among them SOD (EC1.15.1.1) act as a primary line of antioxidant defense and catalyses the dismutation of superoxide radical to hydrogen peroxide (H_2O_2) and molecular oxygen. CAT (EC1.11.1.6) can transform bulk of H_2O_2 to water and scavenges photo respiratory H_2O_2. APX (EC1.11.1.11) use ascorbate as the preferential electron donor and has a higher affinity for H_2O_2, hence APX is involved in scavenging of small amounts of H_2O_2 at more specific locations (Apel and Hirt, 2009). Thus co regulation of different metabolic pathways can contribute to more noticeable reduction of negative effects induced by heavy metals presence.

Vigna radiata belongs to family Fabaceae (Pea family) is the major pulse crop of India and is widely cultivated around the world. Mungbean seeds are sprouted for fresh use or canned for shipment to restaurants. Because of their major use as sprouts and pulses, a high quality seed with excellent germination is required (Somta and Srinives, 2007). The present investigation aims to evaluate the change in the physiological and biochemical responses of *Vigna radiata* in relation to its growth, and antioxidant defense under excessive Cu in order to contribute to an understanding of *V. radiata* adaptation to environmental stress and to identify the biochemical and metabolic limitations for reduced plant performances under stress conditions and ultimately to enhance plant protection under Cu stress conditions.

Materials and Method

Plant Material and Experimental Design

Graded dry seeds of Mung beans (*Vigna radiata* L. Wilkzek) were procured from Krishi Vigyan Kendra, Banasthali University and used as the experimental material. Seeds were surface sterilized with 5 per cent sodium hypochloride for 5 minutes and rinsed five times with autoclaved distilled water. Seeds were germinated in dark for 4 days on blotting paper moistened with distilled water in petridishes. Seedlings were transferred to hoagland nutrient solution (pH- 6.8), for hydroponic cultures and maintained in growth chamber at 25.0 ± 2 °C temperature, 50-55 per cent relative humidity and an illumination of 175 μmol $m^{-2}s^{-1}$ light intensity. Nutrient solutions were renewed weekly to prevent media depletion. Seedlings grown hydroponically for 7 days were used in the experiment and further cultured in Cu enriched solutions containing different concentrations *viz*; 0, 5, 10, 50, 75, 100, 200 μM (added in the form of $CuCl_2.2H_2O$) and maintained in 800 ml of hoagland nutrient solution. Control did not contain any additional Cu. Basic hoagland nutrient solution contained 0.05 μM Cu as micronutrient. All morphological and biochemical parameters were analyzed on fresh root and shoot tissues of *Vigna radiata* seedlings after 72 hrs and 120 hrs of exposure to Cu metal.

Growth Parameters

Growth was measured in terms of root length, shoot length and fresh weight and dry weight of individual seedlings at 72 and 120 hrs of Cu exposure. Dry weights of harvested plants were also recorded after drying the samples at 70.0 °C for 24 hrs.

Protein Extraction and Estimation

Total soluble protein was extracted by homogenizing root tissues in phosphate buffer (pH-7). The extract was centrifuged at 10000 rpm at 4.0 °C. The supernatant was used for protein estimation following the method of Lowry *et al.* (1951).

Chlorophyll Estimation

Chlorophyll was extracted by homogenizing fresh root tissues in chilled acetone on ice-cold mortar pestle according to the method of Arnon (1949). The extract was centrifuged at 10000 rpm for 10 minutes and absorbance of supernatant was recorded at 645 and 663 nm.

Lipid Peroxidation Estimation

The concentration of MDA was estimated using the method of Heath and Packer (1968). Fresh tissues were homogenized in 10 ml solution containing 0.25 per cent w/v thiobarbituric acid in 10 per cent w/v trichloroacetic acid and the resulting mixture was centrifuged at 1000 rpm for 20 minutes after heating at 95 °C for 30 minutes in water bath. The absorbance of resulting supernatant was taken at 532 and 600 nm.

Free Proline Estimation

Proline content was estimated following the method of Bates *et al.* (1973). Fresh tissues were homogenized in 5 ml of 3 per cent sulphosalisylic acid and centrifuged at 12000 rpm for 20 minutes at 4.0 °C. Supernatant was mixed with acid ninhydrin solution (1.25 g ninhydrin, 30 ml of glacial acetic acid and 20 ml of 6 M phosphoric acid) and glacial acetic acid and heated at 100 °C for 1 hour. The reaction was terminated in ice bath and toluene was added to each tube. The absorbance of upper phase was recorded at 520 nm.

Ascorbic Acid Estimation

Ascorbic acid content was measured by the method of Zhang and Kirkham (1996) with little modifications. Tissues were homogenized in chilled TCA. Homogenate was centrifuged at 14000 rpm for 15 min and the extract was neutralized with 5 M sodium hydoxide. A mixture containing 150 mM sodium phosphate buffer (pH-7.4), distilled water, 10.0 per cent TCA, 44.0 per cent phosphoric acid, 4.0 per cent bipyridyl in 70.0 per cent ethanol, 3.0 per cent $FeCl_3$ (freshly prepared) was added to the extract. Samples were incubated at 37.8 °C for 60 minutes in a water bath. Absorbance of the samples was measured at 525 nm. The amount of reduced ascorbic acid was calculated using standard ascorbic acid curve.

Assay of Antioxidant Enzyme Activities

Sample tissue were ground in ice cold mortar in extraction mixture containing 1mM EDTA, 0.5 per cent w/v Triton x100, 2.0 per cent w/v Polyvinyl pyrolidone and 1 mM Ascorbate in 50 mM phosphate buffer (pH-7.2). The homogenate was centrifuged at 10000 rpm for 20 minutes at 4.0 °C. The resulting supernatant was kept at -20.0 °C for enzyme assay.

Assay of APX Activity

APX activity was measured as the rate of hydrogen peroxide dependent oxidation of ascorbic acid, according to Asada (1992). Enzyme extract was added to 2.8 ml of reaction mixture containing, 5 mM ascorbic acid in 50 mM phosphate buffer (pH 7.2).The reaction was initiated by addition of 10 µl of 10 per cent H_2O_2 and the enzyme activity is assayed by measuring the oxidation rate of ascorbic acid at 290 nm for 20 seconds (ε=2.8mM^{-1}cm^{-1}).

Assay of CAT Activity

CAT activity was determined by following the method of Abei (1974). The assay mixture contained 0.1 ml of 3.125 mM H_2O_2 in 2.8 ml 50 mM phosphate buffer (pH 7.2) and 0.1 ml of enzyme extract in total of 3 ml final volume. CAT activity was measured by recording the decrease in absorbance at 240 nm. The molar extinction coefficient at 240 nm is 0.04 µMol^{-1}cm^{-1}.

Assay of SOD Activity

SOD activity was assayed according to the method of Beauchamp and Fridovich (1971) by measuring its ability to inhibit the phytochemical reduction of nitroblue tetrazolium (NBT). 2.5 ml of reaction mixture containing 0.1 mM EDTA, 13 mM methionine and 75 µM NBT was added to a suitable enzyme extract. After the addition of riboflavin (1.67 mM) in dark, the mixture was incubated under fluorescent lamp for 30 minutes. A tube containing enzyme was kept in dark served as a blank while the control tube without enzyme was kept in light as control. The photoreduction of NBT (formazone formation) was measured at 560 nm. Calculations were made by using extinction coefficient of 100 $mM^{-1} cm^{-1}$.

Statistical Analysis

Statistical analysis was performed using one-way analysis of variance (ANOVA) followed by Tukey's post-hoc multiple comparison test using the SPSS 16.0 (Statistical program for Social Sciences) program. All the data are presented as mean ± SEM at $p<0.05$ and exact number of experiments (n) given in figures and tables, denote individual plant measured for each parameter from every Cu concentration.

Results

The effect of Cu on root length, stem height and fresh and dry biomass is shown in Table 20.1. Root length and shoot length decreased throughout at all concentrations of Cu. A significant decrease was observed in root length and shoot length of *Vigna radiata* at high Cu concentrations (50, 75, 100, 200 µM Cu). A marked 3 fold reduction with respect to control was observed in root length at 200 µM Cu treatment while a 1.6 fold decrease with respect to the control was found in shoot length of *Vigna radiata* treated with 200 µM Cu after 120 hrs of treatment (Table 20.1). Correspondingly a significant reduction in total fresh biomass and dry biomass of *Vigna radiata* seedlings was observed at high Cu concentrations (50, 75, 100, 200 µM Cu) indicating the toxic effect of excess Cu on seedling growth. Fresh weight decrease by 1.83 folds (200 µM Cu) and dry weight by 1.6 folds (200 µM Cu) with respect to control after 120 hours of exposure. Similar trends of reduction in root length, shoot length, fresh weight and dry weight were observed after 72 hrs and 120 hrs of exposure. A significant decrease in total

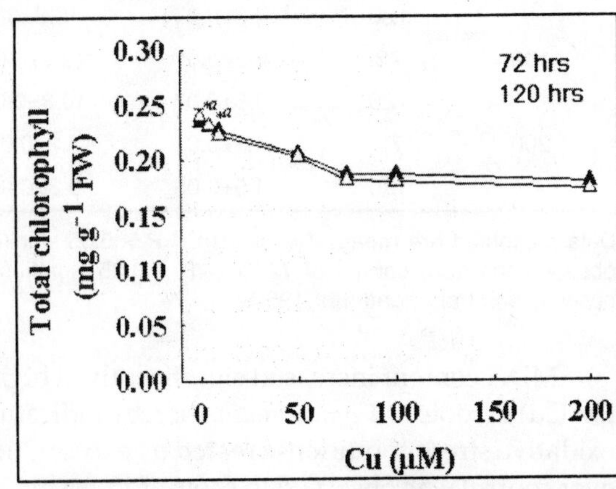

Figure 20.1: Changes in Chlorophyll Content in Leaves of *Vigna radiata* after 72 and 120 hrs of Cu Stress Treatment.

Data represents mean ± SE (n=6). *P > 0.05 signifies insignificant difference between 72 hrs observations from control of 72 hrs, [a]P > 0.05 signifies insignificant difference between 120 hrs observations from control of 120 hrs.

chlorophyll content in comparison to control was observed at 10, 50, 75, 100, 200 µM Cu (Figure 20.1). Total protein content in roots of *Vigna radiata* seedlings increased to 19.8 per cent at 10 µM Cu with respect to control whereas shoot protein content increased to 25.77 per cent at 50 µM Cu in comparison to control after 120 hrs followed by a significant decrease in protein content at high Cu concentrations (Figure 20.2). An apparent difference was observed with increasing exposure time. At lower Cu concentrations total protein content increased with increasing exposure time while it decreased significantly at higher concentrations.

Table 20.1: Root Length (cm), Shoot Length (cm), Fresh Weight (g) and Dry Weight (g) of *Vigna radiata* Seedlings Under Different Concentrations of Cu After 72 and 120 hrs of Treatment

Cu (µM)	*Exposure Time (hrs)*	*Root Length (cm)*	*Shoot Length (cm)*	*Seedling Fresh Weight (g)*	*Seedling Dry Weight (g)*
0	72	12.9±0.05	13. 5±0.05	0.55±0.02	0.24±0.01
	120	13.1±0.07	14.1±0.07	0.59±0.01	0.28±0.01
5	72	12.0±0.07*	13.8±0.07*	0.57±0.03*	0.24±0.01*
	120	12.3±0.06*a*	14.2±0.04*a*	0.59±0.01*a*	0.29±0.02*a*
10	72	9.2 ±0.05	13.7±0.11*	0.56±0.07*	0.24±0.01*
	120	9.8±0.07	14.1±0.05*a*	0.58±0.01*a*	0.27±0.01*a*
50	72	7.3±0.08*	11.5±0.11	0.51±0.09*	0.22±0.01*
	120	7.5±0.07	11.6±0.07	0.51±0.05*a*	0.23±0.02*a*
75	72	7.1±0.09	10.4±0.06	0.45±0.06	0.18±0.03
	120	7.3±0.07	10.6±0.1	0.45±0.02	0.21±0.01
100	72	6.3±0.01	10.8±0.09	0.34±0.06	0.16.±0.01
	120	6.5±0.07	10.8±0.08	0.35±0.01	0.16±0.02
200	72	4.3±0.01	9.2±0.09	0.28±0.07	0.15±0.01
	120	4.5±0.06	9.3±0.02	0.30±0.01	0.15±0.04

Data presented are mean± SE, (n=10). * P > 0.05 signifies insignificant difference between 72 hrs observations from control of 72 hrs, [a]P > 0.05 signifies insignificant difference between 120 hrs observations from control of 120 hrs.

MDA content increased significantly at high Cu concentrations (50, 75, 100, 200 µM Cu) in roots of *Vigna radiata*, thereby indicating membrane leakage caused due to oxidative stress conditions created by excess Cu. A 3.8 fold increase in MDA content over control was observed in roots of *Vigna radiata* at 75 µM Cu concentration while it increased 2.5 folds at 100 µM Cu in shoots after 72 hrs of Cu treatment, which was further increased slightly after 120 hrs of treatment (Figure 20.3). No major difference was observed in MDA contents on increasing exposure time from 72 hrs to 120 hrs in shoots while MDA content increased appreciably at 50,75 µM Cu after 120 hrs in comparison to 72 hrs treatment.

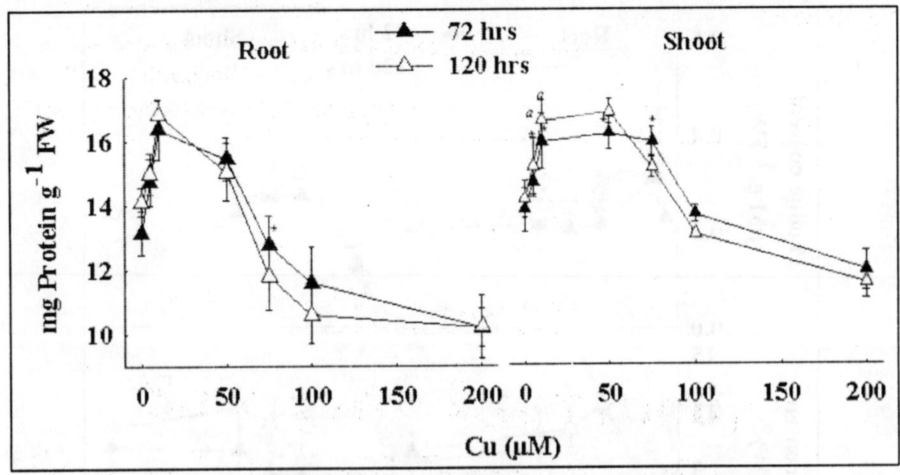

Figure 20.2: Changes in Protein Content in *Vigna radiata* Roots and Shoots after 72 hrs and 120 hrs of Cu Stress Treatment.

Data presented mean ± SE (n=6). *P > 0.05 signifies insignificant difference between 72 hrs observations from control of 72 hrs, [a]P > 0.05 signifies insignificant difference between 120 hrs observations from control of 120 hrs.

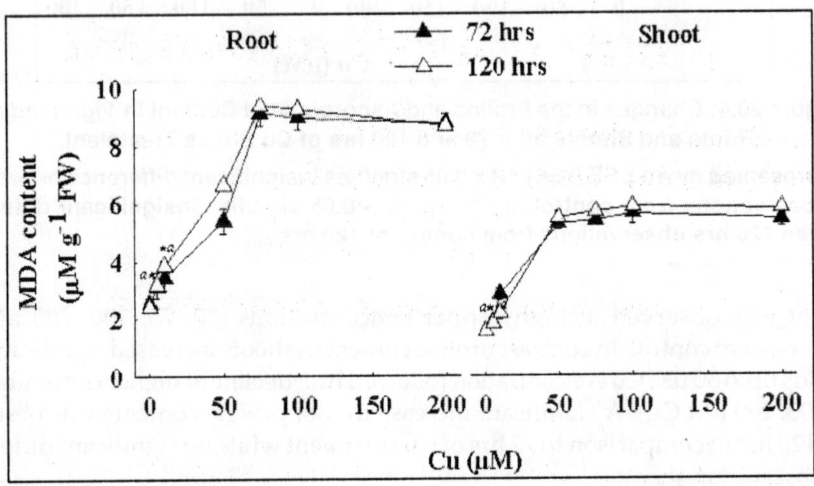

MDA: Malonaldehyde

Figure 20.3: Changes in MDA Content in Roots and Shoots of *Vigna radiata* after 72 and 120 hrs of Cu Stress Treatment.

Data presented mean ± SE (n=6). *P > 0.05 signifies insignificant difference between 72 hrs observations from control of 72 hrs, [a]P > 0.05 signifies insignificant difference between 120 hrs observations from control of 120 hrs.

The stimulatory effect of Cu concentrations on proline synthesis was more in roots of Cu treated seedlings (Figure 20.4). Proline content increased significantly by 2.2 folds at low Cu concentrations (5, 10 μM) in roots while slight increase in proline

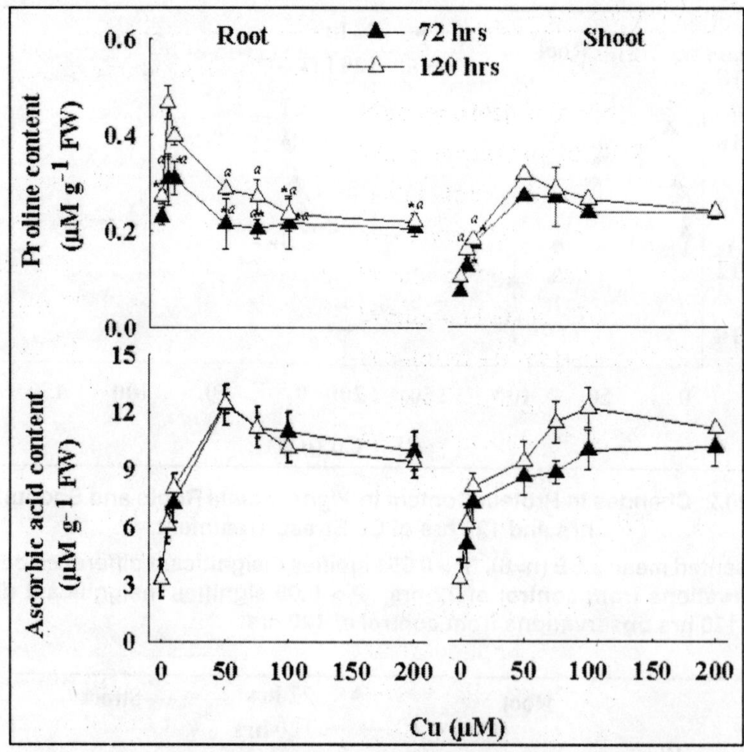

Figure 20.4: Changes in the Proline and Ascorbic Acid Content in *Vigna radiata* Roots and Shoots after 72 and 120 hrs of Cu Stress Treatment.

Data presented mean ± SE (n=6). *P > 0.05 signifies insignificant difference between 72 hrs observations from control of 72 hrs, [a]P > 0.05 signifies insignificant difference between 120 hrs observations from control of 120 hrs.

content was observed at high copper concentrations (50, 75, 100, 200 μM) in comparison to control. In contrast proline content in shoots increased significantly to 1.5 folds upto 50 μM Cu concentration followed by a decline at higher concentrations (75, 100, 200 μM Cu). A significant increase in root proline contents was observed after 120 hrs in comparison to 72 hrs of Cu treatment while no significant difference was observed in shoot.

A significant increase in ascorbic acid level was recorded in both roots and shoots of seedlings (Figure 20.4) at all Cu concentrations after 72 and 120 hrs of exposure. Ascorbic acid content in roots of Cu treated plantlets was found to be 2.8 folds higher in comparison to control at 50 μM Cu after120 hrs. However in shoots a marked 2.5 fold increase was observed at 100 μM Cu after 72 hrs of exposure while it further increased to 3 folds at 100 μM Cu after 120 hrs of Cu treatment. Both root and shoot antioxidants showed a decline in their contents at higher concentrations of Cu (100, 150, 200 μM).

Cu treatment significantly increased the SOD activity in both roots and shoots of *Vigna radiata* seedlings (Figure 20.5). However observed SOD activity was more in root than in shoot tissue. SOD activity increased significantly with maximum upto 4.5 folds of control at 50 μM Cu concentration in root tissue followed by a significant

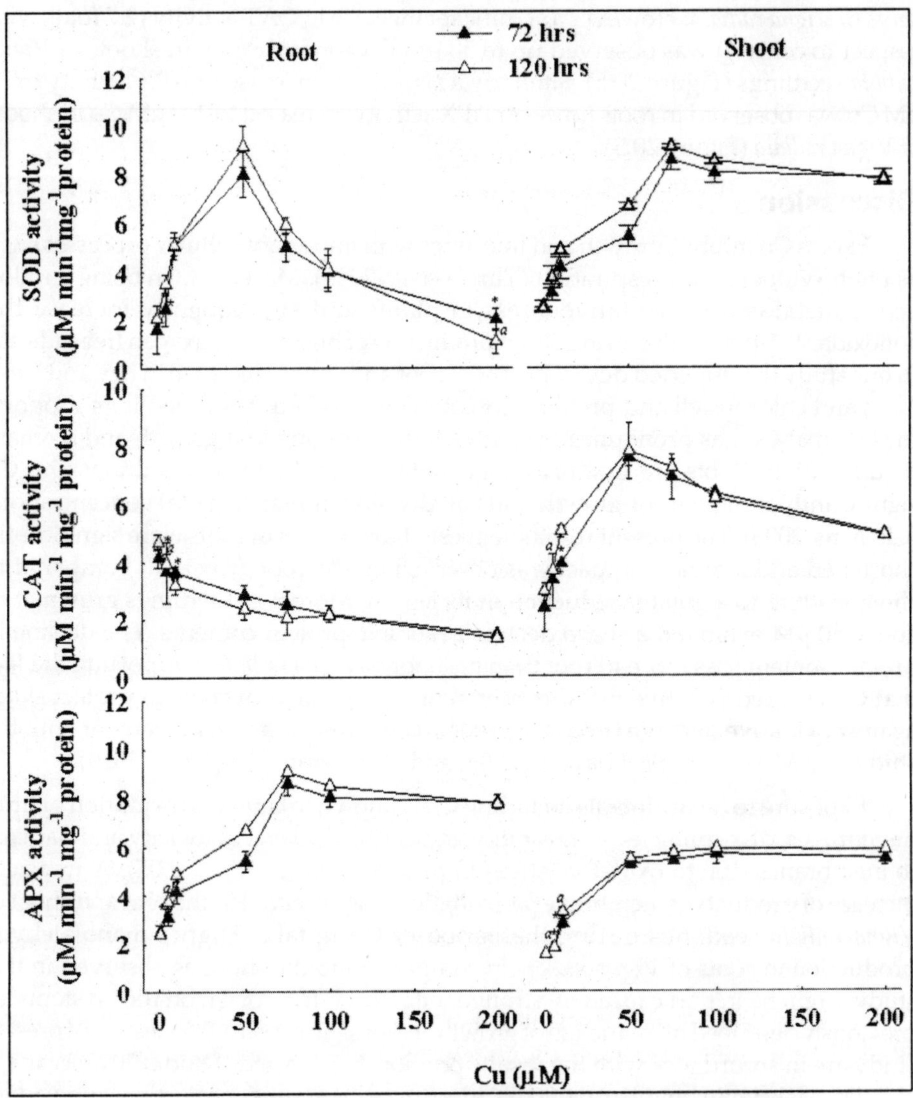

SOD: Superoxide dismutase; APX: Ascorbate peroxidase; CAT: Catalase

Figure 20.5: Changes in SOD, CAT and APX Activities in Roots and Shoots of *Vigna radiata* after 72 and 120 hrs of Cu Stress Treatment.

Data presented mean ± SE (n=6). *P > 0.05 signifies insignificant difference between 72 hrs observations from control of 72 hrs, [a]P > 0.05 signifies insignificant difference between 120 hrs observations from control of 120 hrs.

decrease at 75, 100 and 200 μM Cu concentrations. At the same time in shoot SOD activity increased upto 3.7 folds at 75 μM Cu followed by a slight decrease at 100 and 200 μM Cu concentrations.

In contrast to its effect on SOD, excess Cu also decreased the activity of CAT and APX. CAT activity was significantly reduced throughout all Cu concentrations in roots of *Vigna radiata*. However a significant increase in CAT activity (2.7 folds with respect to control) was observed up to 50 μM Cu concentration in shoots of *Vigna radiata* seedlings (Figure 20.5). Similarly a significant increase in APX activity to 75 μM Cu was observed in roots however APX activity increased to 200 μM Cu in shoots of *Vigna radiata* (Figure 20.5).

Discussion

Excess Cu inhibits growth and interferes with important cellular processes such as photosynthesis and respiration (Yurela *et al.*, 2000). Moreover Cu being a redox active metal causes oxidative stress in plants and subsequently increase the antioxidant responses due to increased production of highly toxic oxygen free radicals. In our study the impeded development of root and shoot, decrease in dry and fresh mass and chlorophyll and protein content in excess Cu treated seedlings supports the fact that Cu has pronounced negative influences on plant growth and biomass accumulation. Roots are first organs in contact with the toxic metal ions and Cu significantly inhibits root growth and cell division at higher metal concentrations (Clemens, 2006). The present results suggested that length of roots were significantly shortened and a greater impact was observed on the root growth as compared to shoot leading to a greater reduction in its length. Moreover seedlings grown at or above 50 μM exhibited a sharp decline in soluble protein contents. The decline in protein content was more in root than in shoot of *Vigna radiata*, supporting the fact that Cu as a redox active metal is preferentially retained in roots to protect shoot against oxidative damage (Lou *et al.*, 2004). These results are in agreement with the studies on *M. sinensis* (Scebba *et al.*, 2006) and *G. uralensis* (Zeng *et al.*, 2010).

Exposure to heavy metals including Cu is known to cause peroxidation of lipid membranes, disrupting its permeability, indicating the level of toxicity and damage to membranes due to oxidative stress in plant tissues (Li *et al.*, 2005). A transient increase of products of membrane peroxidation was observed in the root and shoot of *Vigna radiata* seedlings during the period of Cu uptake. Higher manolalehyde production in roots of *Vigna radiata* in comparison to the shoot as observed in this study might be in part caused by stronger negative effects of Cu on root systems, as previously described by reduction of growth and protein content. The results of present study are in accordance with the results previously reported (Panda, 2008; Wang *et al.*, 2004, Maksymiec and Krupa, 2007). In terms of plant stress protection, antioxidant compounds and enzymes represent important tool enabling them to counteract oxidative stress (Sgherri *et al.*, 2003). Up regulation of proline is often encountered in plants under stress (Verma *et al.*, 2008). Enhanced proline accumulation in root and shoot tissue of *Vigna radiata* signifies its role in plant protection under elevated Cu concentrations. Marked increase in proline activity of roots could to be due to, decrease in activity of electron transport system during stress conditions that leads to accumulation of NADH and H^+. Proline accumulation (presumably through synthesis from glutamic acid) might be an adaptive mechanism for reducing (a) the level of accumulated NADH and (b) the acidity; (2NADH+2H^+) is used for synthesizing each molecule of proline from glutamic acid (Venekemp *et al.*, 1987). At the same time slow increase in proline content at higher concentrations of Cu (50, 75, 100 μM) suggests the low mobility of Cu in shoot tissue. Proline accumulation under heavy metal stress has also been reported by earlier authors (Alia and Saradhi, 1991; Bassi and Sharma, 1993). Ascorbic acid is a key antioxidant and is involved solely in protection of plant cells against oxidative damage. The progressive increase in ascorbic acid content found in Cu treated seedlings indicate plant defense response against oxidative stress. Prakash *et al.* (2007) observed similar results in *Cicer arietinum*. However, a slight decrease in ascorbic acid content at high concentrations of Cu has been observed. It might occur due to reduction of monodehydro-ascorbate produced by APX *via* Fd or by monodehydroreductase or alternatively by dehydroascorbate reductase using reduced glutathione as an electron donor (Mohanty, 2005).

Enhanced CAT activity up to 50 μM was recorded mainly in shoot treated with Cu. However in roots CAT activity decreased continuously on increasing Cu concentration. CAT is sensitive to superoxide radicals thus metal stress leads to inactivation of enzyme (Prakash *et al.*, 2007). The decrease may also be related with degradation caused by peroxisomal proteases or may be due to photo inactivation of enzymes. Decreased CAT activity as observed in the present study is in accordance with the earlier results reported in oat leaves (Luna *et al.*, 1994). Moreover Cu induced inactivation of CAT might be another possible reason of decreased CAT activity. However an enhanced CAT activity in shoot of *Vigna radiata* might be an indication that leaf CAT activity was affected by low mobility and uptake of Cu from roots of *Vigna radiata*. By virtue of their capacity to dismutate superoxide radical, increased SOD activities in root tissues of plant appears to provide a primary line of antioxidant defense and catalyses the dismutation of super oxide radical O^{2-} to H_2O and molecular oxygen. (Chongpraditnum *et al.*, 1992) The Cu mediated increase in SOD activity may possibly be the result of either a direct effect of these metal ions on genes for SOD or an indirect effect, mediated *via* an increase in level of ROS. The decrease in level of SOD activity at higher metal concentration may be due to inactivation of enzyme by H_2O_2, produced in different cell compartments and from a number of non-enzymatic and enzymatic processes in cells. The results of present study are in consent with the results of earlier authors (Okamoto *et al.*, 1996; Gao *et al.*, 2008). H_2O_2 produced by SOD catalyzed reaction is further reduced to water with the help of CAT and APX enzymes. The stimulation of APX activity of Cu treated seedlings appears to be due to increased levels of H_2O_2 generated either by metal toxicity or as the result of increased SOD activity, that induces stress related genes and may play a signaling role during the oxidative stress (Luna *et al.*, 1994). Unrelated activity shown by these enzymes indicates that detoxification in roots mainly occurred through APX (Cuypers *et al.*, 2002). The decrease in CAT activity in plant roots has been compensated by increase in activity of APX which has been used to reduce H_2O_2 to water.

Conclusion

In conclusion our results indicated a substantial reduction in growth of *Vigna radiata* seedlings under elevated Cu concentrations (50, 75,100, 200 µM). Root metabolism and antioxidant system is largely affected under Cu toxicity. However, defense responses in shoot were enhanced transiently on exposure to elevated levels of Cu. At lower concentration of Cu, increase in protein, proline ascorbic acid and enzymes were observed; but at higher concentrations it decreased. A range of antioxidant enzymes (SOD, CAT and APX) and compounds ascorbic acid and proline play a major defensive role against Cu stress in *Vigna radiata*.

Acknowledgement

G.S. Shekhawat acknowledges Professor Aditya Shastri, Vice Chancellor, Banasthali University, Banasthali for his unwavering support to the study.

References

Abei, H. 1974: Catalases. In: Bergmeyer HU (ed) Method of enzymatic analysis, Verlag Chemic, Weiheim. Academic Press. Inc, Newyork, p 680.

Alia, Pardha and Saradhi, P. 1991: Proline accumulation under heavy metal stress. *J. Plant Physiol.* 138, 554–558.

Apel, K. and Hirt, H. 2009: Reactive oxygen species: Metabolism, oxidative stress, and signal transduction. *Annu Rev Plant Biol* 55: 373-399.

Arnon, D.L. 1949: Copper enzymes in isolated chloroplasts, Polyphenol oxidase in *Beta vulgaris*. *Plant Physiol*. 24: 1-15.

Asada, K. 1992: Ascorbate peroxidase—a hydrogen peroxide scavenging enzyme in plants. *Plant Physiol.* 85: 235–241.

Bassi, R. and Sharma, S.S. 1993: Proline accumulation in wheat seedling exposed to zinc and copper. *Phytochemistry*, 33, 1339–1342.

Bates, L.S., Waldren, R.P. and Teare, I.D. 1973: Rapid determination of free proline for water-stress studies. *Plant Soil.* 39: 205–207.

Beauchamp, C. and Fridiovich, I. 1971: Superoxide dismutase: improved assays and an assay applicable to acrylamide gels. *Anal biochem.* 44: 83 - 116.

Cho, M., Chardonnens, A.N. and Dietz, K.J. 2003: Differential heavy metal tolerance of *Arabidopsis halleri* and *Arabidopsis thaliana*: a leaf slice test. *New Phytol.* 158, 287-293.

Chongpraditnum, P., Mori, S. and Chino, M. 1992: Excess copper induces a cytosolic Cu, Zn- superoxide dismutase in soybean roots. *Plant Cell Physiol.* 33: 239-244.

Clemens, S. 2006: Toxic metal accumulation, responses to exposure and mechanisms of tolerance in plants. *Biochimie.* 88: 1707-1719.

Cuypers, A., Vangronsveld, J. and Clijsters, H. 2002: Peroxidases in roots and primary leave of *Phaseolus vulgaris* Copper and Zinc Phytotoxicity: a comparison. *Plant Physiol.* 159: 869-876.

Gao, S., Yan, R., Cao, M., Yang, W., Wang, S. and Chen, F. 2008: Effects of copper on growth, antioxidant enzymes and phenylalanine ammonia-lyase activities in *Jatropha curcas* L. seedling. *Plant soil environ.* 54: 117–122.

Heath, R.L. and Packer, L. 1968: Photoperoxidation in isolated chloroplasts: Kinetics and stoichiometry of fatty acid peroxidation. *Arch Biochem Biophys.* 125: 189-190.

Li, M., Hu, C., Zhu, Q., Chen, L., Kong, Z. and Liu, Z. 2005: Copper and zinc induction of lipid peroxidation and effects on antioxidant enzyme activities in the microalga *Pavlova viridis* (Prymnesiophyceae). *Plant Physiol.* 62: 565-572.

Lou, L-q., Shen, Z-g., Li, X-g. 2004: The Copper tolerance mechanisms of *Elsholtzia haichowensis*, a plant from copper-enriched soils. *Environ Exp Bot.* 51:111-120.

Lowry, O.H., Rosenborough, A.L. and Randall, R.J. 1951: Protein measurement with Folin-phenol reagent. *J. Biol Chem.* 93: 265-272.

Luna, C.M., Gonzalez, C.A. and Trripi, U.S. 1994: Oxidative damage caused by an excess copper in Oat leaves. *Plant Cell Physiol.* 35: 11-15.

Maksymiec, W. and Krupa, Z. 2007: Effects of methyl jasmonate and excess copper on root and leaf growth. *Biol. Plant.* 51: 322-326.

Mohanty, M., Jena, A.K. and Parta, H.K. 2005: Effect of chelated chromium compounds on chlorophyll content and activities of catalase and peroxidase in Wheat seedlings. *Indian J. Agri Biochem.* 18: 25-29.

Okamoto, O.K., Assano, C.S., Adar, E. and Colepicolo, P. 1996: Effects of cadmium on growth and superoxide dismutase activity of the marine microalga *Tetraselmis gracilis* (*Prasino phyceae*). *J Phycol*, 34: 74-79.

Panda, S.K., 2008: Impact of copper on reactive oxygen species, lipid peroxidation and antioxidants in *Lemna minor. Biologia plantarum*, 52: 561-564.

Prakash, V. and Alka, Lawrence, K. 2007: Antioxidative response to copper –induced stress in *Cicer arietinum* L. *Indian J Agric Biochem.* 20: 23-26.

Scebba, F., Arduini, I., Ercoli, L. and Sebastiani, L. 2006: Cadmium effects on growth and antioxidant enzymes activities in Miscanthus sinensis. *Biologia Plantarum*, 50: 688-692.

Sgherri, C., Cossi, E. and Navari-Izzo, F. 2003: Phenols antioxidative status of *Raphanus sativus* grown in copper excess. *Physiol Plant*, 118: 21-28.

Shekhawat, G.S, Kumar, D., Verma, K., Singh, K., Jana, S., Teotia, P. and Yadav, S. 2009: Copper as a Biometal: An insight on transport, toxicity and tolerance in Photosynthetic Organisms. In Biometals: Molecular structures, binding properties and applications, Nova Sciences Publishers, Hauppauge, NY.USA.

Somta, P. Srinives, 2007: Genome Research in Mungbean [*Vigna radiata* (L.) Wilczek] and Blackgram [*V. mungo* (L.) Hepper] Science Asia 33 Supplement 1. doi: 10.2306/scienceasia1513-1874.2007.33(s1).069.

Venekemp, J.H., Lampe, J.E. and Kout, T.M. 1987: Organic acid as a source of drought-induced proline synthesis in field bean plant Vicia faba, L. *J. Plant Physiol.* 133, 654–659.

Verma, K., Shekhawat, G.S., Sharma, A., Mehta, S.K. and Sharma, V. 2008: Cadmium induced oxidative stress and changes in soluble and ionically bound cell wall peroxidase activities in roots of seedling and 3-4 leaf stage plant of *Brassica juncea* (L.) czern. *Plant ceil rep.* 27: 1261-1269.

Wang, S., Yang, Z., Yang, H., Lu, B., Li, S. and Lu, Y. 2004: Copper induced stress and antioxidative responses in roots of *Brassica juncea* L. - *Bot. Bull. Acad. sin.* **45**: 203-212.

Yruela, J., Alfonso, M., Baron, M. and Picorel, R. 2000: Copper effect on the protein composition of photosystem II. *Physiol Plant*. 110: 551-557.

Zeng, G., Lv, H.P., Gao, S. and Wang, S.R. 2010: Effects of cadmium on growth and antioxidant responses in Glycyrrhiza uralensis seedlings. *Plant soil Environ.* 56: 508-515.

Zhang, J.X. and Kirkham, M.B. 1996: Antioxidant responses to drought in flower and sorghum seedlings. *New Phytol.*132: 361-373.

2013, Climate Change Effects on Agriculture and Economy *Pages* ***187–194***
Editors: **D.R. Khanna, A.K. Chopra, Vikas Singh, Rakesh Bhutiani & Gagan Matta**
Published by: **BIOTECH BOOKS, NEW DELHI**

Chapter 21

Haematological Changes Due to High Fluoride Concentration in Drinking Water of Few Villages in District Pali, Rajasthan

A. Choudhary, D. Gujar and M. Kasim
Department of Zoology, Jai Narain Vyas University, Jodhpur, Rajasthan

Groundwater resources, especially in Rajasthan State have high level of fluoride concentration above maximum permissible level suggested by WHO (1971). Such resources causes health hazardous affect *viz;* Dental, skeletal and soft tissue fluorosis. Study revels with the effect of high fluoride concentration on haematological parameter in human being of fluoride endemic area of district Pali, Rajasthan. 32 blood samples of fluorotic patient were collected from endemic area to find out the altered level of blood biochemical parameter *viz;* haemoglobin, RBC's histoarchitecture, WBC's count, acid and alkaline phosphates, serum protein, cholesterol, glucose and electrolyte *etc.* Highly significant decrease (Pd≤0.05) was observed in haemoglobin level mean value 7.6 ± 0.41 gm per cent in fluorotic patient, 13.0 ± 0.35 gm per cent in non fluorotic person, highly significant (Pd≤0.001) decreased value of WBC count was observed in fluorotic patients in compare to non fluorotic persons, mean value of WBC count was 6100.00 ± 368.00 cu/mm in fluorotic patients and 9650.00 ± 35.40 cu/mm observed in non fluorotic persons, mean level of alkaline phosphates 194.71 ± 2.94 IU/l in fluorotic patient and 169.0±0.66 IU/l was observed in non fluorotic person, highly significant (Pd≤0.001) mean level of acid phosphates 4.45±0.05 IU/l and 6.05±.01 IU/l in non fluorotic person. Fluoride concentration in blood serum was found 0.78±0.11 mg/l as mean value maximum permissible level suggested by WHO (1971) was 0.02 mg/l. Significant decrease (Pd≤0.001) in level of serum electrolytes were also observed during the course of study.

Keywords: *Fluoride, Drinking water, Fluorotic patient, Haematology.*

Introduction

Water pollution is inevitable fallout of development in industry, urban centers and most importantly increasing population. Generally water always has some impurities like dissolve solids, ions, minerals and biotic materials as water is a universal solvent. All these elements present in catchment area are within the desirable limit and if there level is above the maximum permissible level (WHO, 1971) then self purifying activities of the aquatic body is lost and water becomes unfit for human consumption and other domestic uses. These elements continuously get dissolved in water aquifers and out of all these elements some are very essential for proper leading of the life supporting system.

The groundwater plays major role as a source of drinking water in the most of the part of our country. Fluoride is one of such element which is found abundantly in the earth crust and through weathering and leaching process of earth crust it dissolve in groundwater resources. It is an essential element and significantly takes part in metabolic activity of living being within desirable limit, but more than desirable limit it create health hazardous effect to living organism.

Present investigation deals with the alteration of blood biochemical parameter *i.e.* haemoglobin, deformation in RBC, WBC count, acid and alkaline phosphate, serum, cholesterol, glucose and protein along with electrolytes in fluorotic patient of fluoride endemic area, results were compared with non fluorotic person.

Materials and Method

To find out the haematological alteration in fluorotic patient and non fluorotic person, a set of 32 blood samples were collected in blood sample testing vial and added EDTA anticoagulant for haemoglobin estimation and WBC count. Other blood parameters were analyzed without EDTA anticoagulant in ice cooled condition, after then carried out to laboratory and centrifuged for separation of serum; centrifuged serum was analyzed for serum biochemistry. Following methods were adapted to analyze haematological parameters:

Blood glucose estimated by Astoor and King method (1954), WBC count by Lynch *et al.* (1969) method, Haemoglobin estimated by Dravkin's cyanmethelogin method, blood cholesterol by Wybenga and Pileggi (1970) method, total protein by Lowry *et al.* (1951) method, a few electrolytes were estimated in the blood serum of the fluorotic patient as well as non fluorotic persons, serum calcium by Clark and Collik, (1925) method, Potassium and sodium by Willard *et al.* (1965) method.

Results and Discussion

It has been evidently proved that continuous intake of long time high fluoride concentration through various sources ultimately caused health hazardous effects and altered the metabolic activities of entire body. Fluoride enters inside the body mainly through drinking water and edibles and ingested fluoride directly reaches into the alimentary canal and gets absorbed by intestinal mucosa through a passive diffusion and then enters into the circulatory system.

Table 21.1 shows comparison between level of haematological parameter of fluorotic patient of endemic area and non fluorotic person. Fluoride concentration in blood serum was observed high in fluorotic patient and mean value was found 0.78±0.11 mg/l with maximum and minimum value 1.11 mg/l and 0.11 mg/l respectively. Maximum permissible value suggested by WHO (1971) was 0.02 mg/l in blood serum. In case of non fluorotic person mean value of fluoride concentration in blood serum was found 0.03±0.01 mg/l. The increase level of fluoride in serum was found highly significant (Pd≤0.001). Bano *et al.* (1996) and Swarup *et al.* (1998) were reported that excessive intake of fluoride contents reduce the level of haemoglobin concentration. Mishra and Mohapatra (1987), Jagdish *et al.* (1998) and Kirstinassan *et al.* (1997) also reported reduced level of haemoglobin in fluorotic individuals. Susheela (2001) stated that ingested fluoride will also accumulate on erythrocyte membrane, fluoride has a negative co relationship with calcium ion so membrane losses its calcium contents. Edward and Hassal (1980) also reported that fluoride is a metabolic inhibitor and also affect the mechanism of sodium pump and stop the transportation of sodium ion, Yingyan and Xiaojic (1994) reported that fluoride reduces the fluidity of human erythrocyte membrane and reduction in fluidity of erythrocyte membrane is time related so fluoride change the conformation of membrane protein to a certain extent. Due to the deficiency of electrolytes, decreased level of protein and changes in conformation of membrane protein, erythrocyte membrane becomes irregular in shape and is thrown into folds and such erythrocyte called as echinocytes, this would lead to a low haemoglobin level in patient chronically ill due to fluoride toxicity resulting in anemia. A highly significant (Pd≤0.001) decrease value of WBC content was observed in fluorotic patients in comparison to non fluorotic persons. In fluorotic patients, the mean value of WBC content was 6100.0 ± 368.0 cu/mm and maximum and minimum WBC content were 7300 cu/mm and 5000 cu/mm observed respectively. But in non fluorotic persons the mean value of WBC content was 9650.0 ± 35.4 cu/mm observed during the course of study. White blood cells count was found decreased in fluorotic patients which was highly significant (Pd≤0.001) in comparison to non fluorotic persons. Jagdish *et al.* (1998) and Kristansson *et al.* (1997) both reported decreased level of WBC count in fluorotic patients. Decrease in WBC's may be due to autolysis caused by the activity of some hydrolytic enzymes released into plasma under stress conditions reported by Reddy *et al.* (1993).

Acid phosphatase level was decreased in fluorotic patients which was highly significant (Pd≤0.001), in comparison to non fluorotic persons. Mean level of acid phosphatase was observed 4.45 ± 0.05 IU/l in fluorotic patients and its maximum and minimum level were 4.79 IU/l and 4.11 IU/l in fluorotic patients, while in non

Table 21.1: Haematological Comparison between Fluorotic Patients and Non-Fluorotic Persons with Level of Fluoride in Serum (Mean ± SE of 5 values)

Sl.No.	Persons	Serum Fluoride mg/l	Serum Calcium mg/l	Hemoglobin gm/ml	W.B.C. cu/mm	Alkaline Phosphate IU/L	Acid Phosphate IU/L	Blood Protein mg/dl	Blood Cholesterol mg/dl	Blood Glucose mg/dl	Serum Sodium mol/l	Serum Potassium mol/l
1.	Pateints (Male and Female)	$0.78^{d}\pm$ 0.11	$2.65^{d}\pm$ 0.60	$7.6^{d}\pm$ 0.11	$6100^{d}\pm$ 0.368	$194.71^{d}\pm$ 2.24	$4.45^{d}\pm$ 0.05	$6.21^{d}\pm$ 0.05	$129.65^{d}\pm$ 3.37	$101.0^{b}\pm$ 5.2	$117.2^{d}\pm$ 1.45	$4.22^{b}\pm$ 0.16
2.	Maximum	1.11	6.19	9.8	7300	210	4.79	6.8	141.2	109.0	122.0	4.78
3.	Minimum	0.11	1.16	6.0	5000	183	4.11	6.0	115.2	88.0	111.5	3.42
4.	Control Non-fluorotic persons	0.03± 0.01	12.07± 0.15	13.0± 0.35	9650± 35.4	169.0± 0.66	6.05± 0.01	7.8± 0.14	172.5± 1.77	85.0± 0.70	142.29± 0.70	5.45± 0.02

Normal limit of fluoride in serum = 0.02 mg/l; b = p non significant; d = $p \leq 0.01$ significant.

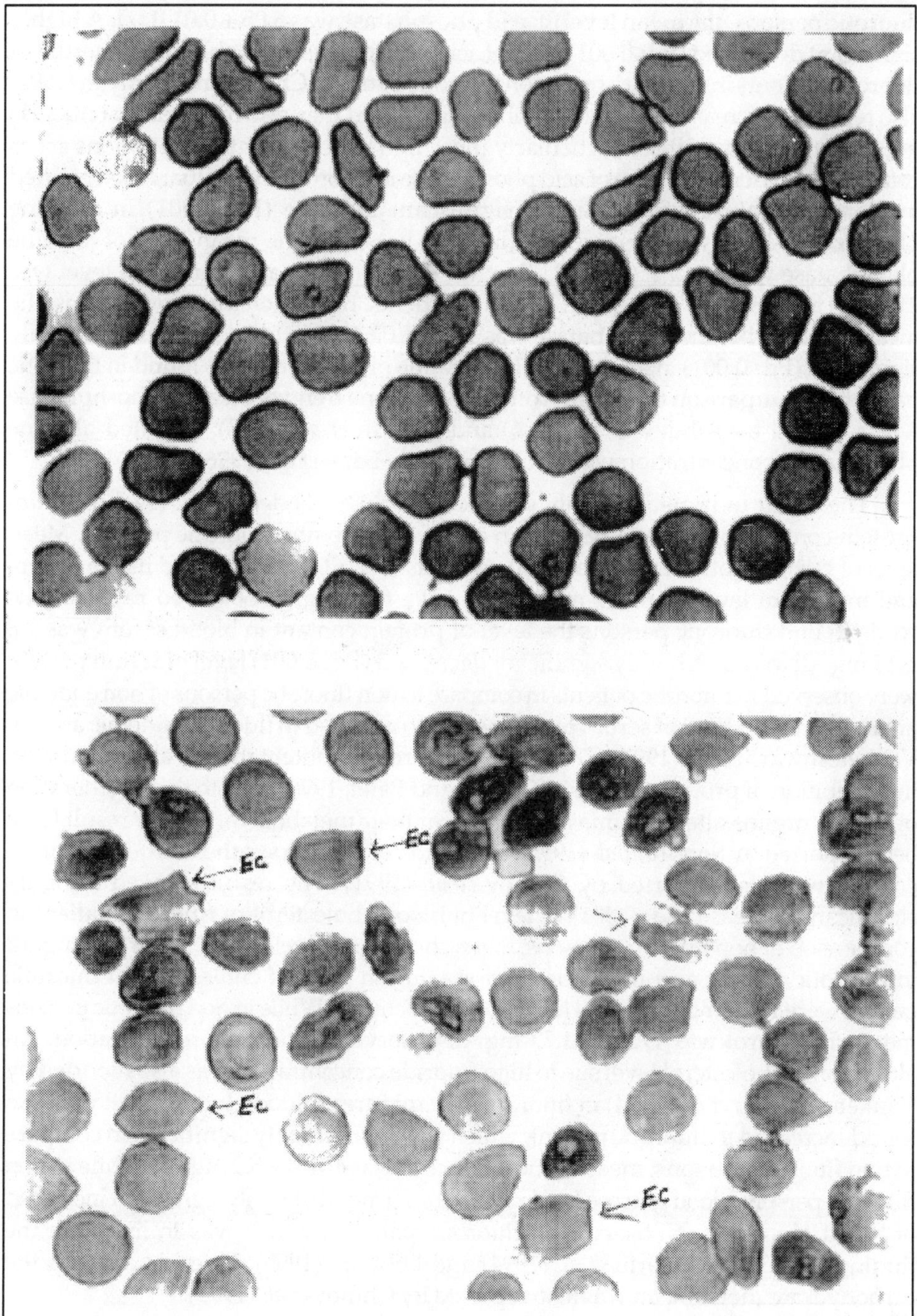

Figure 21.1

fluorotic persons, the mean level of acid phosphatase was 6.05 ± 0.01 IU/l. A highly significant decreased (Pd≤0.001) level of acid phosphatase was observed in serum of fluorotic patients in comparison to non fluorotic persons. Chinoy and Narayan (1994) also reported decreased level of acid phosphatase and they also reported that fluoride inhibits enzyme activities, particularly those in which bivalent metal cations act as cofactors. But increased level of acid phosphatase in fluorotic individuals was reported by Sondhi *et al.* (1995). A highly significant increase (Pd≤0.001) in alkaline phosphatase level was observed in fluorotic patients. The mean level of alkaline phosphatase was 194.71 ± 2.24 IU/l, while its maximum and minimum level was 210 IU/l and 183.0 IU/l found in fluorotic patients. But in non fluorotic persons, the mean level of alkaline phosphatase was 169.0 ± 0.66 IU/l. In present study a highly significant (Pd≤0.001) increased level of alkaline phosphatase was found in fluorotic patients in comparison to non fluorotic persons same trend of alkaline phoshphatase was reported by Sahdev *et al.* (1996) and Jagdish *et al.* (1998). Elevated alkaline phosphatase concentration was also reported by Bano *et al.* (1996).

The result indicated a highly significant (Pd≤0.001) decreased level of serum protein content in fluorotic patients in comparison to non fluorotic persons. Mean level of serum protein in fluorotic patients was 6.12 ± 0.10 mg/dl and its maximum and minimum level were 7.2 mg/dl and 6.0 ± 0.01 mg/dl observed respectively, while in non fluorotic persons the level of protein content in blood serum was 7.8 ±.14 mg/dl found. A highly significant decreased (Pd≤0.001) level of serum protein were observed in fluorotic patients in compare to non fluorotic persons of non endemic areas. Decreased level of serum protein was also reported in fluoride endemic area by Venketeshwarlu *et al.* (1994). The decrease in protein content mainly attributed with the inhibition of protein synthesis (Chinoy and Patel, 1996). Due to the degeneration of cells in organs alteration may take place in their metabolism, similar result have been reported by Sondhi *et al.* (1995). Inhibition of protein synthesis due to fluoride content was also reported by Chinoy *et al.* (1991). The result showed a highly significant decrease (Pd≤0.001) in level of blood cholesterol of fluorotic patient in comparison of non fluorotic persons, mean cholesterol level was 129.65 ± 3.27 mg/dl in fluorotic patient and minimum and maximum level of cholesterol in fluorotic patient were 115.2 mg/dl and 141.2 mg/dl respectively. While in non fluorotic persons mean cholesterol was 172.5 ± 1.77 mg/dl observed during the investigation. An alteration in cholesterol level sue to high fluoride concentration was also reported by Venketeshwarlu *et al.* (1994) in fluoride endemic area. Blood glucose level was also found increased in fluorotic patient, which was found highly significant in compare to non fluorotic persons, mean value of glucose was 101.0 ± 5.2 mg/dl, while in non fluorotic persons blood glucose level was 85.0 ±.70 mg/dl. Highly significant increased of blood glucose was observed in fluorotic patients, which was in line with the finding of Venketeshwarlu *et al.* (1994) and Drobnik (1999). Fluoride disturbs the carbohydrate metabolism was also reported by Chinoy *et al.* (1994).

Haematological investigation clearly indicated a highly significant decreased (Pd≤0.001) in level of calcium, sodium and potassium content. The mean level of calcium content was 2.65 ± 0.60 mg/dl, sodium content was 117.2 ± 1.45 mol/l and potassium electrolyte 4.22 ±.16 mol/l respectively in comparison to non fluorotic persons. Electrolytic imbalance was also reported by Chinoy (1996). Alteration in sodium electrolyte in serum was also reported by Drobink (1999). Jackson *et al.* (1994) also observed decreased trend of sodium electrolyte in serum sample of osteoporatic females. Zhai *et al.* (1985), Magre *et al.* (1997) and Jagdish *et al.* (1998) reported low concentration of serum calcium in endemic area. Ream *et al.* (1983) also reported decreased level of serum calcium concentration due to high concentration of fluoride in serum. Richard (1994) also reported low level of calcium in human serum due to fluoride poisoning.

References

Astoor, A.M. and King E.J. 1954: Simplified calorimetric blood sugar method. *Biochem.* J.56 XLIV.

Bano, R., Sahdev, S. and Lal, S.B. (1996) Biochemical changes in the testes of Swiss albino mice exposed to chronic ingestion of sodium fluoride. *Ind. J. Environ. Toxicon.* 6(1): 19-21.

Chinoy, N.J., Jospeh, R., Sequerira, E. and Narayan, M.V. 1991: Effect of sodium fluoride on the muscle and liver of albino rats. *Ind. J. Environ. Toxicol.* 1(2): 129-134.

Chinoy, N.J. and Narayan M.V. 1994: In- vitro fluoride toxicity in human spermatozoa. *Reproductive Toxicol.* 8(2) 155-159.

Chinoy, N.J. and Patel, D. 1996: Ameliorative role of amino acid on fluoride induced alterations in uterine carbohydrate metabolism in mice. Fluoride 29 (4) 217-26.

Clark, E.P. and Collip, J.B. 1925: Estimation of calcium in biological fluid. *J. Biochem.* 63, 461.

Drobinik, M. 1999: Evaluation of the biological action of fluoride water supplied in the form of the potable cure in experimental animal. *Rocznikai-Panstu wego-zak-ladu-Higieny.* 50 (4): 403-408.

Edward, N.A. and Hassal, K.A. 1980: Biochemistry and physiology of the cell. An introductory text book, McGraw –Hill book company Ltd (UK) second edition 590.

Jackson, R., Kelly, S., Noblitt, T., Zhang, W., Dunipace, A., Li, Y., Stooky, G., Katz, B., Brizendine, E. and Farley, S. 1994: Effect of fluoride therapy on blood chemistry parameters in osteoporatic female. Bone mineral 27 (1) 13-23.

Jagdish, S., Singh, L.K., Samad, A. and Bhalrao, D.P. 1998: Bovine fluorosis I: clinical pathology and investigation into source of fluorine. *Indian Vet. J.* 75 (4) 335-337.

Kristansson, J., Gunnarsson, E., Johannesson, P., Parsson, P.A. and Thormar, H. (1997) Experiomental fluoride poisoning in ice landic sheep. Buvisindi 1 (1): 107-112.

Lowry, O.H., Rosenbrough, M.J., Farr, A.L. and Rawdall, R.J. 1951: Protein measurement with folin phenol reagent. *J.Biol. Chem.* 193: 265-275.

Lynch, J.M., Raphel, S.S., Mellor, L.D., Spare, D.D. and Inwoof, M.J.H. 1969: Medical laboratory technology and clinical pathology. W.B. Sounders company, Igaku shoin ltd. Tokyo, 626, 647, 662.

Magare, S.R., Ahirrao, D.V. and Suryawanshi, R.V. 1997: Alterations due to fluoride on protein metabolism in the crab *Barytelphusa cunicularis. Environ. and Ecology.* 15(1) : 74-78

Mishra, P.C. and Mohapatra, A.K. 1987: Haematological characteristics and bone fluoride content in *bufo melenostictus* from an aluminum industrial site., *Environ. Toxicol. Chem.* 6 (7) 529-533.

Ream, L.J., Pendergrass, P.B. and Scott, J.N. 1983: Fluoride ingestion during pregnancy and lactation: Morphological effects on maternal rat bone. Fluoride 16 (1) 23-33.

Reddy, K.P.O., Latha, E.S., Muthuraj, D. and Neeraja, P. 1993: Effect *In vivo* of ammonium sulphate haematology and oxygen consumption in fish, *labeo rohita., J. Inland Fish Soc. India.* 25 (1): 51-54.

Richard, G.F. 1994: More fluoride poisoning in Hooper Bay, Alaska- A case report. Fluoride 27 (1) 31-35.

Sahdev, S., Bano, R. and Lal, S.B. 1996: Effect of sodium fluoride on testicular enzymes in the sexually mature Swiss albino mice. *Indian J. Environ. And Toxicol.* 6(1) : 14

Sondhi, H., Gupta, M.L. and Gupta G.L. 1995: Effect of sodium fluoride on Swiss albino mice. *Geobios*, 22 (1) 18-21.

Susheela, A.K. 2001: A treatise on fluorosis. Rajiv Gandhi National drinking water mission, research and rural development foundation, ministry of rural development. New Delhi 92, 1-119.

Swarup, D., Dwivedi, S.K., Dey, S. and Ray, S.K. 1998: Fluoride intoxication in Bovines due to industrial pollution. *Indian J. Animal Science* 68 (7) : 605-608.

Venkateshwarlu, M., Reddy V.S.C., Sharma, G.P., Reddy, V.S., Yadagirl, B. and Reddy, C.E. 1994: Studies on certain biochemical constituents of blood in cyclic and true and estrus buffaloes in fluoride endemic areas. *Indian Vet.* J. 71 (12) : 1190-1194.

Willard, H.H., Merritt, L.L. and Dean, J.A. 1965: Instrumental methods of analysis. Forth edition, Venwostrand, New York 590.

World Health Organization, 1971: Fluoride and human health McGraw, Hill Encyclopedia of Science and Technology, Vol. 5, McGraw-Hill book company, New York.

Wybenga, D.R. and Pilleggi, V.J. 1970: *Invitro* estimation of cholesterol. *Clini. Chem.* 16: 980.

Yingyan, W. and Xiaojic, Li. 1994: ESR spectrum studies of the influence of fluoride on the human erythrocyte membrane protein SH binding site property. Fluoride, 27 (3), 129-135.

Zhai X. J., Cheng, Y.H., Qri, Z.M., Wang, Y.W., Yan, X.F., Cai, Y., Zang, K.R. and Ji, Nm. 1985: Reproduction of chronic fluorosis in sheep Chinese. *J. of Vet. Science and Techno.* Zhonggio- Shouyi- Keji. 9: 8-12.

2013, Climate Change Effects on Agriculture and Economy *Pages* **195–200**
Editors: **D.R. Khanna, A.K. Chopra, Vikas Singh, Rakesh Bhutiani & Gagan Matta**
Published by: **BIOTECH BOOKS, NEW DELHI**

Chapter 22

Analysis of Some Physico-chemical Parameters and Heavy Metals of Dhamola River at Distt. Saharanpur, Uttar Pradesh

Priyanka Rajvanshi
Department of Chemistry, M.S. College, Saharanpur, Uttar Pradesh (INDIA)

In the present work, seasonal variation in physico-chemical parameters of river Dhamola at Dist. Saharanpur was studied (2010-2011). For this study three study sites were selected for water sampling seasonally for a period of one year. Samples were analyzed for various physico-chemical parameters like Temperature, Turbidity (JTU), Total solids (mg/l), Total dissolved solids (mg/l), Total suspended solids (mg/l), pH, BOD (mg/l), COD (mg/l), DO (mg/l) and Total heavy metals (mg/l). Some parameters were analyzed at the sampling sites and others were in the laboratory. The less concentration of total heavy metals was observed at sampling station A (near Numaish camp Bridge), while at sampling station B (Near Rakesh Cinema Bridge) and C (Near Vishwa Karma Chowk Bridge) greater concentration were observed.

Keywords: *Water pollution, River Dhamola, Physico-chemical parameters, Water quality.*

Introduction

In most of the rivers the water is polluted naturally as well as through anthropogenic processes. In natural processes, water pollution is mainly caused by the decomposition of vegetables, animals and weather products. It affects the colour, odour and biological properties of water. In anthropogenic processes, water pollution is caused by industrial, agricultural, domestic, radioactive, mining sources, use of fertilizer and pesticides by human being *etc*. These pollutants are regularly poured into water deterioration. These pollutants contain simple nutrients as well as heavy metals.

Many rivers flow through the district Saharanpur as Krishna, Kalindi, Yamuna, Hindon, Maskara, Panvdhoi, Dhamola *etc*. River Dhamola originates from Muzzafrabad and then it goes to Saharanpur and mix with river Panvdhoi. It is a tributary of river Hindon. In district Saharanpur rapid industrialization is taking place day by day, so most peaceful area is changing in industrialized and urbanized area. Most drains of these industries carry effluents from factories and also from adjacent residential colonies with their domestic sewage which finally poured into river Dhamola. Thus river water is polluted. As river passes through Saharanpur its water turns brownish and blackish as small scale industrial effluents and house hold and municipal wastes are discharged in it. Dhamola river confluence with Panvdhoi near Rakesh Cinema and thereafter it continues up to Tapri, thereafter it confluence with Hindon River. Its water is polluted by Dhamola River and is used for agriculture purposes causing bad effect on human being and cattle. So it can be well thought that polluted river water carries a variety of pollutants of equally different physico-chemical nature. Therefore, it was of interest to carry such work.

Materials and Method

For physico-chemical study of river Dhamola at Saharanpur the water samples were collected seasonally from different sampling stations *i.e.* (A) Near Numaish Camp Bridge (B) Near Rakesh Cinema Bridge (C) Near Vishwa Karma Chowk Bridge.

The study was carried out from March, 2010 to February, 2011. The samples were taken in morning hours from 7.00 AM to 10.00 AM in borosil glass bottles of 300 ml, plastic cans of 1 liter from each site. The analysis of samples was done with the standard methods suggested by APHA (1998), Khanna and Bhutiani (2005). Some parameter like Temperature and pH were analyzed at the site and other mentioned parameters were analyzed in the laboratory.

Results and Discussion

The value of water temperature was observed 15.0 °C ± 0.80 to 16.0 °C ± 0.86 in winter season, 24.7 °C ± 0.59 to 25.7 °C ± 0.68 in summer season and 25.2 °C ± 0.37 to 26.1 °C ± 0.26 in monsoon season in all three sampling stations. The seasonal average value of temperature varied between 21.63 °C ± 5.75 (at sampling station A) to 22.60 °C ± 5.72 (at sampling station C). The temperature showed an upward trend from winter to summer season followed by downward trend from monsoon season onwards. More or less similar studies have been observed by Khanna (1993) in river Ganga and Gautam *et al.* (2000) in river Ganga at Rishikesh. Same trend of temperature was observed by Khanna *et al.* (2003, 2005) of various bathing Ghats of river Ganga at Bulandshahar and in river Panvdhoi at Saharanpur respectively. Dalal and Arora (2008) observed the similar trend in river Hindon at Ghaziabad. The seasonal average values of Turbidity in all three sampling stations varied between 290.67 JTU ± 62.61 and 576.67 JTU ± 85.20 in which maximum value at station C and minimum value at sampling station A. Water of river is found turbid throughout the year but goes highest in 665 JTU ± 110 in monsoon (sampling station C) and minimum 226 JTU ± 6.1 in winter season at sampling station A. Same pattern was reported by Khanna (1993), Seth *et al.* (2000). Khanna *et al.* (2005) also found similar trend in river Panvdhoi at Saharanpur.

Table 22.1: Seasonal Variation in Physico-chemical Parameters of Dhamola River at Saharanpur at Sampling Station A

Physico-chemical Parameters	*Summer*	*Monsoon*	*Winter*	*Average*
Temperature (C)	24.7±0.59	25.2±0.37	15.0±0.80	21.63±5.75
Turbidity (JTU)	295.0±7.2	351.0±3.00	226.0±6.1	290.67±62.61
Total Solids (mg/l)	951.0±32.00	1350.0±110.00	740.0±60.12	1013.67±309.79
TDS (mg/l)	485.0±25.00	595.0±63.81	328.0±35.18	469.33±134.19
TSS (mg/l)	466.0±23.35	755.0±90.25	412.0±50.12	544.33±184.43
pH	8.90±0.17	8.64±0.18	8.27±0.25	8.60±0.32
BOD (mg/l)	300.0±10.23	330.0±20.15	217.0±7.21	282.33±58.53
COD (mg/l)	895.0±50.13	1105.0±200.10	685.0±32.14	895.0±210.0
DO (mg/l)	6.05±0.11	4.02±0.51	6.75±0.01	5.61±1.42
Total heavy metals (mg/l)	9.97±0.62	10.24±0.5	9.76±0.15	9.97±0.24

Table 22.2: Seasonal Variation in Physico-chemical Parameters of Dhamola River at Saharanpur at Sampling Station B

Physico-chemical Parameters	*Summer*	*Monsoon*	*Winter*	*Average*
Temperature (C)	25.1±0.54	25.7±0.76	15.30±0.32	22.03±5.84
Turbidity (JTU)	350.0±6.8	536.0±14.50	283.0±7.21	389.67±131.08
Total Solids (mg/l)	1130.0±145.10	1420.0±192 21	875.0±120.00	1141.67±272.69
TDS (mg/l)	595.0±16.28	626.0±128.18	403.0±20.21	541.33±120.80
TSS (mg/l)	535.0±92.00	794.0±78.10	472.0±50.10	600.33±170.65
pH	8.95±0.10	8.72±0.14	8.32±0.12	8.66±0.32
BOD (mg/l)	436.0±32.60	477.0±28.10	402.0±16.15	438.33±37.55
COD (mg/l)	1020.0±126.10	1138.0±62.10	927.0±78.18	1028.33±105.75
DO (mg/l)	3.07±0.80	2.79±0.08	3.68±0.65	3.18±0.46
Total heavy metals (mg/l)	11.09±0.18	12.68±0.41	10.79±0.08	11.52±1.02

Total solids were found maximum 1790.0 mg/l ± 276.0 at station C and minimum 742.0 mg/l ± 60.12 sampling station A. Seasonal average value of total solids varies between 1015.0 mg/l ± 309.9 and 1464.0 mg/l ± 289.71. Minimum average value of total solid during investigation found at sampling station A and maximum at sampling station C. Total dissolved solids were found maximum 784.0 mg/l ± 130.0 at station C and minimum 485.0 mg/l ± 25.0 at sampling station A. Seasonal average value of TDS varies between 469.33 mg/l ± 134.19 and 673.33 mg/l ± 133.55. Total suspended solids were maximum 1014.0 mg/l ± 110.92 at station C and minimum 466 mg/l ± 23.35 at sampling station A. Seasonal average value of TSS varies between 544.33 mg/l ± 184.43 and 786.67 mg/l ± 197.70. Same conditions were shown by Khanna (1993), Khanna *et al.* (1997) and Seth *et al.* (2000) also found the same observation. Similar trends were shown by Khanna and Chugh (2004) during study of water quality of river Ganga at Haridwar. Khanna *et al.* (2005) and Dalal and Arora (2008) also found similar pattern during the physico-chemical study of river Panvdhoi at Saharanpur and river Hindon at Ghaziabad, respectively.

Table 22.3: Seasonal Variation in Physico-chemical Parameters of Dhamola River at Saharanpur at Sampling Station C

Physico-chemical Parameters	*Summer*	*Monsoon*	*Winter*	*Average*
Temperature (C)	25.7±0.68	26.1±0.26	16.0±0.86	22.60±5.72
Turbidity (JTU)	570.0±80.0	665.0±110.0	495.0±45.0	576.67±85.20
Total Solids (mg/l)	1366.0±226.0	1790.0±276.0	1236.0±167.10	1464.0±289.71
TDS (mg/l)	711.0±42.18	784.0±130.0	525.0±25.25	673.33±133.55
TSS (mg/l)	655.0±167.01	1014.0±110.92	691.0±21.12	786.67±197.70
pH	8.98±0.10	8.81±0.16	8.48±0.16	8.76±0.25
BOD (mg/l)	502.0±45.10	536.0±11.10	488.0±27.24	508.67±24.68
COD (mg/l)	1172.0±130.0	1290.0±55.20	1027.0±71.10	1163.0±131.73
DO (mg/l)	1.37±0.52	1.01±0.06	2.9±0.90	1.76±1.00
Total heavy metals (mg/l)	13.78±0.10	15.5±0.02	13.20±0.08	14.16±1.20

During the course of study it was recorded that pH was always alkaline at all the sampling stations. The seasonal average value of pH varied between 8.60 ± 0.32 to 8.76 ± 0.25. The highest pH value 8.98 ± 0.10 was found in monsoon season at sampling station C and lowest pH value 8.27 ± 0.25 was observed in winter season at sampling station A. More or less similar results were reported Sangu and Sharma (1985) in river Yamuna and by Khanna (1993) in river Ganga and Khanna *et al.* (2007) in river Ganga from Rishikesh to Haridwar. Dalal and Arora (2008) found that pH of water Hindon remains slightly alkaline throughout the study period.

The maximum value of BOD was observed 536.0 mg/l ± 11.10 in monsoon season and minimum 217.0 mg/l ± 7.21 in winter season. The seasonal average value of BOD ranged between 282.33 mg/l ± 58.53 to 508.67 mg/l ± 24.68. Khanna *et al.* (1997, 2005) observed peak values in monsoon season in river Ganga at Laljiwala Haridwar and river Panvdhoi at Saharanpur. Dalal and Arora (2008) also observed similar trends in river Ganga at Bulandshahar and in river Hindon at Ghaziabad, respectively.

Chemical Oxygen Demand (COD) was noted maximum 1290.0 mg/l ± 55.20 in monsoon season and minimum 685.0 mg/l ± 32.14 in winter season. The seasonal average value of COD ranged between 895.0 mg/l ± 210 to 1163.0 mg/l ± 131.73. Similar trend of COD have shown by Khanna *et al.* (2003, 2005) in river Ganga and Panvdhoi and Dalal and Arora (2008) in river Hindon at Ghaziabad.

Maximum DO was recorded 6.75 mg/l ± 0.01 in the winter season and minimum 1.01 mg/l ± 0.06 in monsoon season. The seasonal average value varies between 1.76 mg/l ± 1.00 (at sampling station C) and 5.61 mg/l ± 1.42 (at sampling station A). Khanna (1993) has reported the same trend in the river Ganga. Same observation was made by Pandey *et al.* (2003) in Ganga canal at Haridwar, Khanna *et al.* (2005) in river Panvdhoi at Saharanpur and Singh *et al.* (2006) in river Ganga at Bulandshahar (U.P.).

In the present study total heavy metals were taken for observations. The seasonal average value of total heavy metals range between 14.16 mg/l ± 1.20 (at sampling station C) to 9.97 mg/l ± 0.24 (at sampling station A). Results obtained shows that heavy metal accumulation is more in downstream study site C as compared to upstream study site A. Khanna *et al.* (2003, 2005) reported heavy metals in water of Ganga at Bulandshahar and river Panvdhoi at Saharanpur. Singh *et al.* (2006) also reported heavy metals in river Ganga at Anupshahar.

Acknowledgement

I am greatly obliged to Prof D.R.Khanna, Dept. of Zoology and Env. Science Gurukul Kangri University, Haridwar and my guide Dr. S.C. Bhatia, Dept. of Chemistry, Saharanpur for their kind co-operation, continuous inspiration and precious advice.

References

APHA. 1998: Standard methods for the examination of water and wastewater. *American Public Health Association*, 1015 fifteenth St, NW Wahington. 15: 1–1134.

Dalal, Anupama and Arora, M.P. 2008: Seasonal Physico – Chemical fluctuations of water quality of river Hindon. *J. Exp. Zool. India* 11(1): 125–131.

Gautam, A., Khanna, D.K. and Sarkar, Praveen. 2000: Diurnal variation in the Physico – Chemical characteristics of Ganga water at Rishikesh during winter season. *Indian J. Environ. and Ecoplan* 3(2) : 369–371.

Khanna, D. R. 1993: Ecology and pollution of Ganga River, *Ashish Publishing House*, Delhi 1–241.

Khanna, D.R., Asharaf, J., Chauhan, Beena and Bhutiani, R. 2005: Physico – Chemical Anayasis of River Panvdhoi at Saharanpur (U.P.). *Environment Conservation Journal*. 6(2): 89–94.

Khanna, D.R., Bhutani, R.C. and Bhutiani, R. 2007: Limnological Modeling – A case study of river Suswa. *Daya Publishing House*. New Delhi. pp 1–297.

Khanna, D.R., Badola, S.P. and Malik, D.S. 1997: Population of green algae in relation to Physico – Chemical factor of the river Ganga at LaljiWala, Haridwar. *U. P. J. Zool.* 17(3): 237–240.

Khanna, D.R. and Bhutiani, R. 2005: Water analysis at a glance. Published by ASEA Rishikesh pp 1–108.

Khanna, D.R. and Chugh, T. 2004: Microbial Ecology, A study of river Ganga at Haridwar 1–273.

Khanna, D.R., Singh, S., Gautam, A. and Singh, J.P. 2003: Assessment of water quality of river Ganga in District-Bulandshahar, (U.P.). India. *J. Nat. Conservation* 15(1): 167–175.

Pandey, S. Kaushik, P. and Parashar, N.B. 2003: Physico-chemical and Microbiological studies of Ganga Canal at Haridwar during Kumbh Period 1998. *Him. J. Env.Zool.* Vol 17(2) : 167–171.

Sangu, R.P.S. and Sharma, K.D. 1985: Studies on water pollution on Yamuna River at Agra. *Indian. J. Environ. Hlth.* 27(3): 257–261.

Seth, T.R., Khanna, D.R., Gautam, Ashutosh, Chugh, Tarun and Sarkar, Praveen. 2000: Temporal Trends of phytoplanktonic diversity in the river Ganga at Haridwar. *Him. J. Env. Zool.* Vol- 14, pp 129–134.

Singh, J.P., Shakun, Singh and Khanna, D.R. 2006: Water quality of river Ganges in respect of Physico-chemical And Microbial Characteristics at Anupshahar, District- Bulandshahar (India) *Environment Conservation Journal* 7(1-2): 29–34.

2013, Climate Change Effects on Agriculture and Economy *Pages* ***201–212***
Editors: **D.R. Khanna, A.K. Chopra, Vikas Singh, Rakesh Bhutiani & Gagan Matta**
Published by: **BIOTECH BOOKS, NEW DELHI**

Chapter 23

Environmental Pollutants Causes Male Infertility and Tumorigenesis in Swiss Albino Mice

A. Nath, J.K.Singh, Preety Sinha, Manisha Lall, Anamika and Arti Kumari

Research Centre, Mahavir Cancer Institute and Research Centre, Phulwarisharif, Patna, Bihar

Environmental pollutants in the 21st century have increased many folds leading to various health hazards. Endosulfan an organochlorine pesticide is widely used by farmers in India to kill the pests for better yield of crops. But, these pesticides through the food chain have lead to various dysfunctions of the body's metabolic activity causing reproductive health problems in males and females.

In the present investigation endosulfan was administered to mice @ 3 mg/kg b. w. for 4 weeks to observe the toxicity induced in the body especially the male and female reproductive system. In parallel, arsenic was also administered to another group of mice @ 3 mg/kg b. w. for 4 weeks to observe the toxicity induced in the body especially the male and female reproductive system. The observations showed the testicular and ovarian damages both due to endosulfan toxicity as well as arsenic toxicity is at such an extent that it is causing infertility in them.

In both the treatment group the light microscopic study showed testicular dystrophy especially the abruptions in the spermatogenesis with vacuolations in them while in ovary follicular damage quiet prominent. The ultrastructural study strongly confirmed the damage with changes in cytoarchitecture especially in the spermatozoa tending towards the tumorogenesis while in female adenoid cystic carcinoma was very common.

The hormonal assessment also showed declination in the levels of Testosterone and Estrogen in both the treatment groups.

Thus, from the entire study it reveals that Endosulfan and Arsenic which are environmental pollutants cause high degree of degeneration in Swiss albino mice leading to infertility and tumorigenesis in mice.

Keywords: *Pollutants, Infertility, Tumorigenesis, Albino mice, Toxicity.*

Introduction

Environmental pollutants in the 21st century have increased many folds leading to various health hazards. Endosulfan an organochlorine pesticide (Insecticide and Acaricide) is widely used by farmers in India to kill the pests for better yield of crops. But, these pesticides through the natural food chain are contaminating every layer in the ecosystem. Human beings, at the top of the food chain are most vulnerable to the health risks, as ingestion of toxic contaminants is several-fold higher through the process of bio-magnification and lead to various dysfunctions metabolic activity of the body causing reproductive health problems in males and females.

Arsenic (heavy metal) is also an environmental pollutant. It is found commonly in water and soil. It's primarily use in agriculture, forestry, animal husbandry, wood preservation, pigments, semiconductor industry (Arsine gas). Arsenic can be inorganic or organic and trivalent (Arsenite) or pentavalent (Arsenate). Generally the inorganic forms and the trivalent forms are more harmful. Ingestion is the main route because arsenic is present in water and on many food products, especially marine food products. Inhalation and dermal absorption are typically related to occupational exposures. Once arsenic has been absorbed (Intestine, Lung, Skin) it is quickly transported in the blood through the body. In case of inorganic arsenic (arsenic trioxide), 80-90 per cent is absorbed from intestine while organic arsenic (seafood) is poorly absorbed from intestine.

Many attempts have been made to study the toxic effects of pesticide on the reproductive organs of mice (Russel, 1995; Kumar and Nath, 1995) especially testicular atrophy. Endosulfan may also increase defects in male sex organs (Solo, 1993). It has been well established that Endosulfan causes deleterious effect on reproductive physiology of the body (Nath *et al.*, 2007). Similar observation has been found by Arumugamn (1992).

The present study is an attempt to demonstrate the changes in the testicular and ovarian cells due to Endosulfan as well as by Arsenic.

Materials and Method

Pesticide

Endosulfan (EC = 35 per cent, w/v) was obtained from local market (manufactured by Excel Industries Ltd., Mumbai, India).

Experimental Model

In the present investigation, experiments were performed on 3 months old, sexually matured healthy female Swiss albino mice, *Mus musculus* weighing 25-35g b. wt. The animals were housed at controlled environmental conditions 22.0 ± 2 C, relative humidity 50.0 ± 10 per cent, and 12 h dark-light cycle. Animals were housed and allowed to free access to food and water. All experimental procedures were conducted as per the guide lines of CPCSEA (Committee for the Purpose of Control and Supervision of Experiments on Animals).

Grouping

- ✰ **GROUP- A**: Control; Normal male mice received distilled water only for 4 weeks (n = 10)
- ✰ **GROUP- B**: Control; Normal female mice received distilled water only for 4 weeks (n = 10)
- ✰ **GROUP- C**: Male Mice administered orally with Endosulfan @ 3mg/kg b.w. for 4 weeks (n = 10)
- ✰ **GROUP- D**: Female Mice administered orally with Endosulfan @ 3mg/kg b.w. for 4 weeks (n = 10)
- ✰ **GROUP- E**: Male Mice administered orally with Arsenic @ 3mg/kg b.w. for 4 weeks (n = 10)
- ✰ **GROUP- F**: Female Mice administered orally with Arsenic @ 3mg/kg b.w. for 4 weeks (n = 10)

Control, Endosulfan and Arsenic treated male and female mice were sacrificed at the end of treatment.

Methodology

Collection of Blood

Blood sample was obtained from mice by orbital sinus puncture and then separation of the serum was done by centrifugation @ 3000 rpm for 15 minutes.

Hormonal Assessment

Serum was analyzed for Testosterone and Estrogen by ELISA kits.

Epididymal Sperm Count

Cauda epididymis were dissected out and washed thoroughly in normal Saline (0.85 per cent). Cauda were incised in 1 ml distilled water so as to allow the sperm to ooze out. After that, two drops of Eosin Y is mixed well. Then sperm counts were made using an improved Neubauer's chamber taking a drop of above preparation in it and observed at 400 X magnification.

Histopathological Study

Fixations of the tissues (Ovary and Testis) were done in formalin for Light microscopy and in Gluteraldehyde for Electron microscopy.

Results and Discussion

The most fascinating result was that the female mice showed skin tumors after 4 weeks of endosulfan post treatment. One case of uterus tumor (1 per cent) was also found in 3.0 mg/kg b.w. female mice. Ovarian cysts were observed in 3 per cent of female mice. But, male mice showed no such signs (Figure 23.1).

Light Microscopic Study

Control Testis and Epididymis of Mice

The seminiferous tubules were round or oval with different stages of spermatogenesis. Seminiferous epithelium is normal. Spermatogonia were nearly round and walls were smooth. Spermatids were clearly visible under the 40X magnification of light microscope (Figure 23.2).

In Epididymis, Spermatozoa are seen in lumen. Principle cells are normal. Basal cells are also well arranged (Figure 23.4).

Endosulfan Treated Testis and Epididymis of Mice

In Testis, many changes are noticeable after the administration of endosulfan. The diameter of the seminiferous tubule decreased. Interspaces between seminiferous tubule increased. Seminiferous epithelium is very thin and looks serrated. Spermatids tail in lumen and agglutination of sperm tail was also observed (Figure 23.3).

In Epididymis, multiple vacuoles are seen in epithelium. Lumen is filled with spermatozoa. Basal cells and apical cells are lacking (Figure 23.5).

Arsenic Treated Testis and Epididymis of Mice

In testis, degeneration of seminiferous tubules was observed. Vacuolated spaces were observed inside seminiferous tubules. Seminiferous tubule epithelium is very thin. Clustered nucleus are seen in a layer with agglutinated sperm tail in lumen (Figure 23.6).

In epididymis, different stages of degeneration of caput epithelium were observed. Principle cells are not normal. Multiple vacuoles were observed in epithelium.

Control Ovary of Mice

Histopathological study showed the stages of maturation of follicles which was regular in all female of control group (Figure 23.8(a).

Endosulfan Treated Ovary of Mice

After administration of Endosulfan, some degenerations are seen in ovarian follicles. Spaces between zona pellucida and multiple layers of follicle cells were increase (Figure 23.8a-b).

Arsenic Treated Ovary of Mice

(Figure 23.8c-d).

Electron Microscopic Study

The control group showed normal structure of spermatozoa with nuclear membrane, chromatin, plasma membrane (PM), acrosome (AC) and head cap (HC) (Figure 23.7a).

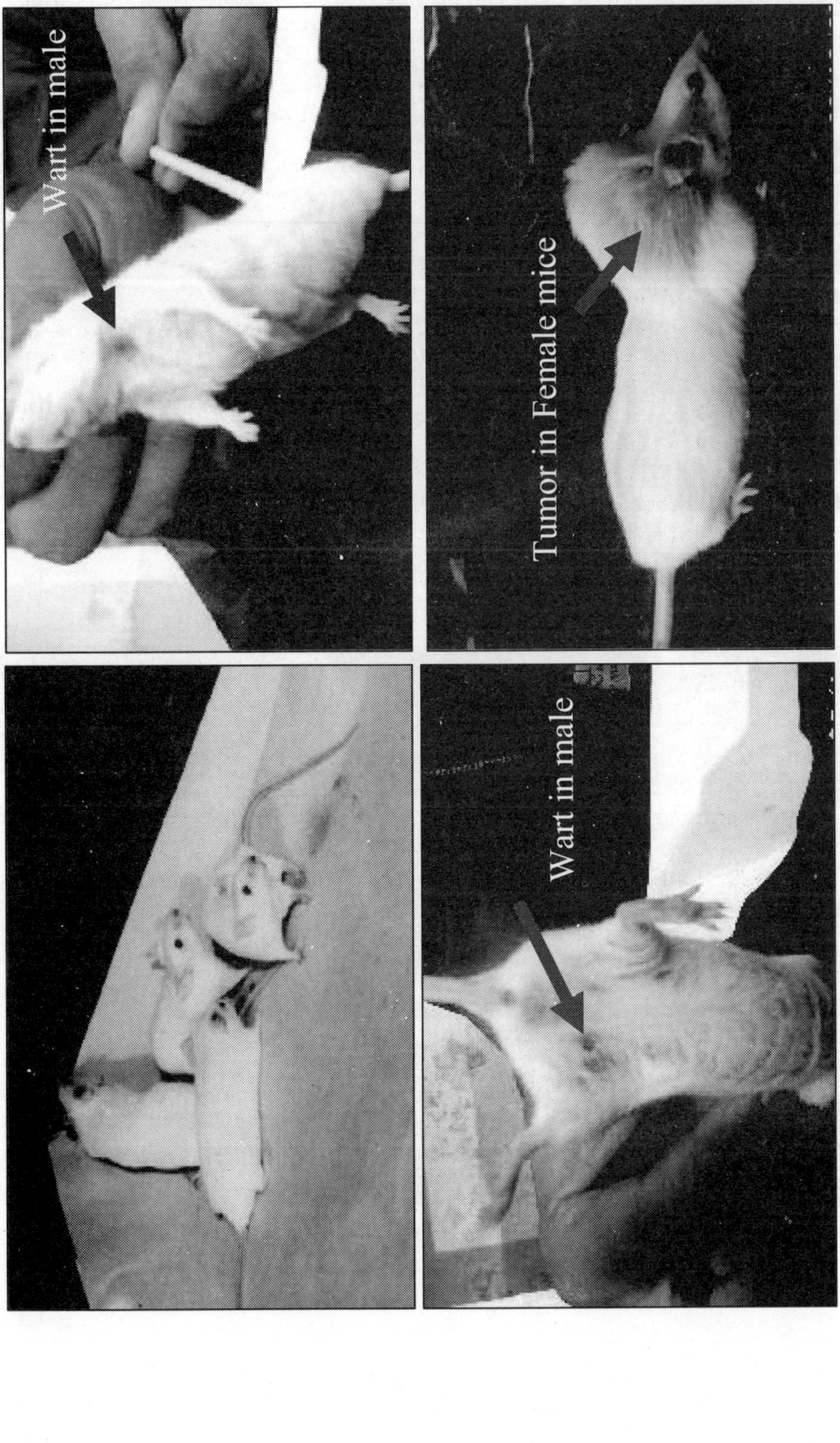

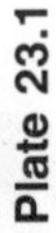

Plate 23.1

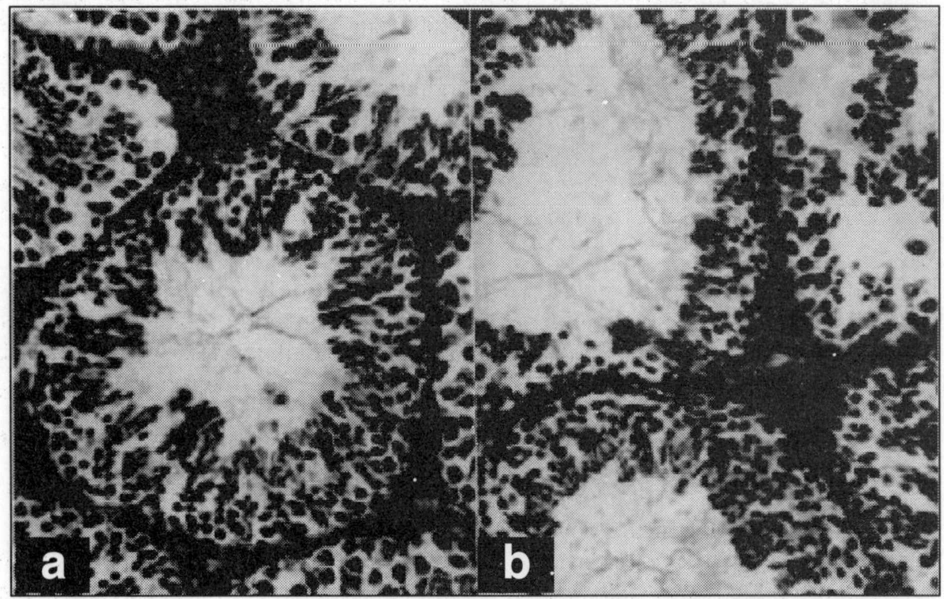

Plate 23.2a-b

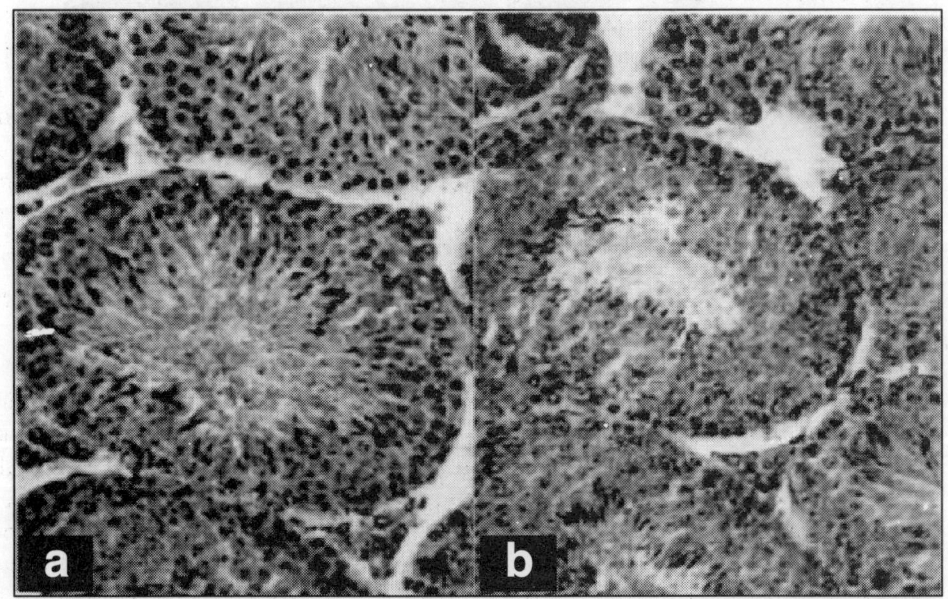

Plate 23.4a-b

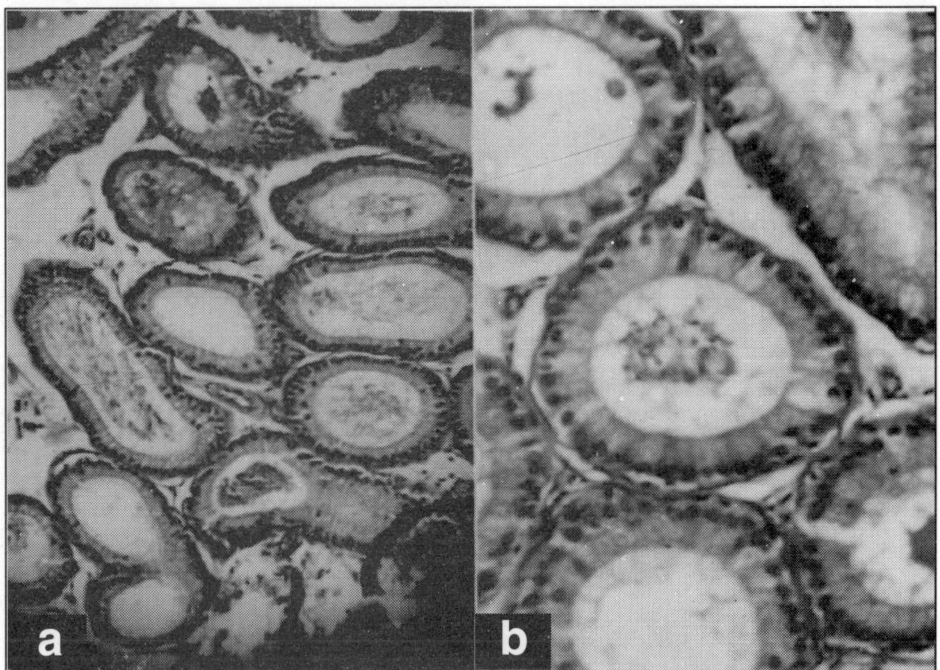

Plate 23.5a-b

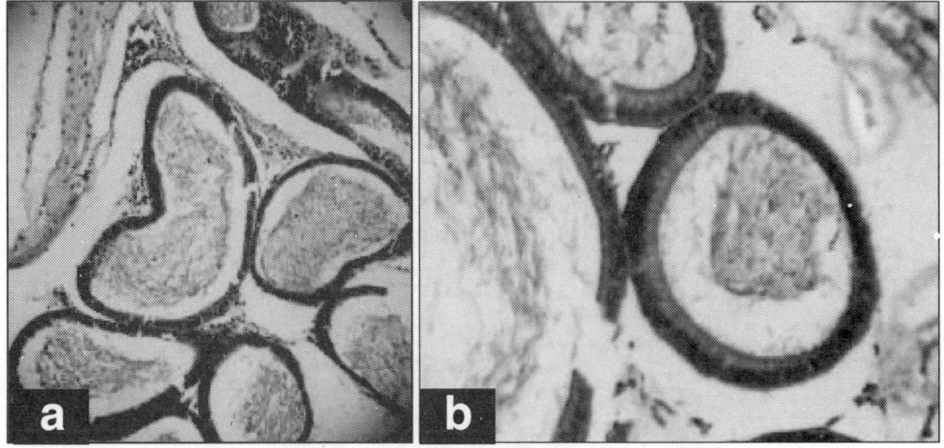

Plate 23.6a-b

Endosulfan treated group showed degeneration in apical acrosome region (AC) with vacuolation in the cytoplasm. Degenerated head cap (HC) and heterochromatization (HCR) were clearly observed (Figure 23.7b)

After administration of Endosulfan, electron microscopic study showed the shape of nucleus of ovarian cells which was found abnormal as compared to the control.

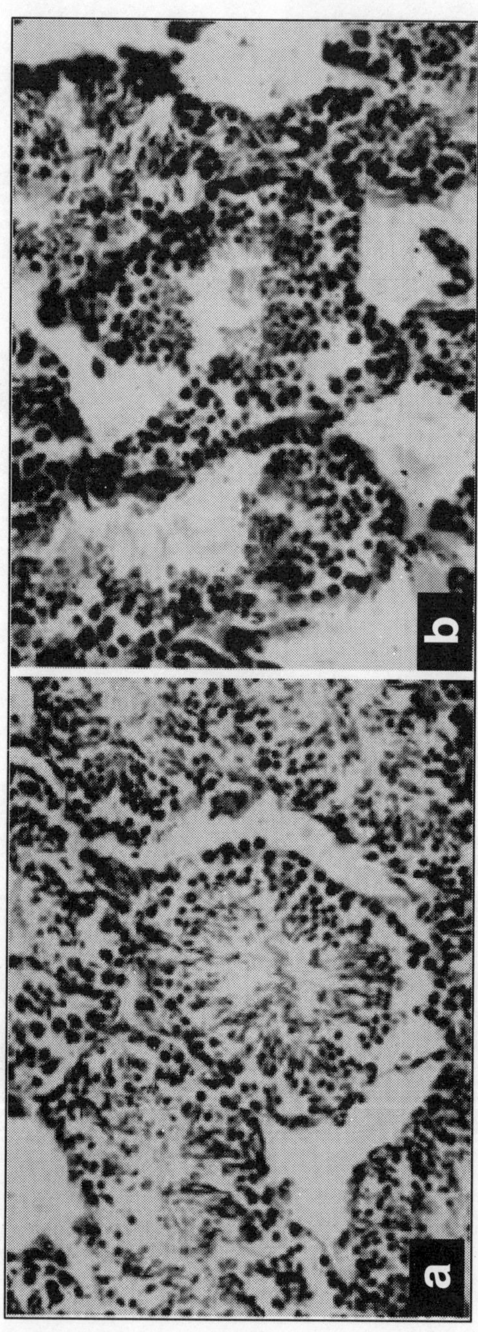

Plate 23.7a-b

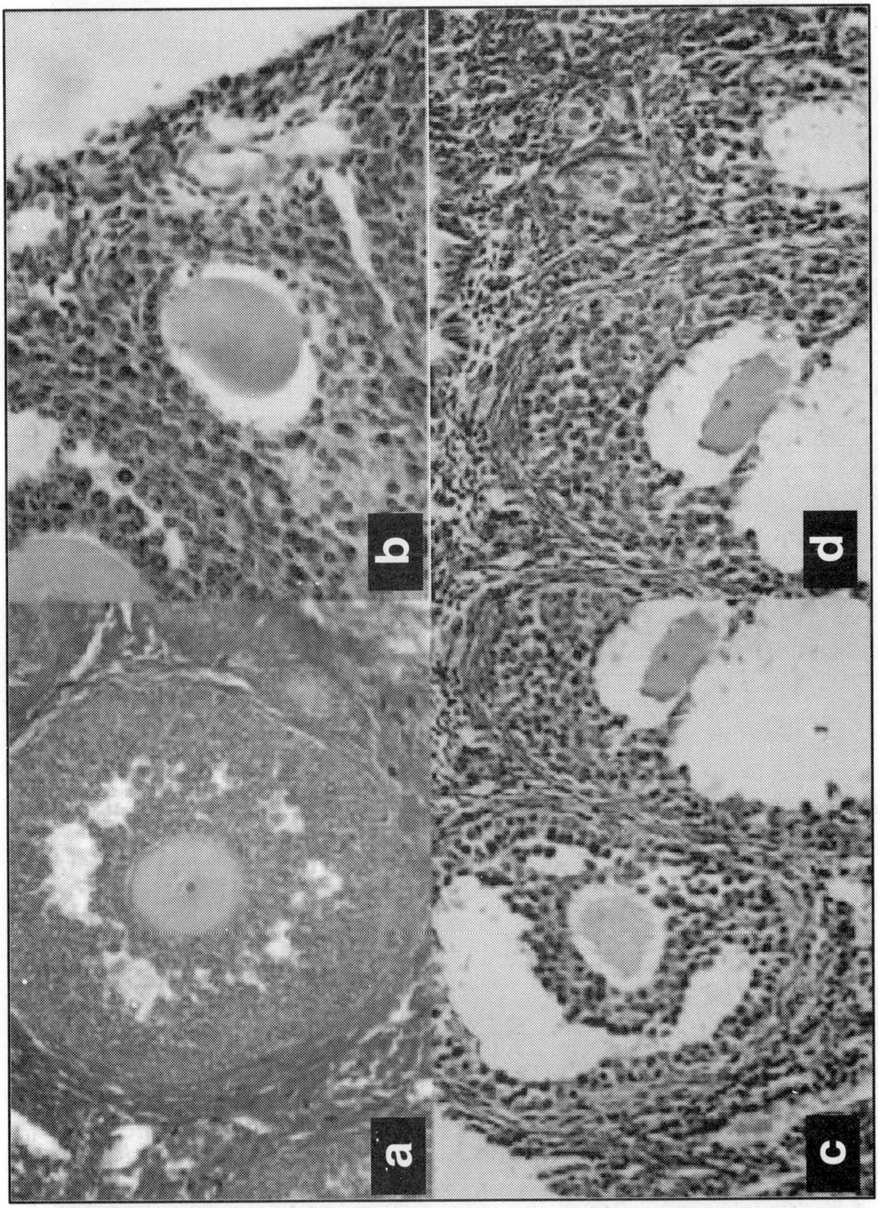

Plate 28.7a-d

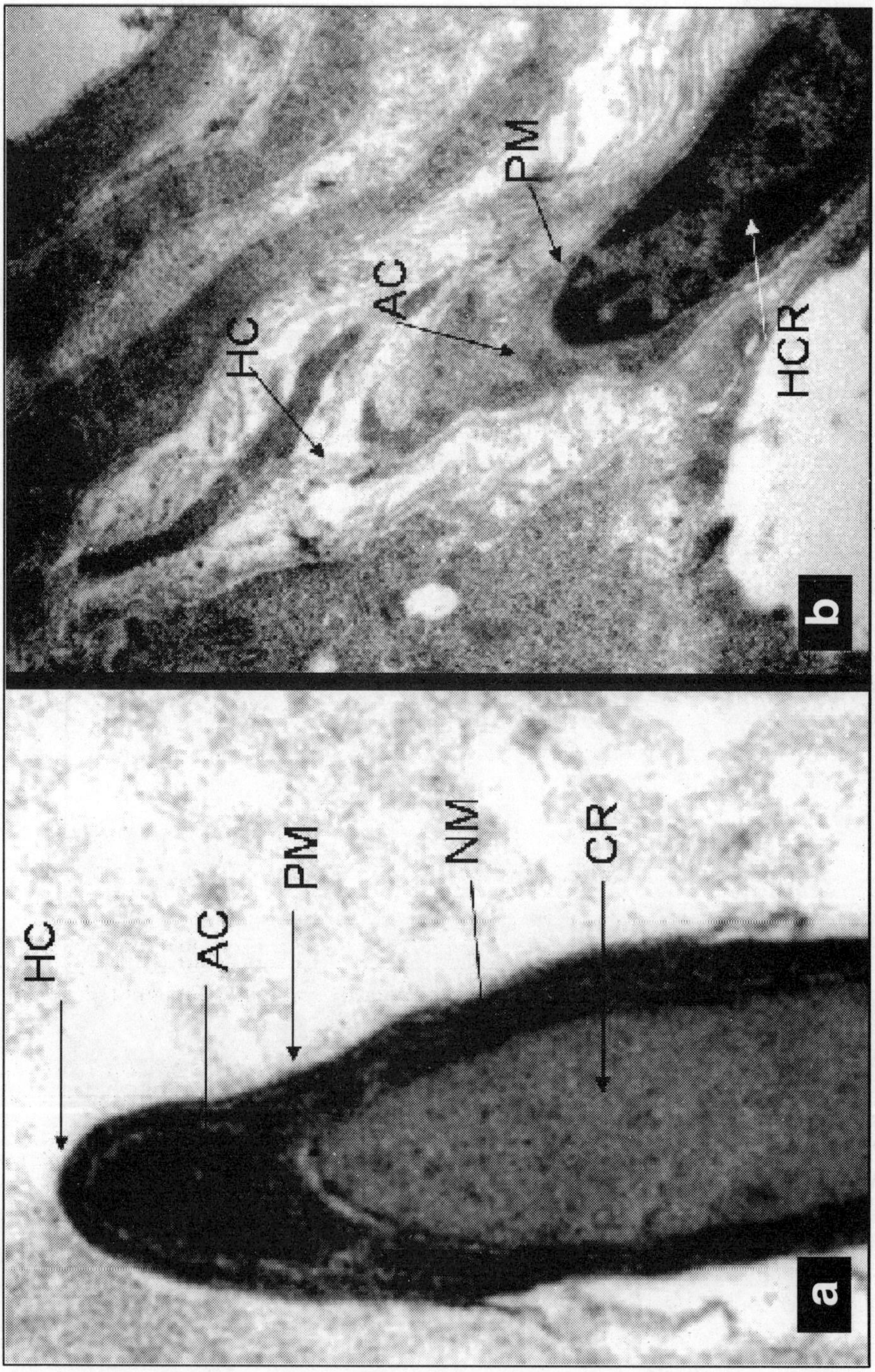

Plate 23.9a-b

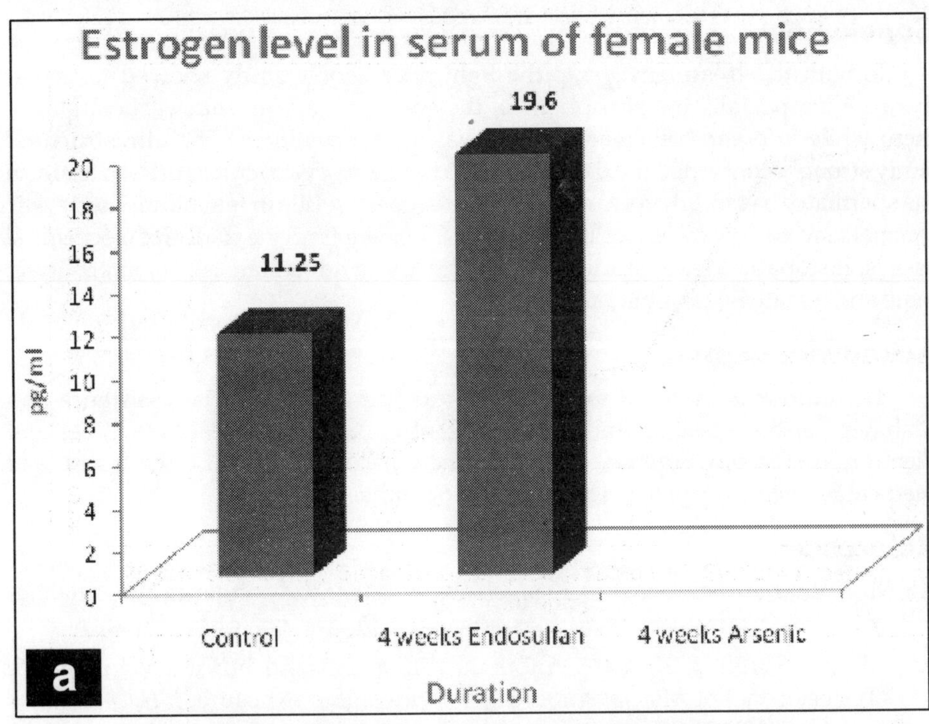

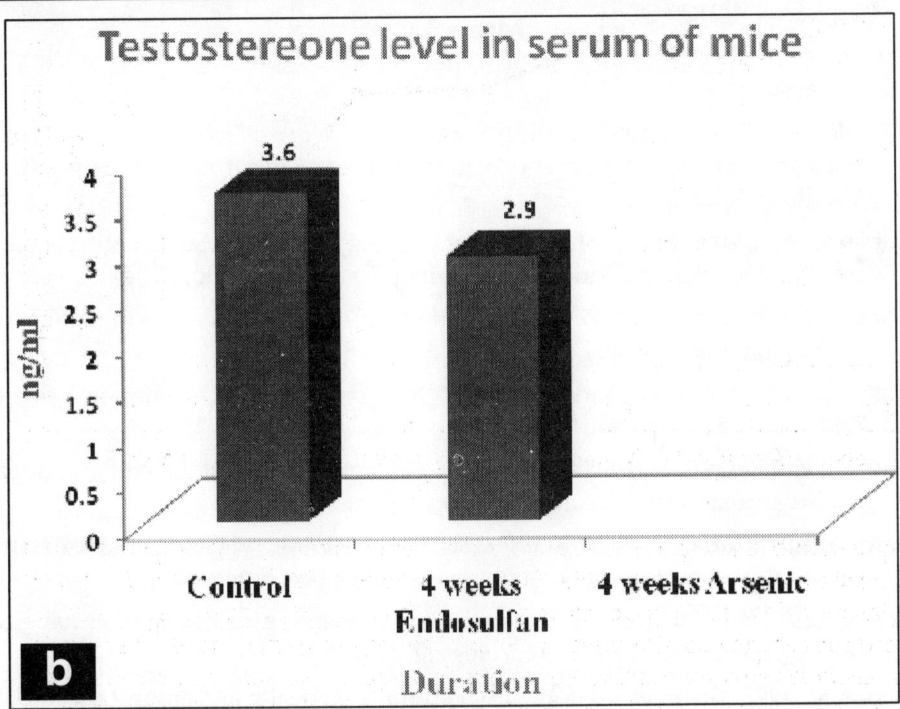

Figure 23.10a-b

Conclusions

In both the treatment group the light microscopic study showed testicular dystrophy especially the abruptions in the spermatogenesis with vacuolations in them while in ovary follicular damage was quiet prominent. The ultrastructural study strongly confirmed the damages with changes in cytoarchitecture especially in the spermatozoa tending towards the tumorogenesis while in female adenoid cystic carcinoma were very common. That is why, it is evident from the studies of biochemical assays, histopathological studies by LM and EM, Endosulfan causes infertility in male and female Swiss albino mice.

Acknowledgement

The author is thankful to ICMR for providing the financial assistance and Mahavir Cancer Sansthan and Research Centre, Patna for research facilities and Electron Microscopy Sophisticated Instrument Facility Unit, All India Institute of Medical Sciences, New Delhi for microphotographs.

References

Ali, M., Kumar, A., Srivastava, A.K. and Nath, A. 2006: Study of sperm abnormalities of Endosulfan induced mice. *Mus musculus. J. Biospectra*; Vol. (1): 59-62.

Ali, M. and Singh, J. K. 2009: Anomalies in sperm and hormonal imbalance (Testosterone) of *Mus musculus* due to Endosulfan exposure. *J. Haematol. and Ecotoxicol.* 4(2).30-35

Arumugam, V. 1992: Victims without a voice: a study of women pesticide workers in Malaysia.

ATSDR. 2000: Toxicological profile for Endosulfan. United States Agency for Toxic Substances and Disease Registry, Atlanta, GA Available from NTIS, Springfield, VA. PB93-182558.

Kumar R., *et al.*, 2009: Endosulfan exposure Causing Acrosome Malformation leading infertility in Swiss Albino Mice. J. Haematol. and Ecotoxiclo.4 (2) 51.55.

Kumar, S. and Nath, A. 1995: Study of effects of sublethal doses of malathion on spermatogonia of mice. *J. Trend in Life Sci.* 10, 53-58.

Nath, A., Sahay, R. and Kumar, R. 2007: Ovarian Follicular Degeneration due to Endosulfan Treatment in Swiss Albino Mice. Embryo Talk. Vol 2.2.

Nath, A. and Kumar, A. 2007: Occurrence of apoptosis due to Endosulfan induced testicular toxicity in Mus musculus. *Int. J. Toxicology*.

Nath, A. and Kumar, R. 2007: Interference of Endosulfan on Golgi complex action during Acrosome formation in spermatozoa of mice. Embryotalk.

Russel, L.D. 1995: The Role of histopathology of testes in short term toxicological testing protocol. Inhalation toxicology, 7, 961.

Solo, A.M. 1993: Testimony before the sub comitte on health and environment, U. S. house of representative.

2013, Climate Change Effects on Agriculture and Economy *Pages* ***213–222***
Editors: **D.R. Khanna, A.K. Chopra, Vikas Singh, Rakesh Bhutiani & Gagan Matta**
Published by: **BIOTECH BOOKS, NEW DELHI**

Chapter 24
Importance of Metal Speciation Study in Sediment in Detection of Pollution

Ruchi Raghuwanshi[1], M.K. Sharma[2], Ritu Sharma[2] and Prashant Singh[1]
[1]Department of Chemistry, DAV (PG) College, Dehradun, Uttarakhand
[2]National Institute of Hydrology, Roorkee, Uttarakhand

The sediments existing at the bottom of the water column play a major role in the pollution scheme of the river systems by heavy metals. The sediments, act as traps for numerous compounds, thus allowing a control of pollution of overlying water to some extent. However, they still remain less variable in sediment phase as compared to the water in equilibrium. Therefore sediments are the better indicator of pollution in pollution scheme of a surface water body. Speciation study of metals in sediment provides information on the potential availability of metals (toxic) to biota under various environmental conditions. The term chemical speciation was described as the determination of individual concentration of the various chemical species of an element which covers the total concentration of that element in a given sample. The behaviour of a particular matter depends on its chemical form. In the present paper, metal concentration bound to different forms of sediment *viz*; (i) adsorptive and exchangeable, (ii) bound to carbonate phases, (iii) bound to reducible phases (iron and manganese oxides), (iv) bound to organic matter and sulphides and (v) detrital or lattice metals have been studied These forms have different behaviour with respect to remobilisation under changing environmental conditions. Their toxicity depends especially on their chemical forms rather than their total metal content and has been illustrated by few case studies carried out by different workers in India.

***Keywords**: Pollution, Sediments, Metal speciation, Toxicity.*

Introduction

Water pollution is contamination of water by foreign matter that deteriorates the quality of the water. Water pollution covers pollutions in liquid forms like ocean pollution and river pollution. As the term applies, liquid pollution occurs in the oceans, lakes, streams, rivers, undergroundwater and bays, in short liquid-containing areas. It involves the release of toxic substances, pathogenic germs, substances that require much oxygen to decompose, easy-soluble substances, radioactivity etc. that becomes deposited upon the bottom and their accumulations will interfere with the condition of aquatic ecosystems. For example, eutrophication is lack of oxygen in a water body caused by excessive algae growths because of enrichment of pollutants.

Heavy metals are not biodegradable and enter the food chain through a number of pathways causing progressive toxicity due to the accumulation in human and animal organs during their life span on long term exposure to contaminated environments. Despite the presence of trace concentrations of Cr, Mn, Co, Cu and Zn in the aquatic environment, which is essential to a number of life processes, high concentrations of these metals become toxic.

In the pollution study of riverine systems heavy metal pollution assume a great significant. Unlike the organic pollutants, heavy metals are not removed by natural processes of decomposition. On the contrary, they may be enriched by organism (Bio-magnification) and can be converted to organic complexes which may be toxic. A typical example is the transformation of elementary and divalent ionic mercury into methyl mercury (Forstner and Muller, 1973).

The sediments existing at the bottom of the water column play a major role in the pollution scheme of the river systems by heavy metals. They reflect the current quality of the water system and can be used to detect the presence of contaminants that do not remain soluble after discharge into surface waters. The sediments, act as traps for numerous compounds, thus allowing a control of pollution of overlying water to some extent. However, they still remain less variable as compared to the water in equilibrium (Forstner and Muller, 1973; Harding and Whiton, 1978; Forstner and Wittmann, 1983).

The sediments play an important role in determining concentration, distribution and the fate of several pollutants in a riverine system. The concentration of heavy metals in sediments and their characteristic can act as indicator of the state of the aquatic environment. The total metal will only be able to provide information about the pollution if the background level or geochemical composition is known: metal origin (natural or anthropogenic) is rather difficult to be determined. Thus to assess the environmental impact of sediments, the determination of trace metal is not sufficient in itself (Forstner and Salomons, 1980). The chemical form of metal in the sediment ultimately determines the behaviour and mobilisation ability of the metal in the environment.

Chemical Speciation

The term chemical speciation was described (Ure, 1990; Ure *et al.*, 1993) as the determination of individual concentration of the various chemical species of an element which covers the total concentration of that element in a given sample. On the basis of this principle, there are two approaches as organometallic speciation and inorganic speciation. In organometallic speciation, the inorganic and organic forms are specified by their different toxicities and mobilities while in inorganic speciation different oxidation states of a metal are specified by their different toxicities and mobilities. A few (Quevauviller, 1998; Kramer and Allen, 1988; Brinckman and Salder, 1986; Patterson and Passino, 1990; Ali and Aboul-Enein, 2006), among several publications on metal speciation of environmental samples are available and helpful in this area. Scientists have defined the elemental speciation as an analytical activity to identify the actual chemical species of an element and also their quantity (Caruso and Bayon, 2003; Montes-Bayon, 2003; B'Hymer *et al.*, 2000). Metal ion speciation analysis is necessary and need of the day in order to know very low concentration of a particular metal species, which has adverse effects on human health.

The total concentration of metal does not reliably predict its bioavailability, toxicity and the distribution between water-sediment. The behaviour of a particular matter depends on its chemical form. The major mechanism of accumulation of heavy metal in sediment lead to the existence of five categories (Gibbs, 1977) adsorptive and exchangeable, bound to carbonate phases, bound to reducible phases (iron and manganese oxides), bound to organic matter and sulphides and detrital or lattice metals. These categories have different behaviour with respect to remobilisation under changing environmental condition.

Determination of only total element in sediments does not give an accurate estimate of the likely environmental impacts. Speciation study of metals in sediment provides information on the potential availability of metals (toxic) to biota under various environmental conditions. In water, the toxic metal specie is the free hydrated metal ion. The toxicity of metals depends especially on their chemical forms rather than their total metal content (Korfali and Jurdi, 2011).

Speciation Studies

Jain *et al.* (2003) studied the metal pollution in water and sediment of the River Hindon in western Uttar Pradesh (India) for Cd, Cr, Cu, Fe, Mn, Ni, Pb and Zn. The ratio of heavy metals to conservative elements (Fe, Al *etc.*) may reveal the geochemical imbalances due to the elevated metal concentrations normally attributed to anthropogenic sources. Metal/Al ratios for the bed sediments of the river Hindon were used to determine the relative mobility and general trend of relative mobility occurred Fe > Mn > Zn > Cr > Ni > Pb > Cu > Cd.

Singh *et al.* (2003) studied the concentrations of cadmium, chromium, copper, iron, lead, manganese, nickel, and zinc in bed sediments of river Gomti from Neemsar to Jaunpur. The river water and sediment samples were processed and analyzed for heavy metals *viz.*, Cd, Cr, Cu, Fe, Pb, Mn, Ni, and Zn using ICP-AES. In the sediments, the same were found in the range: Cd (0.70–7.90 μg/g); Cr (6.105–20.595 μg/g); Cu (3.735–35.68 μg/g); Fe (5051.485–8291.485 μg/g); Mn (134.915–320.45 μg/g); Ni (13.905–37.370 μg/g); Pb (21.25–92.15 μg/g); and Zn (15.72–99.35 μg/g) of dry weight respectively.

Gaur *et al.* (2005) studied water and sediment of the river Gomti contaminated due to heavy metals and other pollutants. Impacts of domestic/industrial wastes on the water and sediment chemistry of river Gomti with special reference to heavy metals have been investigated in different seasons (summer, winter and rainy). For this, seven sampling sites were selected and six metals (Cd, Cr, Cu, Ni, Pb and Zn) were analyzed by ICP-AES in the river Gomti. High concentrations of all the metals were noticed in water and sediment in rainy season compared to summer and winter.

Jain *et al.* (2010) studied the metal fractionation of bed sediments of Hussainsagar Lake, Hyderabad, India. Heavy metals entering the lake get adsorbed onto the suspended sediments, which eventually settle down in the bottom of the lake. Comparison of sediments with average shale values indicated anthropogenic enrichment with copper, nickel, lead, cadmium, and zinc. The risk assessment code as applied to the study revealed that 10–17 per cent of manganese, 10–18 per cent of nickel, 14–24 per cent of chromium, 10–19 per cent of lead, 21–30 per cent of cadmium and 18–28 per cent of zinc exist in exchangeable fraction and therefore comes under medium risk category and may enter into food chain.

Sakan *et al.* (2010) studied trace elements as tracers of environmental pollution in the canal of the Danube River, Serbia by collecting the alluvial sediments samples in August and September of 2001 along the canal. The most significant fractions binding all of the studied elements were oxides and silicates. The high proportion of elements in the residual fraction and the generally low contents of extractable elements reflected the background geochemical conditions. The contents of trace elements (except Hg and As) were dominantly controlled by the presence of Fe and Mn oxides as well as by the grain size and the geochemical composition of the sediment. The alluvial sediments were contaminated only at few sites where an anthropogenic influence on the concentrations of some metals (Cu, Cd, Zn, and Pb) was noticed. The results revealed that impact of Panicevo industrial zone on the quality of the Danube River and its contamination with trace element is minor.

Nasrabadi *et al.* (2010) studied the efficiency of sediment metal pollution indices in interpreting the pollution of Haraz River sediments, Southern Caspian Sea Basin. The concentrations of Cu, Zn, As, Cd, Pb, Fe, Ni, Cr, Co and Sr in Haraz River (Iran) bed sediments were measured to determine the proportions of metals in different forms. The order of metals introduced by anthropogenic activities was as follows: Sr > Pb > Co > Cd > Zn > Cu > Ni > As > Cr > Fe. The results revealed that relatively higher concentrations of Cd, As, Sr, and Pb in comparison with other metal were observed. However, based on the chemical partitioning of metals, it was found that Sr, Pb, Co and Cd were the most mobile metals. Also, Fe, Cr, and Ni were present in the greatest percentages in the residual fraction, which implies that these metals are strongly linked to the sediments.

Wang *et al.* (2010) analysed the heavy metal (copper, lead, zinc, and cadmium) concentrations in water and sediment from the Lanzhou section of the Yellow River, China and compared with the threshold values in the guidelines of water, sediment and food given by the National Environmental Protection Agency of China, the National Oceanic and Atmospheric Administration of America, and the National

Standards Management Department of China. The results suggested that (1) Cu, Pb, Zn, and Cd did not contaminate the aquatic ecosystem severely and did not threaten the safety of human consumption in the Yellow River, Lanzhou section, and (2) organs that are sensitive to accumulating heavy metals may be useful to develop bioindicators for monitoring metal contamination.

Alam *et al.* (2011) studied the prevalence of heavy metal resistance in bacteria isolated from tannery effluents and affected soil and reported 110 samples of agricultural soil irrigated with the tannery effluents were collected. Tannery effluents and soils were analyzed for metal concentrations by atomic absorption spectrophotometer. The tannery effluents and soil samples were found to be contaminated with chromium, nickel, zinc, copper, and cadmium.

Cui and Chen (2011) studied the lead (Pb) and arsenic (As) bioaccessibility in various soils from South China. Seventeen soil samples with various concentrations of lead (Pb) and arsenic (As) were collected from five provinces of South China, and bioaccessibility of Pb and As in the soils were examined using a physiologically based extraction test. The results showed that the bioaccessibility ranged from 24.6 per cent to 82.5 per cent and 2.3 per cent to 57.5 per cent for Pb, 2.5 per cent to 65.5 per cent and 1.2 per cent to 31.8 per cent for As in the gastric and small intestinal phases, respectively.

Das and Chakrapani (2011) investigated trace metal contamination from mining activity in soil, rock and water samples collected from India's oldest coalfield Raniganj. Data revealed that trace metal concentration in soil samples lies above the average world soil composition; especially, Cr, Cu, Ni and Zn concentrations exceed the maximum allowable concentration proposed by the European Commission for agricultural soils. In particular, Cr, Cu and Ni exceeded the ecotoxicological limit and Ni exceeds the typical value for cultivated soils. Mineral dissolution from overburden material and high adsorption capacity of laterite soil were responsible for the elevated concentrations.

Hoque *et al.* (2011) studied heavy metal (Fe, Mn, Zn, Cu, Ni, Pb and Cd) concentrations and their chemical speciations for the first time in bed sediments of Bharali River, a major tributary of the Brahmaputra River of the Eastern Himalayas, India. Levels of Fe, Mn, Pb and Cd in the bed sediments were much below the average Indian rivers; however, Cu and Zn exhibit levels on the higher side. The metals speciations, done by a sequential extraction regime, showed that Cd, Cu and Pb exhibit considerable presence in the exchangeable and carbonate fraction, thereby showing higher mobility and bioavailability. On the other hand, Ni, Mn, and Fe exhibit greater presence in the residual fraction and Zn was dominant in the Fe–Mn oxide phase.

Mahanta and Bhattacharyya (2011) studied the total concentrations, fractionation and mobility of heavy metals Cd, Co, Cr, Cu, Mn, Ni, Pb, and Zn in soils of urban area of Guwahati, India. Sequential extraction procedure was used to evaluate the relative distribution of the eight metals in exchangeable, carbonate, reducible (Fe–Mn oxide), organic and sulfide and residual fractions. The two metals, Mn and Zn have domination over the other metals and the values are e"300 mg/kg. Industrial and roadside soil contains much more Mn, while commercial soil is most enriched with Zn. Of the metals, Ni has the largest proportion (~42 per cent) bound to the exchangeable fraction and Co, Cr and Pb also have appreciable proportion bound to the same fraction. A significant amount of Co is associated with carbonates. The reducible fraction has bound considerable quantity of Mn and Zn, while most of Cu is associated with the organic and sulfide fraction. Both Cd and Pb are dominantly associated with the residual fraction. Computation of the mobility factor of the metals indicates Mn to be the most mobile metal present in the soil samples.

Abdallah (2007) studied the sediments of El-Mex Bay estuary on the southern Mediterranean Sea for the analysis of trace metals (Mn, Cu, Cd, Cr, Zn and Fe). Extracted concentrations of trace metals analyzed after all five steps, were found to be (ìg/g) for Mn: 1930.2, Cu: 165.3, Cd: 60.9, Cr: 386.3, Zn: 2351.3 and Fe: 10895. Labile (non-residual) fractions of trace elements (sum of the first four fractions) were analyzed because they are more bioavailable than the residual amount. since Fe-and Mn-oxides play an important role in trace metals sorption within aquatic systems, especially within El-Mex Bay sediments that characterized by varying metal bioavailability.

Alomary and Belhadj (2007) studied the heavy metals (Cd, Cr, Cu, Fe, Ni, Pb, Zn) by ICP-OES and their speciation in Algerian Mediterranean Sea sediments in 18 samples to assess the distribution and bioavailability of metals and to study the anthropogenic factors affecting their concentrations. The metals were sequentially extracted into five phases namely exchangeable (P1), carbonates (P2), Fe–Mn oxides (P3), organic (P4) and residual (P5). The average total extractable metal concentrations for Cd, Cr, Cu, Fe, Ni, Pb and Zn were 1.1, 8.8, 4.7, 1,291.3, 13.9, 5.7 and 20.4 ìg/g, respectively. The P5 phase had the highest percents of Cr, Cu, Fe, Ni, and Zn. Cadmium and lead were predominant in the P4 phase, while Cu, Fe and Zn were distributed in the order P5 > P3 > P4 > P2 > P1. The following order of bioavailability was found with the heavy metals Pb > Cr > Cd > Ni > Zn.

Dessai and Nayak (2009) studied the distribution and speciation of selected metals Fe, Mn, Cu, Zn, Cr and Co to determine their distribution in five geochemical phases [Exchangeable, carbonate, Fe–Mn oxide (reducible) organic bound (oxidisable) and residual] in surface sediments, from the tropical Zuari estuary, central west coast of India. The results obtained from sequential procedure revealed that among the studied elements, Mn and Co are potentially available in the bioavailable fractions (exchangeable, carbonate and Fe–Mn oxide bound fractions) indicating their importance in toxicity whereas rest of the metals *viz.* Fe, Cu, Zn and to some extent Cr are largely available in residual phase although they are available in other fractions.

Ratuzny *et al.* (2009) studied the total concentrations and speciation of heavy metals in soils of the Shenyang Zhangshi Irrigation Area, China. The study revealed that the most predominant metal was cadmium which was mainly associated with mobile, easily soluble and easily reducible fractions. Lead was bound to organic matter and poorly crystalline Fe-oxides. Nickel and zinc were mainly associated with the residual and strongly reducible fractions. Although copper, lead, nickel and zinc concentrations were of minor importance mobile metal concentrations of these metals as well as of cadmium exceeded German trigger values for plant production and plant growth.

Reis *et al.* (2009) studied the total-recoverable metals (Al, Cr, Cu, Fe, Mn, Ni, Pb and Zn) in sediments from Minho estuary salt marsh to evaluate possible increase in anthropogenic contamination by metals. This study revealed that the spatial distribution of metals in the salt marsh sediments was not homogeneous and that two sampling sites (sites 5 and 7) had indications of anthropogenic contamination. However, metal levels in these salt marsh sediments were lower than those observed in the wetlands of the main Portuguese estuaries. In addition, metal levels in the sediments were in chemical forms that were not easily available to organisms, indicating that these sediments probably will not have negative influences in the organisms living in the salt marsh.

Jernstorm *et al.* (2010) studied the heavy metals (Co, Cu, Fe, Mn, Ni, Pb, U and Zn.) in bottom sediments of Lake Umbozero in Murmansk Region, Russia. Metal distribution in sediments was found to vary with respect to the cores, although in general the concentrations were at the same level throughout the lake. The result revealed that the concentrations of most of the metals studied were significantly lower. Sediment layers were subjected to four-step sequential extraction procedure to reveal the metal distribution in soluble, exchangeable, acid-soluble and residual fractions. Indicative of their potential higher lability, Mn, U, and Zn were generally found in exchangeable fraction; as also Mn and U extensively in the acid-soluble fraction.

Villalobos-Castañeda *et al.* (2010) studied the distribution and partitioning of iron, zinc and arsenic in surface sediments in the Grande River mouth to Cuitzeo Lake, Mexico. A sequential extraction procedure was carried out to determine the concentrations of these elements in different geochemical phases of the sediments. Total metal concentrations were evaluated by using the enrichment factor and the geoaccumulation index. A comparison with sediment quality guidelines and shale values has also been made. The results indicate that sediments are considered unpolluted by iron and moderately polluted by zinc and arsenic. However, fractionation studies showed that significant amounts of Zn and As could be released to the lake ecosystem depending on the environmental conditions, representing a medium risk potential of bioavailability to the biota.

Wang *et al.* (2010) studied the concentration and speciation of heavy metals (Cr, Ni, Cu, Zn, Cd, Pb) in surface sediments (d"} pKš 63 ìm) of Jinjiang River tidal reach. A modified BCR three-step sequential extraction procedure is carried out, and the residual fraction is undertaken by microwave-assisted acid digestion. The index of geo-accumulation indicates that Cd appeared highest among all heavy metals in surface sediments; Cr, Cu, Zn lower and Ni, Pb the least. The percentage of Zn, Cd is comparatively higher in the acid soluble fraction, Pb and Cu higher in the reductive fraction, indicating larger potential danger to the environment. So it is essential for developing the future remediation plans and pollution control strategies.

Yu *et al.* (2010) studied the chemical speciation and ecological risk of nine heavy metals in intertidal sediments of Quanzhou Bay, China, The results revealed that Mn presents the highest percentage in the acid-soluble fraction and Pb and Cu present the highest percentages in the reducible fraction. The highest percentages of Fe, V, Cr,

Ni, Zn, and Co were found in the residual fraction. The mobility order of the heavy metals studied on the basis of the nonresidual content of the elements is Mn > Pb > Cu > Co > Zn > Ni > Cr > V > Fe. On the whole, the comprehensive potential ecological risk index of Cu, Zn, Ni, Cr and Pb in the sediments presents moderate degree.

Conclusion

The importance of speciation study of metals in sediment can not be ignored as this aspect provides information on the potential availability of metals (toxic) to biota under various environmental conditions. Therefore the toxicity of individual metal should be studied in different components of the sediment which will provide the actual picture of the toxicity.

References

Abdallah, M.A.M. 2007: Speciation of Trace Metals in Coastal Sediments of El-Mex Bay South Mediterranean Sea–West of Alexandria (Egypt). *Environment Monitoring and Assessment*, 131 (1-3), 111-123.

Alomary, A.A. and Belhadj, S. 2007: Determination of heavy metals (Cd, Cr, Cu, Fe, Ni, Pb, Zn) by ICP-OES and their speciation in Algerian Mediterranean Sea sediments after a five-stage sequential extraction procedure. *Environmental Monitoring and Assessment*, 135(1-3), 265-280.

Alam Md. Z., Ahmad, S. and Malik, A. 2011: Prevalence of heavy metal resistance in bacteria isolated from tannery effluents and affected soil. *Environmental Monitoring and Assessment*, 178 (1-4), 281-291.

Ali, I. and Aboul-Enein, H.Y. 2006: Instrumental methods in metal ions speciation: Chromatography, Capillary Electrophoresis and Electrochemistry. *Taylor and Francis Ltd.*, New York, USA, ISBN: 0-8493-3736-4.

B'Hymer, C., Brisbin, J.A., Sutton, K.L. and Caruso, J.A. 2000: New approaches for elemental speciation using plasma mass spectrometry. *American Laboratory*, 32(3), 17-39.

Brinckman, F. and Salder, P. 1986: The importance of Chemical Speciation in Environmental Processes. *Springer-Verlag, New York.*

Caruso, J.A. and Montes-Bayon, M. 2003: Elemental speciation studies–new directions for trace metal analysis. *Ecotoxicology and Environmental Safety*, 56 (1), 148-163.

Cui, Y. and Chen, X. 2011: Lead (Pb) and arsenic (As) bioaccessibility in various soils from south China. *Environmental Monitoring and Assessment*, 177(1-4), 481-492.

Das, S.K. and Chakrapani, G.J. 2011: Assessment of trace metal toxicity in soils of Raniganj Coalfield, India. *Environmental Monitoring and Assessment*, 177(1-4), 63-71.

Dessai, D.V.G. and Nayak, G.N. 2009: Distribution and speciation of selected metals in surface sediments, from the tropical Zuari estuary, central west coast of India. *Environmental Monitoring and Assessment*, 158(1-4), 117-137.

Forstner, U. and Muller, G. 1973: Heavy metal accumulation in river sediments: a response to environment pollution. *Geoforum*, 14, 53-61.

Forstner, U. and Salomons, W. 1980: Trace metal analysis on polluted sediments, Part I, Assessment of sources and intensities. *Environ. Technol. Lett.*, 1, 494-504.

Forstner, U. and Wittmenn, G.T.W. 1983: Metal pollution in the Aquatic Environment. 2nd ed., *springer Verlag.*, 160.

Gaur, V.K., Gupta, S.K., Pandey, S.D., Gopal, K. and Misra, V. 2005: Distribution of heavy metals in sediment and water of river Gomti. *Environmental Monitoring and Assessment*, 102(1-3), 419-433.

Gibbs, R.J. 1977: Transport phases of transition metals in the Amazon and Yukon Rivers. *Geological society of America bulletin*, 88, 829-843.

Harding, J.P.C. and Whiton, B.A. 1978: Zinc, cadmium and lead in water, sediments and submerged plants of the Derwent reservoir, Northern England. *Water Reearch*, 12, 307-316.

Hoque, R.R., Goswami, K.G., Kusre, B.C. and Sarma, K.P. 2011: Distribution and solid-phase speciation of toxic heavy metals of bed sediments of Bharali tributary of Brahmaputra River. *Environmental Monitoring and Assessment*, 177(1-4), 457-466.

Jain, C.K., Rao, V.V.S.G., Prakash, B.A., Kumar, K.M. and Yoshida, M. 2010: Metal fractionation study on bed sediments of Hussainsagar Lake, Hyderabad, India. *Environmental Monitoring and Assessment*, 166 (1-4), 57-67.

Jain, C.K., Singhal, D.C. and Sharma, M.K. 2003: Metal Pollution Assessment of Sediment and Water in the River Hindon, India. *Environmental Monitoring and Assessment*, 105 (1-3), 193-207.

Jernstrom, J., Lehto, J., Dauvalter, V.A., Hatakka A., Leskinen, A. and Paatero, J. 2010: Heavy metals in bottom sediments of Lake Umbozero in Murmansk Region, Russia. *Environmental Monitoring and Assessment*, 161(1-4), 93-105.

Kramer, J. and Allen, H. 1988: Metal Speciation Theory, Analysis and Application, *Lewis Publishers, Chelsea*, MI.

Korfali, S.I. and Jurdi, M.S. 2011: Speciation of metals in bed sediments and water of Qaraaoun Reservoir, Lebanon. *Environmental Monitoring and Assessment*, 178 (1-4), 563-579.

Mahanta, Mayur J. and Bhattacharyya, Krishna G. 2011: Total concentrations, fractionation and mobility of heavy metals in soils of urban area of Guwahati, India, *Environmental Monitoring and Assessment* 173, 221–240.

Montes-Bayon, M., DeNicola, K. and Caruso, J. A. 2003: Liquid Chromatography – Inductively Coupled plasma mass spectrometry. *Journal of Chromatography* A, 1000 (1-2) 457-476.

Nasrabadi, T., Bidhendi, G. N., Karbassi, A. and Mehrdadi, N. 2010: Evaluating the efficiency of sediment metal pollution indices in interpreting the pollution of Haraz River sediments, southern Caspian Sea basin. *Environmental Monitoring and Assessment*, 171(1-4), 395-410.

Patterson, J. and Passino, R. 1990: Metal Speciation: Separation and Recovery, Vol.II, *Lewis Publisshers*, Chelsea, MI.

Quevauviller, P. 1998 Methods Performance Studies for Speciation Analysis. *Royal Society of Chemistry Publishers, Cambridge, UK.*

Ratuzny, T., Gong, Z. and Wilke, B.M. 2009: Total concentrations and speciation of heavy metals in soils of the Shenyang Zhangshi Irrigation Area, China. *Environmental Monitoring and Assessment*, 156 (1-4), 171-180.

Reis, P.A., Antunes, J.C. and Almeida C.M.R. 2009: Metal levels in sediments from the Minho estuary salt marsh: a metal clean area. *Environmental Monitoring and Assessment*, 159 (1-4), 191-205.

Singh, V.K., Singh, K.P. and Mohan, D. 2003: Status of Heavy Metals in Water and Bed Sediments of River Gomti–A Tributary of the Ganga River, India. *Environmental Monitoring and Assessment*, 105 (1-3), 43-67.

Ure, A. 1990: Trace elements in soil: Their determination and speciation. Fresenius' *Journal of Analytical Chemistry*. 337 (5), 567–581.

Ure, A., Quevauviller, P., Muntau, H. and Grieping, B. 1993: Speciation of Heavy Metals in Soils and Sediments. An Account of the Improvement and Harmonization of Extraction Techniques Undertaken Under the Auspices of the BCR of the Commission of the European Communities. *International Journal of Environmental Analytical Chemistry*. 51 (1-4), 135–151.

Villalobos-Castaneda B., Alfaro-Cuevas, R., Cortes-Martinez, R., Martinez-Miranda, V. and Márquez-Benavides, L. 2010: Distribution and partitioning of iron, zinc, and arsenic in surface sediments in the Grande River mouth to Cuitzeo Lake, Mexico. *Environmental Monitoring and Assessment*, 166 (1-4) 331-346.

Wang, L., Yu, R., Hu, G. and Tu, X. 2010: Speciation and assessment of heavy metals in surface sediments of Jinjiang River tidal reach, southeast of China. *Environmental Monitoring and Assessment*, 165 (1-4), 491-499.

Wang, Y., Chen, P., Cui, R., Si, W., Zhang, Y. and Ji, W. 2010: Heavy metal concentrations in water, sediment, and tissues of two fish species (*Triplohysa pappenheimi, Gobio hwanghensis*) from the Lanzhou section of the Yellow River, China. *Environmental Monitoring and Assessment*, 165 (1-4), 97-102.

Yu, R., Hu, G. and Wang, L. 2010: Speciation and ecological risk of heavy metals in intertidal sediments of Quanzhou Bay, China. *Environmental Monitoring and Assessment*, 163 (1-4), 241-252.

2013, Climate Change Effects on Agriculture and Economy *Pages* ***223–228***
Editors: **D.R. Khanna, A.K. Chopra, Vikas Singh, Rakesh Bhutiani & Gagan Matta**
Published by: **BIOTECH BOOKS, NEW DELHI**

Chapter 25

River Water Pollution and its Removal by Bed Sediment

M.K. Sharma, V.K. Choubey and R.D. Singh
National Institute of Hydrology, Roorkee, Uttarakhand

River serves as a source of water supply for domestic and industrial purposes and also for agriculture, fisheries and power development. The same water resources are also utilized for the disposal of industrial wastes and sewage leading to water pollution. In recent years, with the rapid development of industries, water pollution especially metal pollution is assuming serious concern. A number of technologies have been developed to remove toxic matter from water. The most important of these include chemical precipitation, ion exchange, reverse osmosis and adsorption. Adsorption is one of important phenomena in water quality control, which may determine the fate and transport of pollutants in the aquatic environment. Heavy metals have a considerable significance due to their toxicity and adsorption characteristics. The sediment at the bottom of the water column also plays a major role in pollution studies due to their specific adsorption capacity. Adsorption behaviour of different metals on bed sediments of different rivers has been studied by different workers under different operating variables *viz;* contact time, pH, initial concentrations, adsorbent sizes and doses. In the present paper, few adsorption studies and role of bed sediment in different rivers have been discussed. Results of adsorption studies of different metals *viz;* cadmium, lead, chromium and zinc in different rivers Kali and Hindon have been compared and concluded that the adsorption phenomena plays an important role in pollution removing scheme in aquatic environment.

Keywords: *Rivers, Metal concentrations, Bed sediments, Adsorption, Pollution.*

Introduction

Numerous studies have demonstrated that trace metal concentrations in suspended and bed sediments are better indicators of contamination than the elements in solutions in hydrologic systems (Salomons and Forstner, 1984; Luoma, 1990). Heavy metals added to a river system by natural or manmade sources during their transport are distributed between the aqueous phase, suspended and bed sediments. The fraction in the sediment is expected not to have direct adverse effects, if the metal ions are tightly bound to it and subsequently get settled at the bottom in course of time. This state of affairs is maintained until there is remobilization from the sediment due to changing conditions in the system. Thus, in the natural conditions of river water, suspended load and sediments have the important function of buffering higher metal concentrations of water particularly by adsorption or precipitation. Therefore, the study of sediments and their sorptive properties can provide valuable information relating to the tolerance of the system to the added heavy metal load and may determine the fate and transport of pollutants in the aquatic environment. Adsorption is an important phenomena in water quality control, which may determine the fate and transport of pollutants in the aquatic environment. Heavy metals have a considerable significance due to their toxicity and adsorption characteristics. The sediment at the bottom of the water column also play a major role in pollution studies due to their specific adsorption capacity.

Heavy Metal Pollution

Pollution from industrial and agricultural sources, to a great extent, is responsible for high concentration of heavy metals in river water. This may also be contributed by water distribution system due to leaching from galvanized pipes. Domestic wastewater constitutes the single largest source of heavy metals in river water. The elevated levels of Cu, Pb, Zn and to a lesser degree Cd, can be due to corrosion within the urban water supply network (Preuss and Kollmann, 1974). The use of detergents also creates a possible pollution hazard, since common household detergent products can affect the water quality. Angino *et al.* (1970) also reported presence of various elements *viz.*, Fe, Mn, Cr, Co, Zn, Sr and B in most detergents.

Lead has been known for its toxicity for a long time. It is used to manufacture of acid accumulators, alkyl lead compounds for gasoline, tetraethyl lead as anti knock compounds, solders, pigments and paints, ammunition, cable sheathing, roofing material, piping material, including for drinking water, in manufacture of sulfuric acid, lead arsenate, insecticides, rubber *etc.* Its presence in drinking water tends to cause various types of serious health problems leading to death in extreme exposure case (Volesky, 1990). It has cumulative property in human system and exhibits some symptoms of acute poisoning like tiredness, lassitude, abdominal discomfort, irritability and behavioral changes. Lead at high level can reduce the activity of enzyme to synthesize hemoglobin resulting in anemia (WHO, 1977).

Cadmium finds its way to the water bodies through wastewater from metal plating industries and industries of cadmium nickel batteries, phosphate fertilizer, mining, pigments, stabilizers and alloys (Low and Lee, 1991). The effects of acute cadmium poisoning in humans are very serious, among them are high blood pressure, kidney damage and destruction of testicular tissue and red blood cells (Encyclopaedia of Environmental Sciences, 1980). Even low concentrations of cadmium are known to have detrimental effects on man, having been associated with hypertensive diseases. The international Agency for Research on Cancer classified cadmium among the chemicals which are probably carcinogenic to man (IARC, 1976). Therefore, an understanding of the conditions and mechanisms that regulate the mobilization and distribution of cadmium in the environment is essential.

Zinc is an essential element for both animals and man and is necessary for the functioning of various enzyme systems, deficiency of which leads to growth retardation. Low intake of zinc results in retardation of growth, immaturity and anemia, condition known as 'zinc deficiency syndrome'. Symptoms of zinc toxicity in humans include vomiting, dehydration, electrolyte imbalance, abdominal pain, nausea lethargyness, dizziness and lack of muscular coordination. Zinc imparts undesirable, bitter astringent taste to water at levels above 5.0 mg/l (WHO, 1996). Water containing zinc at concentrations in excess of 5.0 mg/l may appear opalescent and develops a greasy film on boiling. Toxic concentrations of zinc above recommended value cause adverse effect in the morphology of fish by inducing cellular breakdown of gills. Zinc deficiency in human body may result in infantilism, impaired wound healing and several other diseases.

Chromium is least toxic of the trace elements on the basis of its oversupply and essentiality. Generally, the mammalian body can tolerate 100-200 times its total body content of chromium without harmful effects. Chromium is introduced into the natural bodies of water from industries like electroplating, leather tanning, cement industries, steel industries and photography. Chromium (VI) compounds are approximately 100 times more toxic than chromium (III) salts. The stomach acidity leads to reduction of Cr (VI) to Cr (III) of which gastrointestinal absorption is less than 1 per cent. Cr (VI) causes skin irritation resulting in ulcer formation. Over exposure of Cr (VI) leads to liver damage, pulmonary congestion and oedema. The recommended limit of chromium in potable water is 0.05 mg/L.

Adsorption Studies

A number of technologies have been developed to remove toxic metals from water. The most important of these include chemical precipitation, ion exchange, reverse osmosis and adsorption. The suspended and riverbed sediments play an important role in the transport of metals in aquatic systems. The important components of the suspended load for geochemical transport are silt, clay, hydrous iron and manganese oxides and organic matter. Generally, the adsorption study for sediment less than 50 μm has been made greatly, while that for the sediment more than 75 μm are few because of little adsorption action. Adsorption properties of the sediments provide valuable information relating to the tolerance of the system to the added heavy metal load.

Despite the apparent wealth of information on adsorption processes, little is known about quantitatively describing adsorption process by coarser sediment. Although clay and silt component adsorb metal ions much better than coarser fraction of sediment, one should take into account that most river sediments contain 90-95 per cent sand and only 0-10 per cent clay and silt. Therefore, in river systems with high sand percentage and low clay and silt content, the overall contribution of sand content to adsorption of metal ions could be comparable to or even higher than that of clay and silt fraction (Jain and Ram, 1997a,b; Jain and Ali, 2000).

Sharma (2001) studied the potentiality of freshly deposited sediments of river Hindon, a tributary of river Yamuna to adsorb lead, cadmium, zinc and chromium ions which, may enter the river system through the disposal of municipal and industrial effluents or by biological and chemical degradation. The study indicates, though these metal ions have more affinity for the clay and silt fraction of the sediment, but the overall contribution of coarser fraction to adsorption is more as compared to clay and silt fraction. The adsorption data further suggests that the pH of the solution is the most important parameter in the control of these metals. The percentage adsorption increases with increasing adsorbent doses, and as such removal increases with decreasing size of the adsorbent material. The two important geochemical phases iron and manganese oxide also play an important role in the adsorption process. The manganese and iron contents in the various fractions of the sediment decreases with increasing particle size, *i.e.*, the clay and silt fraction contain more iron and manganese content than that of 210-250 μm fraction indicating possibility of association of these substrate with clay and silt particles. These findings illustrate the possible role of clay and silt components as substrate for the adsorption of lead, cadmium, zinc and chromium. The content of iron in the sediment fractions was found to be relatively higher indicating the possibility of presence of iron minerals other than hydroxides. The equilibrium data describe the Langmuir and Freundlich isotherm models satisfactorily and given in Table 25.1 and 25.2.

Table 25.1: Langmuir Parameters for Adsorption of Lead, Cadmium, Zinc and Chromium Ions on the Bed Sediments of River Hindon

Metal Ion	*Sediment Fraction (mm)*	*Adsorption Maxima Q^o (mg/g)*	*Binding Energy Constant b (mg/L)$^{-1}$*
Pb	0-75	5.0	8.403
	210-250	5.0	5.900
Cd	0-75	5.0	1.919
	210-250	5.0	1.467
Zn	0-75	5.0	4.338
	210-250	12.5	0.974
Cr	0-75	1.7	0.949
	210-250	1.1	1.077

Jain and Ram (1995) studied the adsorption of zinc and lead on the bed sediments (size fraction 210-250 μm) of river Kali, a tributary of river Hindon and evaluated the mechanistic parameters associated with the adsorption process and given in Table 25.3.

Table 25.2: Freundlich Parameters for Adsorption of Lead, Cadmium, Zinc and Chromium Ions on the Bed Sediments of River Hindon

Metal ion	Sediment Fraction (mm)	Adsorption Capacity K_F (mg/g)	Adsorption Intensity 1/n
Pb	0-75	0.164	0.514
	210-250	0.118	0.556
Cd	0-75	0.043	0.634
	210-250	0.034	0.642
Zn	0-75	0.088	0.593
	210-250	0.046	0.690
Cr	0-75	0.012	0.605
	210-250	0.010	0.580

Table 25.3: Langmuir and Freundlich Parameters for Adsorption of Lead and Zinc Ions on the Bed Sediments of River Kali

Metal ion	Langmuir Parameters		Freundlich Parameters	
	Adsorption Maxima Q^o (mg/g)	Binding Energy Constant b $(mg/L)^{-1}$	Adsorption Capacity K_F (mg/g)	Adsorption Intensity 1/n
Pb	0.613	0.639	0.259	0.309
Zn	0.384	0.118	0.053	0.558

Jain and Ali (1998) studied the adsorption of cadmium on the bed sediments (size fractions 0-75 and 210-250 μm) of river Kali and compared the role of coarser sediment fraction to those of clay and silt fractions. Further, competitive experiments conducted in the presence of lead and zinc ions revealed that both the ions suppress the ability of cadmium to adsorb onto sediments. The mechanistic parameters associated with the adsorption process are given in Table 25.4.

Table 25.4: Langmuir and Freundlich Parameters for Adsorption of Cadmium Ion on the Bed Sediments of River Kali

Particle Size Fraction	Langmuir Parameters		Freundlich Parameters	
	Adsorption Maxima Q^o (mg/g)	Binding Energy Constant b $(mg/L)^{-1}$	Adsorption Capacity K_F (mg/g)	Adsorption Intensity 1/n
0-75 μm	0.125	32.0	0.05	0.20
210-250 μm	0.143	55.9	0.04	0.20

From above data, it is revealed that the adsorption capacity of bed sediment of river Kali is more as compared to bed sediment of river Hindon which may attributed higher content iron and manganese oxide and organic matter present in bed sediment of river Kali, which act as active support material for adsorption phenomena.

Conclusion

Adsorption is a special characteristic of the bed sediment of a riverine system which may determine the fate and transport of the pollutant in the system. The mechanistic parameters related to adsorption behaviour can be used to characterise the sediment of different riverine system and their capacity to adsorb pollutants specially metals.

References

Angino, E.E., Magnuson, L.M., Waugh, T.C., Galle, O.K. and Bredfeldt, J. 1970: Arsenic in detergents – possible danger and pollution hazard, Science, 168, 389-390.

Encyclopedia of Environmental Sciences, 1980: 2nd edn., McGraw-Hill, Newyork, 354.

International Agency for Research on Cancer, 1976: Monographs on the evaluation of carcinogenic risk of chemicals to man. In: Cadmium and Cadmium Compounds, 11, 3974.

Jain, C.K. and Ram, D. 1995: Adsorption of metal ions on sediments, National Institute of Hydrology, Roorkee, Technical Report No. CS (AR) 173.

Jain, C.K. and Ram, D. 1997a: Adsorption of lead and zinc on bed sediments of the river Kali, Wat. Res., 31(1), 154-162.

Jain, C.K. and Ram, D. 1997b: Adsorption of metal ions on bed sediments, *Hydrol. Sci. Journal*, 42(5), 713-723.

Jain, C.K. and Ali, I. 1998: Adsorption of cadmium on bed sediment, National Institute of Hydrology, Roorkee, Technical Report No. TR (BR)-10/97-98.

Jain, C.K. and Ali, I. 2000: Adsorption of cadmium on riverine sediments: Quantitative treatment of the large particles, *Hydrol. Process.*, 14, 261-270.

Low, K.S. and Lee, C.K. 1991: Cadmium uptake by the moss calympers delesertii, besch., *Bioresour. Technol.*, 38, 1-6.

Luoma, S.N. 1990: Processes affecting metal concentrations in estuarine and coastal marine sediments, In: Heavy Metals in the Marine Environment, (edited by Rainbow, P. S. and Furness, R. W.), CRC Press, Cleveland, Ohio.

Preuss, E. and Kollmann, H. 1974: Metallgehalte in Klarschlammen, Naturwissenschaften, 61, 270-271.

Salomons, W. and Forstner, U. 1984: Metals in the Hydrocycle, Springer, Berlin.

Sharma, M.K. 2001: Studies of Hindon river system with special reference to heavy metals, Ph D Thesis, University of Rookee, Roorkee.

Volesky, B. 1990: Biosorption and biosorbents, In: Biosorption of heavy metal (edited by Volesky, B.), CRC Press, Boca, Raton, FL., pp. 3-6,.

WHO. 1977: Environmental Health Criteria– 3, WHO, Geneva.

WHO. 1996: Guidelines for drinking water quality, WHO, Geneva, II nd Edition, Vol. 2.

2013, Climate Change Effects on Agriculture and Economy *Pages* ***229–233***
Editors: **D.R. Khanna, A.K. Chopra, Vikas Singh, Rakesh Bhutiani & Gagan Matta**
Published by: **BIOTECH BOOKS, NEW DELHI**

Chapter 26

Effect of Sulphur Dioxide on *Cicer Arietinum* with Special Reference to the Morphological and Biochemical Aspects

Madhu Sharma

Department of Botany, S.M.J.N. (P.G.) College, Govindpuri, Haridwar, Uttarakhand

The effect of SO_2 on chlorophyll and total free amino acids and Growth of Gram (*Cicer arietinum* L.) were investigated by exposing them to 1ppm of SO_2. The experiments were carried out at Central Institute of Agricultural Engineering, Nabibagh, Bhopal (M.P.). 1ppm of SO_2 was given to twenty days old plants for thirty days. Results were noted for 30, 45, 60, 75, 90 day old plants. Total free amino acids and chlorophyll were quantified. Exposure of plants, developed visible injury symptoms and suffered growth reductions.

Amount of chlorophyll depressed in the leaves due to SO_2 fumigation. There is an increase in total free amino acids in the leaves of fumigated plants over control, respectively.

Keywords: Chlorophyll, Amino acid, SO_2, Plants, Pollution.

Introduction

Green Plants serve as a natural sink for air pollutants by providing an enormous surface of expanded leaves for the absorption and setting of gases and particulate matter. They express their suffering in the form of foliar injury and reduction in growth and yield. Crop growing around industrial areas are often subjected to stress of pollution and subsequently show alterations in various levels. With this purpose in mind, the present investigation was designed. *Cicer arietinum* L. variety JG – 315 was selected and effect of 1 ppm concentration of SO_2 on them was studied across a number of parameters *viz;* morphological, growth and biochemical aspects.

Materials and Method

The leaves of test plant *Cicer arietinum* L. was collected from both control and fumigated plants for biochemical analysis. The following parameters like chlorophyll were determined using spectrophotometer (Duxbury and Yentsch, 1956; Maclachlan and Zalik, 1983). Total free amino acids were determined with use of Ninhydrin (Lee and Sashi, 1966). Leaf area dry weight ratio (LADWR): (Rao and Dubey, 1988) and Leaf injury index (Dubey, 1986) were also recorded.

Results and Discussion

The results are tabulated in Tables 26.1 to 26.4.

Leaf Injury Index

The foliar injury caused by the exposure of Sulphur dioxide on Gram included chlorosis, necrosis, bleaching and leaf fall. The degree of foliar injury was directly proportional to the concentration of pollutant. Initially only chlorotic patches on leaves were observed but in later stages of treatment necrosis, bleaching were also recorded. Leaf injury index of fumigated plants of *Cicer arietinum* after 30 and 90 days was 8.29 per cent and 69.91 per cent at 1ppm of SO_2, respectively.

Leaf Area Dry Weight Ratio

It has been observed in the present study that the LADW Ratio of the plants enhanced when exposed to SO_2 as compared with control plants indicating their decreasing trend of reduction in foliar biomass or dry matter accumulation. The observations made in the present study confirm those results since the enhanced LADW Ratio indicates reduction in biomass or dry weight accumulation Mc Leod *et al.* (1988) noted that the dry weight of winter barley increased when exposed to low concentrations of SO_2 (0.016 µg). The reduction in dry matter production in plants by SO_2 exposure could be due to alternations of chlorophyll pigments (Shimazaki and Sugahara, 1980); which in turn affects photosynthesis (Hallgren, 1984; Gupta, 1986).

Chlorophyll

The chlorophyll of leaves of SO_2 exposed plants reduced in all the age groups of plants. Chlorophyll – a was found to be more sensitive than chlorophyll– b and carotenoids. This seems to be the loss of structural organization of chloroplast (Rabe and Kreeb, 1980; Tiwari and Bansal 1994). Chlorophyll is the index of productivity in plants (Madison and Anderson, 1963). Gaseous air pollutions can inhibit both light and dark reactions of photosynthesis. The inhibitory effect of SO_2 on photosynthetic CO_2 exchange has been well documented (Zeigler, 1975; Hallgren, 1976; Heath, 1980)

Table 26.1: Effect of Sulphur Dioxide on Chlorophyll Contents in *Cicer arietinum* L.

Experimental Condition	*Age of Plant in Days*	*Chl. a*	*Chl. b*	*Carotenoids*
		(mg/g)	(mg/g)	(mg/g)
	30	1.23	0.81	1.62
	45	1.28	0.98	1.68
Control	60	1.31	1	1.72
	75	1.35	1.03	1.83
	90	1.39	1.09	1.89
	30	0.91	0.79	1.59
	45	0.89	0.72	1.52
Fumigated 1 ppm of SO_2	60	0.85	0.7	1.49
	75	0.79	0.68	1.43
	90	0.68	0.65	1.41

Values are in mean n=10

Table 26.2: Effect of Sulphur Dioxide on Total Free Amino Acids in *Cicer arietinum* L.

Experimental Condition	*Age of Plant in Days*	*Total Free Amino Acids (Mg/g)*
	30	14.97
	45	20.11
Control	60	21.1
	75	17.02
	90	10.72
	30	17.61
	45	25.41
Fumigated 1 ppm of SO_2	60	26.55
	75	22.01
	90	15.51

Values are in mean (n=5).

Total Free Amino Acids

The total free amino acids of leaves in the fumigated plants were increased. Generally the SO_2 exposed plants show increase in levels of free amino acids prior to appearance of visual injury symptoms (Jagar *et al.*, 1972; Pierre and Quiroz, 1981). SO_2 interferes with enzymes regulating amino acid metabolism (Jager and Pahlich, 1976; Wellburn *et al.*, 1976) causing qualitative and quantitative change in item (Godzik and Linskens, 1974).

Table 26.3: Periodic Alterations in Foliar Injury Parameters of *Cicer arietinum* (variety J-315) to Sulphur Dioxide

Experimental Condition	*Age of Plant in Days*	*Leaf Injury Index*
	30	NIL
	45	NIL
Control	60	NIL
	75	NIL
	90	NIL
	30	8.29
	45	11.46
Fumigated 1 ppm of SO_2	60	14.67
	75	16.12
	90	19.41

Values are mean of 10 Plants.

Table 26.4: Effect of Sulphur Dioxide on LADW Ratio of Gram at Different Growth Stages

Experimental Condition	*Age of Plant in Days*	*Leaf Area Dry Weight Ratio*
	30	15.41
	45	18.28
Control	60	25.31
	75	28.11
	90	31.13
	30	19.21
	45	25.68
Fumigated 1 ppm of SO_2	60	32.94
	75	39.11
	90	45.02

Values are mean of 10 Plants.

References

Dubey, P.S. 1986: Procedure Manual. All India Cordinated Programme. Air Pollution and Plants. Deptt. Of Environment, Govt. of India, New Delhi (India).

Duxbury, A.C. and Yentsch. C.S. 1956: *J. Marine Res*, 15, 19-101.

Godzik, S. and Linskens, H.F. 1974: Concentration changes of free amino acids in Primary bean leaves after continuous and interrupted SO_2 fumigation and recovery. *Environ. Pollut.* 7: 25-38.

Gupta, M.C. 1986: Morphological and Cuticular responses of Ocimum to Coal – Smoke Pollutants. *Ind. Polln. Contl.*, 2 (1 and 2), 63-68.

Hallgren, J.E. 1976: Physiological and biochemical effects of Sulphur – dioxide on plants. In: J.O. Nariagu (Ed.), Sulphur in the environment. Part –II. Ecological Impact. 163-209, New York, John Wiley and Sons.

Hallgren, J.E. 1984: Photosynthetic gas exchange in leaves affected by air pollutants. In: Gaseous air pollutants and plant metabolism (eds. M.J. Koziol and F.R. Whatley), Butter Worths, London, 147-160.

Heath, R.L. 1980: Initial events in injury to Plants by air pollutants. *Ann. Rev. Pl. Physiol.* 31: 395-431.

Jager, H.J. Pahlich E. and Steubing L. 1972: Effect of Sulphur dioxide on amino acid and protein content of Pea Seedlings. Angew. Bot. 46. 199-211

Jager, H.J. and Pahlich, E. 1976: Influence of SO_2 on the amino acids metabolism of Pea Seedlings. Oecologia. 9: 135-140.

Lee, Y.P. and Takahashi, T. 1966: An improved calorimetric determination of amino acids with use of ninhydrin. *Anal. Biochem.* 193: 265-275.

Maclachlan, A. and Zalik, S. 1983: *Can. J. Bot.* 41, 1053-1062.

Mc Leod, A.R., Roberts, T.M., Alexender, Kand Cribb, D.M. 1988: Effects of open air fumigation with SO_2 on the growth and yield of winter barley. *New phytol.*, 109, 67-78.

Peirre, M. and Quiroz, O. 1988: Air pollution by SO_2 amplifies the effects of water stress on enzymes and total soluble proteins of spruce needles. *Physiol. Plant.* 73(3): 412-417.

Rabe, R. and Kreeb, K.H. 1980: Effects of SO_{2-}, upon enzyme activity in plant leaves. *Inter. J. Plant Physiol.* 97: 215-226.

Rao, M.V. and Dubey, P.S. 1988: Plant response against Sulphur-dioxide in field conditions. *Asian Environ.*, 3 (10): 1-9.

Shimazaki, K. and Sugahara, K. 1980: Inhibition site of electron transport system in lettuce chloroplasts by fumigation of Lettuce leaves with SO_2. *Plant Cellphysiol.* 21(1): 125-135.

Tiwari, S., Bansal, S. 1994: Responses of *Ficus religiosa* exposed to NO_2. *India J. Env. Toxico*, 2 (2) : 21-26.

Wellburn, A.R., Capron, T.M., Chan, H.S. and Hrosman, D.C. 1976: Biochemical effects of atmospheric pollutants on Plant (Ed. T.A. Mansfield). Cambridge University Press. Cambridge. pp. 105-114.

Ziegler, I. 1975: The effect of SO_2 Pollution on plant metabolism. *Res. Rev.* 56: 79-105.

2013, Climate Change Effects on Agriculture and Economy *Pages* **235–244**
Editors: **D.R. Khanna, A.K. Chopra, Vikas Singh, Rakesh Bhutiani & Gagan Matta**
Published by: **BIOTECH BOOKS, NEW DELHI**

Chapter 27

Bioethanol Production Paper Industry Waste–Bagasse Pith

Arun Kumar, J. Harini Banda, P.S.V. Madhuri and M. Kavya Anusha

Department of Environment and Civil Engineering, University of Petroleum and Energy Studies, Dehradun, Uttarakhand

Biofuels produced from various lignocellulosic biomasses such as agro forest residue have been recognized to have potential, to be available substitute for or compliment to gasoline. These biofuels derived from plant materials have the potential to address the two main issues carbon neutral and renewable. In addition they are an integral part of the emerging Bio-economy where the plant material is used to produce specific chemicals and bulk industrial products.

Various institutes and industries have been working and exploring the possibility of utilization of pith, available as a solid waste biomass in paper and sugar industry, as a source of bioethanol. Efforts were made to achieve maximum prehydrolysis and sachharification efficiency to obtain higher yields of fermentable sugars and ethanol.

The present paper describes the preliminary findings of the study for exploring the possibility of conversion of bagasse pith into Bio-Ethanol. There by, helping in meeting out the ever increasing demand of much needed ethanol needed for blending the gasoline. This should also help in addressing the problem of global warming and concern form of GHG emissions thereby improving the energy and environment status of our country.

Keywords: *Bioethanol, Gasoline, Biomass, Bagasse, Hydrolysis.*

Introduction

Ethanol has attracted worldwide attention because of its potential use as an alternative automotive fuel. It has immense importance for countries such as India which depends heavily on import of crude oil, spending a huge sum of its annual budget on import of crude oil. There are numerous advantages of blending ethanol with gasoline that emits lower quantities of carbon monoxide (CO), nitrogen oxides (NO_x) and hydrocarbon after combustion compared to that of gasoline alone because ethanol acts as oxidizing agent (Kasey, 1961; Kadam and Mcmillan, 2003; Knauf and Moniruzzaman, 2004).

Long term economical and environmental concerns have resulted in a great amount of research in the past couple of decades on renewable sources of liquid fuels to replace fossil fuels. The traditional feed stocks like molasses, sugarcane juice, corn *etc.* are used for ethanol production but have social and economical barriers. Due to their primary utility as food and feed, these crops cannot meet the global demand for ethanol production as an alternative transportation fuel.

Apart from these feed stocks, lignocellulosic biomass, which is the most abundant on earth, may prove to be a potential alternative feed stock for bio ethanol production. These biomass including forest residues such as wood; agricultural residues sugarcane baggase, corn cob, corn stover, wheat and rice straw; industrial residues such as pulp and paper processing waste and municipal solid wastes; energy crops such as switch grass are the most potential feed stocks for fuel ethanol (Knauf and Moniruzzaman, 2004; Badger, 2002; Alexandria *et al.*, 2007; McMillan, 1994). The main bottleneck so far has been the technology y concerns, which do not support cost-effective and competitive production of lignocellulosic bioethanol.

Conversion of abundant lignocellulosic biomass to biofuels as transportation fuels presents a viable option for improving energy security and reducing greenhouse emissions. Unlike fossil fuels, which come from plants that grew millions of years ago, biofuels are produced form plants grown today. They are cleaner-burning than fossil fuels and the short cycle of growing plants and burning fuel made from them does not add CO_2 to the atmosphere. It has been reported that cellulosic ethanol and ethanol produced from other biomass resources have the potential to cut greenhouse gas emissions by 86 per cent. Lignocellulosic materials such as agricultural residues (*e.g.*, wheat straw, sugarcane bagasse, corn Stover), forest products (hardwood and softwood), and dedicated crops (Switch grass, sllix) are renewable source of energy.

Despite possessing several advantages over conventional ethanol, lignocellulosic bioethanol has not yet been produced on a commercial scale because of the low ethanol yields and high production costs. Economic analyses (US scenario) indicate that the process of sugar release from lignocellulosic biomass, which includes pre-treatment, enzyme production, and enzymatic hydrolysis, contributes to as much as 40-45 per cent of the projected total cost.

India is the world's largest producer of sugar cane producing nearly 357 million tons per annum and nearly 60 per cent of the cane is utilized for the production of sugar of which about 1/3rd of the total sugar cane is obtained as residue so called bagasse containing appreciable quantity (up to 35 per cent w/w) of the pith. Presently the pith which is considered to be the undesirable entity from view point of papermaking is separated and pith used as a source for fuel of biomass. Nearly 160 t/day of pith is generated in a 300 tpd of Pulp and Paper mill. Due to its inherent chemical composition, pith contains up to 70 per cent of the carbohydrates contents and may become a potential source for Bio-ethanol and other value added products.

Pretreatment of Lignocellulosic Biomass (Pith): Technology Bottlenecks and Possible Remedies

With diminishing oil supplies and growing political instability in oil-producing nations, the world is facing a major energy threat which needs to be solved by virtue of alternative energy source. Bioethanol has received considerable attention in the transportation sector because of its utility as an octane booster, fuel additive, and even as neat fuel. Brazil and the USA have been producing ethanol on a large scale from sugarcane and corn, respectively. However, due to their primary utility as food and feed, these crops cannot meet the global demand for ethanol production as an alternative transportation fuel.

Lignocellulosic biomass is projected as a virtually eternal raw material for fuel ethanol production. The main bottleneck so far has been the technology concern, which do not support cost-effective and competitive production of lignocellulosic bioethanol.

Pretreatment of biomass is requisite to the biological conversion of lignocellulosic biomass to ethanol is the release of the cellulose portion (and subsequently glucose) from the tightly woven lignocellulosic structure. For this, the biomass needs to be pre-treated, either physically, chemically and/or biologically to make pre-treated biomass more amenable to subsequent cellulose hydrolysis. Pretreatment is required to alter the size and structure of the biomass, as well as its chemical composition, so that the hydrolysis process can be significantly improved by removal of lignin and hemicellulose, reduction of cellulose crystalline and increase of porosity through pretreatment processes.

About 18-20 per cent of the total projected cost for biological production of cellulosic ethanol can be attributed to pre-treatment, more than for any other single step. Reducing ethanol cost requires optimizing pre-treatment strategies and condition to the most economical possible. However, not every kind of lignocellulosic biomass can be pre-treated in the same way.

The recovery of fuels from the fermentation broth is achieved by distillation or a combination of distillation and adsorption. The other components, including residual lignin, unreacted cellulose and hemicellulose and enzymes, accumulate at the bottom of the distillation column.

The economical production of any microbial metabolite is governed by three main factors: (1) the initial cost of raw materials, transportation, and processing; (2) the availability of suitable production and recovery procedures; and (3) the final product yield. The lignocellulosic bioethanol process can be made cost-competitive by: (1) minimizing handling and transportation costs, and incorporating the most efficient and cost-effective pre-treatment and hydrolysis strategy;

(2) Developing efficient bioprocesses by optimally integrating the preceding and subsequent process steps; and (3) applying over producing and recombinant strains for enhanced ethanol productivity.

University of Petroleum and Energy studies, as a part of its activity to promote clean and green process technologies in Indian Paper Industry, has been engaged in exploring the potential of utilizing bagasse pith a waste lignocellulosic biomass rich in carbohydrate as a source of bioethanol adopting a biorefinery approach.

The present paper describes the studies on optimization of the process conditions for pretreatment of biomass (prehydrolysis) and sachharification of the prehydrolysed pith through chemical and enzymatic route to produce maximum quantity of fermentable sugars to convert into the bio ethanol helping in making the process, techno economically viable.

Materials and Method

Raw Material

Fresh bagasse was collected from sugar mill, Deoband and depithed using CPPRI patented depithing method. Pith was collected, analyzed for chemical composition and used for the present study.

Microorganism

Ethanol producing thermophilic yeast

Enzymes

Enzymes used in the present study were procured from national commercial enzyme supplier (Enzyme- 1) and R and D Institute (Enzyme- 2).

Methods

Chemical composition (Lignin, holocellulose, pentosan, ash) of the pith samples after pre hydrolysis and sachharification were analyzed by standard TAPPI methods. Total reducing sugars and total sugars was analyzed by Miller's (1959) method (16) and Dubois (1956) method (17). Sugar analysis of hydrolysates was done by high performance liquid chromatography (HPLC). Estimation of ethanol was done by both chromic acid method Caputi *et al.* (1968) and HPLC.

Results and Discussion

Structure and Composition of various Lignocelluloses Biomass (Agro Residues and Waste)

Lignocellulose is the primary building of plant cell walls. Plant biomass is mainly composed of cellulose, hemicelluloses and lignin, along with smaller amounts of pectin, protein, extractives (soluble non structural materials such as nonstructural sugars, nitrogenous material, chlorophyll, and waxes) and ash. The composition of these constituents can vary from one plant species to another. Bagasse pith was collected and subjected to physical chemical characterization like lignin holocellulose *etc.* Table 27.1 shows the chemical composition of bagasse pith and another various lignocellulosic biomasses indicating the potential for use as a source of bioethanol.

Table 27.1: Chemical Composition of Common Agricultural Residues and Wastes

Lignocellulosic Material	*Cellulose (per cent)*	*Hemicellulose (per cent)*	*Lignin (per cent)*
Hardwood stems	40-45	24-40	18-25
Softwood stems	45-50	25-35	25-35
Nut shells	25-30	25-30	30-40
Corn cobs	45	35	15
Grasses	25-40	35-50	10-30
Paper	85-99	0	0-15
Wheat straw	30	50	15
News paper	40-55	25-40	18-30
Waste paper from chemical pulps	60-70	10-20	5-15
Bagasse	36.4	18.2	22.8
Bagasse Pith	46.0	22.0	21.8
Rice straw	26.0	13.0	4.1

Conversion of Biomass (Bagasse Pith) to Biofuel (Ethanol)

Action of microorganisms and enzymes of biological sources can lead to the production of mostly ethanol and less commonly, propanol and butanol.These agents carry out the fermentation being the most difficult. The conversion includes the hydrolysis of various components in the lignocellulosic materials to fermentable reducing sugar and the fermentation of the sugars to fuels such as ethanol and butanol.The pretreatment step is mainly required to for efficient hydrolysis of cellulose to its constituent sugar. The hydrolysis is usually catalyzed by acids or cellulose enzymes and the fermentation is carried out by yeasts or bacteria. The factors affecting the hydrolysis of cellulose include porosity (accessible surface area) of the biomass materials, cellulose fiber crystallinity and content of both lignin and hemicelluloses. The presence of lignin and hemicelluloses makes the accessibility of cellulose enzymes and acids to cellulose more difficult, thus reducing the efficiency of the hydrolysis process.

Prehydrolysis of Bagasse Pith

Sugarcane bagasse was depithed by CPPRI method. Pith was collected and analysed for its chemical composition and it was subjected to prehydrolysis. Bagasse pith was subjected to prehydrolysis with an aim to maximum degradation of hemicelluloses (xylan) to sugars (xylose) with less or no sachharification of cellulose. Optimization of process parameters like acid concentration, temperature and retention time were studied for the maximum production of reducing sugars in terms of xylose by hydrolyzing the pith in series digester. After treatment, pith was filtered, hydrolysates were collected and the solid residue was subjected to yield analysis. The hydrolysates were analyzed for estimation of total reducing sugars (TRS), total solids, phenolics and lignin content.

Bagasse pith was hydrolyzed by dilute acid concentration of 2,4,6 and 8 per cent at 120.0 °C for 90 min. Effect of temperature on reducing sugar production from pith was determined with various temperatures such as 80.0 °C,100.0 °C,120.0 °C and 140.0 °C. Treatment time was optimized so that maximum sugar production was achieved without loosing much lignin. Pith was hydrolyzed with 8 per cent acid concentration for different time periods 60 mins, 75 mins, 90 mins, 120 mins. Based on the results, process parameters for prehydrolysis of pith for maximum sugar production(TRS) without losing much lignin were optimized *i.e.* Acid concentration 8 per cent temperature 120 °C and retention time 90 min shown in Table 27.2.

Table 27.2: Process Parameters for Prehydrolysis of Pith

Conditions: *Acid – 8 per cent*
Treatment time – 90 mins
Temperature –120.0 °C
Material to Liquor : 1:8

Parameters	*Per cent*
Residual biomass	
Biomass	68.29
Klason Lignin	51.96
Ash	4.96
Unhydrolysed biomass	11.8
Hydrolysate	
Total sugars	28.6
Total reducing sugars	38.3
Soluble lignin	3.33

Saccharification of Bagasse Pith Chemical and Enzymatic Routes

Prehydrolyzed solid residue pith subjected to saccharification with an aim to maximum degradation of cellulose to sugars (glucose)

Saccharification of prehydrolyzed pith was done through chemical and enzymatic routes.

Chemical Saccharification

Chemical saccharification of pre hydrolyzed Bagasse was pith was done using the optimized acid concentration and process conditions (Table 27.3).

Enzymatic Saccharification

Enzymatic hydrolysis of lignocellulosíc is achieved by employing hydrolytic enzymes like cellulases laccases and xylanase enzymatic hydrolysis is usually done after a pretreatment step. This step is required to open up the bundles of cellulose and hemicelluloses fibers and hence to facilitate and increase the efficiency of enzymatic hydrolysis. Enzymatic saccharification was done using Enzymes 1 and 2 by following the process conditions given in Table- 4. The samples were filtered and filtrates were subjected to analysis.

Table- 3: Process conditions for chemical saccharification of prehydrolyzed bagasse

Conditions: *Acid – 72 per cent for 2 hrs (at room temperature)*
15 per cent for 4 hrs (Digestion at 120.0 °C)

Parameters	*Per cent*
Residual biomass	
Biomass	27.13
Klason Lignin	22.29
Ash	4.2
Unhydrolysed biomass	0.64
Hydrolysate	
Total sugars	55.2
Total reducing sugars	42.1
Soluble lignin	4.88
pH	Below 1.0

Table 27.4: Process Conditions for Enzyme Saccharification

Conditions: *Enzyme Dose : Enzyme 1: 4 per cent, Enzyme 2 : 2 per cent*
Treatment time – 72 hrs
Temperature – 50.0 °C
Material to Liquor: 1:10
pH – 5-6

Parameters	*Per cent*
Residual biomass	
Biomass	**46.7**
Klason Lignin	**35.5**
Ash	**5.19**
Unhydrolysed biomass	**5.64**
Hydrolysate	
Total sugars	**53.2**
Total reducing sugars	**42.7**
Soluble lignin	**4.0**
pH	**6-7**

Analysis of prehydrolysates showed that both acid and enzymatic hydrolysates yielded same amount of reducing sugars. Data is shown in Table- 3 and 4. As the pH of the acid hydrolysate is acidic it is to be neutralized and detoxified before subjecting it to fermentation which may affect the efficiency and economics of the process. Enzymatic hydrolosate having pH of 6–7 which is suitable for fermentation looks more attractive for further fermentation process.

Fermentation of Bagasse Pith Hydrolysates

Acid pretreated and enzymatic saccharified pith hydrolyzed samples was fermented in 500 ml bottles using thermophillic ethanologen, yeast strain at 50.0 °C.

Acid pretreated and enzymatic saccharified pith hydrolyzed samples with initial total sugar concentration 36.64 g/l was fermented in 500 ml bottles using thermophillic yeast strain (Fig.- 1 and 2). Almost 80 per cent glucose, 30 g/l of enzyme hydrolysate was fermented to ethanol (9.5 mg/l) within 40 hr and there after some of the oligosaccharides also converted to ethanol bringing the final ethanol concentration 11.5 g/l in the broth.

Possible Utilization of Lignin Rich Residual Biomass

This is based on the nature of the residual biomass shown in Table 27.5 which is rich in lignin. It was found that it has reasonably high calorific value both obtained from the chemical and enzymatic saccharification. This will provide an opportunity to use it as a source of fuel for sustaining the energy requirement of the process. Further the purity of lignin after chemical saccharification was found to be more than 83 per cent which may find application in rubber and plywood industry.

Table 27.5: Parameters of Residual Biomass after Treatment

Parameters	*Residual Biomass after Chemical*	*Saccaharification Enzymatic*
Residual biomass (per cent)	27.1	46.7
Silica (per cent) w/w	13.5	6.9
Lignin purity (per cent) w/w	83.28	50.23
Ash (per cent) w/w	4.2	5.2
Gross Calorific value (cals/gm)	4650	4260
Chlorides (per cent) w/w as Cl^-	0.41	0.49
Carbon (per cent) w/w	54.7	47.9
Hydrogen (per cent) w/w	3.34	3.89
Nitrogen (per cent) w/w	0.13	0.54
Sulphur (per cent) w/w	0	0.04

Conclusion

1. The study indicates that the bagasse pith available as a waste biomass may proved to be a potential substrate alternate to the convectional used raw materials *i.e.* molasses, feedstock like corn, maize, starch *etc.* for the production of bioethanol.

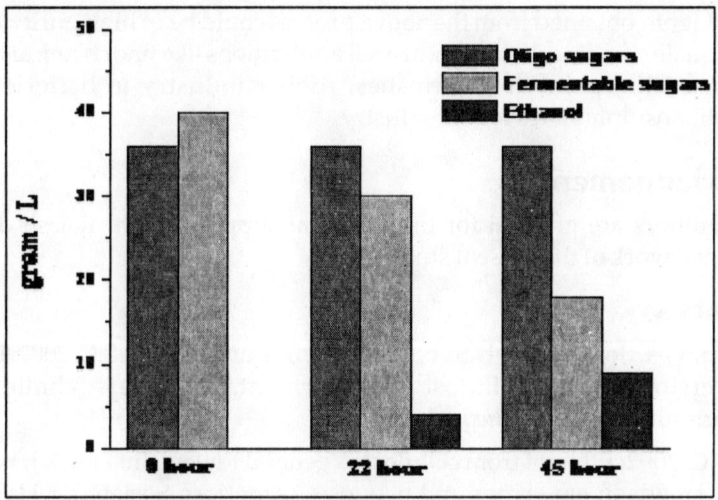

Figure 27.2: Chemical Hydrolysate of Prehydrolysed Bagasse Pith

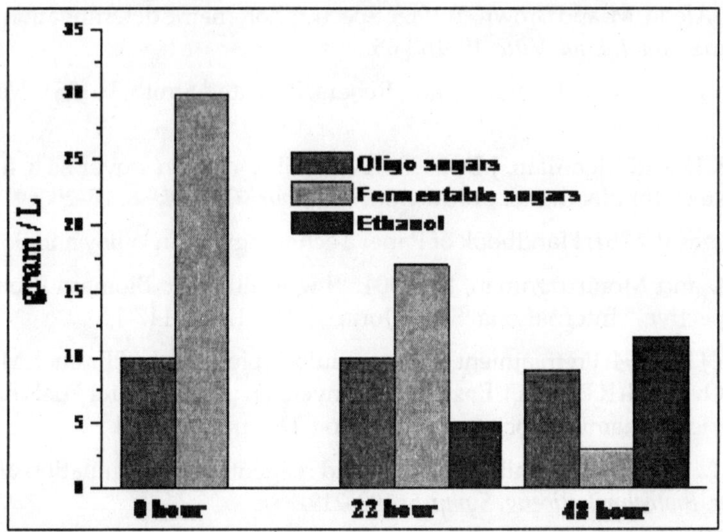

Figure 27.3: Enzymatic Hydrolysate of Prehydrolysed Bagasse Pith

2. It is important to prehydrolysed biomass before saccharification to convert the cellulose ion to fermentable sugars under given optimized process conditions.
3. Enzymatic saccharification route results it to higher yield of fermentable sugars thereby providing opportunity for higher bioethanol yield.
4. High ethanol conversion was found with enzyme saccharified pith hydrolysate when compared with the acid hydrolysate.

5. Lignin obtained from the above process could be of high purity and better quality finding various industrial applications like fine chemicals including vanillin, paint and varnishes, rubber industry in batteries and for lignosulphonates for oil industry.

Acknowledgement

The authors are grateful for the technical support of Mr. Rajesh during the experimental work of the present study.

References

Alexandria, V.A., Jorgenesen, U.S.A., Kristensen, H. and Felby, J.B.C. 2007: Enzymatic conversion of lignocelluloses into fermentable sugars: challenges and opportunities. *Biofuels, Bioprod, Bioref.*

Badger, P.C. 2002: Ethanol from cellulose: a general review. In: Janick J, Whipkey A (eds) Trends in new crops and new uses. American Society for Horticultural Science (ASHS) Press,

Caputi, A., Ueda, M. and Brown, T. 1968: Spectrophotometric determination of ethanol in wine. *Am. J. Enol. Vitic.* 19: 160-65.

Dubois, A.I., Gilles, K., Hamilton, J.K., Rebers, P. A. and Smith, F. 1951: *Nature.*, 168, 167.

Kadam, K.L. and Mcmillan, J.D. 2003: Availability of corn stover as a sustainable feedstock for bioethanol production. *Bioresour. Technol.* 88, 17–25.

Kasey, James P. 1961: Handbook of Paper Technology. John Wiley and Sons.

Knauf, M. and Moniruzzaman, M. 2004: "Lignocellulosic Biomass Processing: A Perspective," International Sugar Journal, 106(1263): 147-150.

McMillan, J.D. 1994: Pretreatment of lignocellulosic biomass. In: Himmel, M.E., Baker, J.O., Overend, R.P. (Eds.), Enzymatic Conversion of Biomass for Fuels Production. American Chemical Society, Washington, DC, pp. 292–324.

Miller, G.L. 1959: Use of dinitrosalicyclic acid reagent for determination of reducing sugar. *Biotechnol. Bioeng. Symp.* 5: 193-219.

2013, Climate Change Effects on Agriculture and Economy *Pages* **245–247**
Editors: **D.R. Khanna, A.K. Chopra, Vikas Singh, Rakesh Bhutiani & Gagan Matta**
Published by: **BIOTECH BOOKS, NEW DELHI**

Chapter 28

Bacteriological Examination of Drinking Water in Ashti Lake with Reference to Coliforms

U.W. Fule[1], *S.S. Nimghare*[1], *S.B. Zade*[2] *and P.M. Telkhade*[3]

[1]*Hutatma Rashtriya Arts and Science College, Ashti Dist. Wardha, Maharashtra*

[2]*Department of Zoology, R.T.M Nagpur University, Nagpur, Maharashtra*

[3]*Arts Commerce and Science College, Tukum, Dist. Chandrapur, Maharashtra*

Present study investigates the pollution status of drinking water at different sites of Ashti Lake in Wardha district, Maharashtra. The study has been carried out for four months from June 2011 to September, 2011. The bacteriological analysis performed was in accordance with standard procedure. The investigations revealed that out of six samples from different sites all had counts higher than the Drinking Water Standards and Station E (West) was the most polluted posing a public health hazard.

Keywords: *Drinking water, Coliform, Pollution level, Public health, Hazard.*

Introduction

Water Pollution is the contamination of water caused by human activities, which is harmful to organisms and plants living in these water bodies or to the humans which use these waters. Organic wastes such as sewage impose high oxygen demand on the receiving water bodies leading to oxygen depletion with potentially severe impacts on the whole ecosystem (Kudesia, 1990; Ao meren and Bordoloi, 2003).

Pathogen's can produce waterborne disease in either human or animal hosts. The coliform group consists of several genera of bacteria in family Enterobacteriaceae that included *E. coli*, which is a normal inhabitant of the intestinal tract of human and other warm blooded animals and thus, regarded as the feacal type coliform. E. coli is regarded as the most sensitive indicator of feacal pollution (APHA, 1998: Bonde, 1977: Edberg *et al.*, 2000). Large numbers of E. coli are present in the gut of human and other warm blooded animals. The fact that they are not generally present in other environment supports their continued use as the most sensitive indicator of faecal pollution (Edberg *et al.*, 2000; Kasuar *et al.*, 2009). In this context the present study was designed to detect the coliform's and to assess the quality of drinking water in various sites of Ashti lake in rainy season.

Materials and Method

The detection and confirmatory test for *E. coli* was made by Most Probable Number (MPN) method using Mac'Conkey's Broth (APHA, 1998; Eresinnies and Schneider, 1988) for the water samples collected from different sites of Ashti Lake in Wardha district from June, 2011 to September, 2011.

Result and Discussion

The detection and confirmatory test for *E. coli* was made by Most Probable Number (MPN) method Mac'Conkey's Broth (APHA, 1998) and the results are presented in Table 28.1. Most Probable Number (MPN) of coliform in case of water samples collected from site E (West) were estimated to be very high 33 per 100 ml, Ashti Lake site A (Center) 22 per 100 ml, site B (North) 24 per 100 ml, Site C (South) 21 per 100 ml, Site D (East) 20 per 100 ml.

Table 28.1: MPN of Coliforms from Different Places

Source of Sample	*Sample Size*		*MPN/100 ml*
	Five of 10 ml	*Five of 1 ml*	
Station A(Center)	05	01	22
Station B (North)	04	02	24
Station C(South)	05	01	21
Station D (East)	03	03	20
Station E (West)	04	00	33

It was found that water of these places is, thus not potable as per WHO Standards (WHO, 1996). According to a regulation published in the federal register (1986), there should be 0 (zero) coliforms per 100 ml of drinking water, as determined by any method for any sampling frequency.

The consumption of drinking water contaminated with pathogens of faecal origin is a significant risk to human health in the developing world, especially in remote rural areas and peri-urban 'shanty' communities. Kasur *et al.* (2009) examined drinking water in different public places with Reference to coliforms and reported unsafe drinking water, as in our present investigation. Over three million deaths per year are attributed to water borne diarrhoeal diseases, especially among infants and young children in poor communities in Africa, Asia and South America (Anon, 1997). Proper management is immediately required to maintain the water quality in public places otherwise it will pose a danger to public at large.

References

Anon, 1997: World Health Report, World Health Forum 97: 181 – 188.

Ao Meren and Bordolor, S. 2003: Amphibian distribution with respect to water chemistry in the wetlands of Kohima district, Nagaland, India. *Aquacult*, 4(2): 259-263

APHA, 1998: Standard Methods for the examinations of water and wastewater, 20th edition. American public health association, Washington DC,

Bonde, G.J. 1977: Bacteriological indication of water pollution in, Advances in Aquatic Microbiology ds. Droops, M.R. and Janvas, H. W. *Academic Press London*, New York, 273-364

Edberg, S.C. Rice, E.W. Karlin, R.J. and Allen, M.J. 2000: *Escherichia Coli* - The best biological drinking water indictor for public health protection. *J. Appl. Microbiol. Symp Supplement*, 88: 160 – 116.

Eresennies, Quentin W.K.L. and Schneider, W. 1988: Water Analysis. A practical guide to analyse physico-chemical and microbiological water examination and quality assurance. Springer verlag berling, Heidelberg, Germany 1-804.

Kasur, R.R. Tidke, R.T., Kore, G.V. and Masarrat, Sultana. 2009: Bacteriological Examination of Drinking water in different public places with Reference to coliforms. *J. Int. Nature Environment and pollution Technology* 8(2): 307-308.

Kudesia, V.P. 1990: Water pollution 3rd revised edn. Pragati Prakashan, Meerut, 84-102.

WHO, 1996: Guidelines for drinking water quality 2nd Edition, Geneva.

2013, Climate Change Effects on Agriculture and Economy *Pages* **249–257**
Editors: **D.R. Khanna, A.K. Chopra, Vikas Singh, Rakesh Bhutiani & Gagan Matta**
Published by: **BIOTECH BOOKS, NEW DELHI**

Chapter 29

Water Quality Assessment of River Narmada using Benthic Macroinvertebrates

***Shailendra Sharma*[1], *L.K.Mudgal*[2], *Kalpana Mimrot*[1], *Taniya Sengupta*[1], *Vishal Shankar*[1], *Tushar Gandhi*[1] *and Vibha Dave*[3]**

[1]*Department of Biotechnology and Zoology, Shari Umiya Girls College, Mandleshwar, Madhya Pradesh*

[2]*Department of Zoology, Govt.Girls P.G. College, Indore, Madhya Pradesh*

[3]*Department of Zoology, P.M.B. Gujarati Science College, Indore, Madhya Pradesh*

Benthic macroinvertebrates are best indicators for bio-assessment. The abiotic environment of the water body directly affects the distribution, population density and diversity of the macro benthic community. Benthic fauna are especially of great significance for fisheries that they themselves act as food of bottom feeder fishes (Walker *et al.*, 1991). The present study deals with the population density and species diversity of aquatic macro invertebrate fauna. The Narmada river of Khandwa district at Omkareshwar dam in Madhya Pradesh was chosen to assess the impact of alteration of the flow regime on water quality, due to the formation of the dam. The present study involved sampling, pre-identification and identification of macro-invertebrates during 2009 and computing the per cent occurrence of families of various taxonomic groups and conducting physiochemical analysis of water samples collected from the selected locations. Macro-invertebrates were identified up to family level and bio assessment at various locations was done using NEPBIOS score system. 29 families of taxonomical group like *Ephemeroptera, Trichoptera, Plecoptera, Coleoptera, Heteroptera, Odonata, Diptera,*

Mollusca, Oligochaetes etc. have been found in different composition inhabiting the river. The results further showed that all the locations assessed for quality using macro-invertebrates and physiochemical analysis are in the range of water quality class II (Good) during Pre-Commissioning stage of the Omkareshwar dam, while during post-commissioning stage of the Omkareshwar dam, the water quality belongs to class III (Moderately Polluted).

Keywords: *Macroinvertebrates, Benthic fauna, Omkareshwar dam, Benthos, Pollution.*

Introduction

Benthos is an important part of the food chain, especially for fish.Many invertebrates feed on algae and bacteria, which are on the lower end of the food chain. Some shred and eat leaves and other organic matter that enters the water. Because of their abundance and position as "middlemen" in the aquatic food chain, benthos plays a critical role in the natural flow of energy and nutrients. As benthic invertebrates die, they decay, leaving behind nutrients that are used by aquatic plants and other animals in the food chain. They also help in assessment of water quality. As like many types of benthos are sensitive to pollutant such as metals and organic wastes, mayflies, stoneflies, and caddisflies are generally intolerant of pollution. If a large number of these insect types are collected in a sample, the water quality in the stream is likely to be good. If only pollution-tolerant organisms such as non-biting midges and worms are found the water is likely to be polluted. They help in water purifier. It is well known that the benthos is the best indicators of water pollution in water bodies.

Materials and Method

About Narmada River

Narmada is the fifth largest river in Indian subcontinent, Narmada a Sanskrit word means 'one who endows with bliss'. It originates from Amarkantak and form the traditional boundary between North India and South India and flows westwards over a length of 1312 km before draining through the Gulf of Combay (Khambat) into the Arabian Sea, 50kms west of Bharuch city of Gujrat. It is one of only three major rivers in peninsular India that runs from east to west (largest west flowing river) along with the Tapti and the Mahi River. It is the only river in India that flows in a rift valley flowing west between the Satpura and Vindhya ranges.

About Omkareshwar

Omkareshwar is a famous place of pilgrimage, situated 77 km from Indore in Khandawa District, Madhya Pradesh. Shaped like the holy Hindu symbol 'OM', this scared island, on the conflux of the river Narmada, Kaveri visit by pilgrims from all over the country to seek blessing at the temple of Shri Omkar Mandhata. It's Latitude (DMS) 22°15'1" N and Longitude (DMS) 76°8' 48" E.

Sampling Site

1. 400 m Upstream from Omkareshwar dam
2. 200 m downstream in Omkareshwar dam.

A sample consists of collection of 20 sub samples each of 0.25 ´ 0.25 m^2 taken from all microhabitat type. This procedure results in sampling of approximately 1.25 m^2 stream bottom area. Net of mesh size 500 μm is used for collecting the macro invertebrates. Every large boulder or cobble in the area is picked up if it could be lifted and organisms vigorously washed by hand into the net. Finally, the substrate with smaller boulders should be distributed by kicking systematically across the area 3-4 times such that invertebrate wash D/S into the net. The organisms are then picked from the net surface and preserved immediately in 80 per cent ethanol or 4 per cent formaldehyde. These samples are returned to the laboratory for processing. Specimen collected are sorted and identified up to family level with the help of regional keys in the laboratory under the dissecting microscope. For identify the fauna, standard literature was consulted.

Water samples were collected in plastic container for different physico-chemical parameters. The chemical characteristics were determined by the standard methods suggested of APHA (2002), Welch (2002) and Golterman (1975).

Table 29.1: Physico-chemical Parameters Analyzed during the Study Period

Sl.No.	*Parameters*	*U/S*	*D/S*
1.	pH	8.2	8.1
2.	Conductivity	366	225
3.	Turbidity	1.42	1.28
4.	Dissolved Oxygen	9.09	8.25
5.	Alkalinity	141	89
6.	Hardness	201	102
7.	Chloride	13.4	7.2
8.	Phosphate	940	1150
9.	BOD	3	3.9
10.	TDS	0.22	0.12
	NSF Index	III	III
	NSF value	69	68
	WQI	50-70	50-70

On the basis of these parameters, the water quality can be determined using NSF water quality index.

NSF Water Quality Index

The National Sanitation Foundation, water quality index (NSFWQI) has long been used for the development of standards, product testing and certification services in the areas of public health, safety and protection of the environment. Calculate a mathematical a single value from multiple test results. The index tells about the status of water quality of a given water body which is scaled from 0 to 100 to rate the quality of the water, with 100 being the highest possible score representing excellent water quality and 0-25 as the bad quality (Table 29.2).

Table 29.2: NSF WQI Scale

Range	WQI Quality	WQ Class
91-100	Excellent water quality	I
71-90	Good water quality	II
51-70	Medium or average water quality	III
26-50	Poor water quality	IV
0-25	Bad water quality	V

The range using NSF water quality index for the site u/s is 69, which is indicative of moderately polluted water quality (Class III). The range using NSF water quality index for the site D/S is 68, which is indicative of moderately polluted water quality (Class III).

Biological Parameters

The common and dominant families of macroinvertebrates of each group encountered are given in (Table 29.3). NEPBIOS score has been calculated using NEPBIOS. The water quality assessment on the basis of macroinvertebrates families is based on the NEPBIOS biotic index system which is based on indices or score system is only when local reference communities are properly scored. Taking this fact into account the Nepalese taxa were scored, the *Average Score Per Taxon* (ASPT) calculated and a different biotic score method for Nepal developed under the name *Nepalese Biotic Score* abbreviated as NEPBIOS.The calculation of the water quality on the basis of the presence of the macroinvertebrates families has been done on the basis of NEPBIOS (Table 29.4) using the pre-classification sheet.

Three procedures were followed in scoring the taxa.

Numerical Procedure

This procedure follows the following formula:

Guide Score: SI/STot × 10 + SI-II/STot × 8.57 + SII/STot × 7.14 + SII-III/STot × 5.71 + SIII/STot × 4.28 + SIII-IV/STot × 2.85 + SIV/STot × 1.43

Where, SI, SI-II, SII, SII-III, SIII, SIII-IV, SIV are the total number of sites representing the pollutional classes I, I-II, II, II-III, III, III-IV and IV, respectively.

Stot = SI+SI-II+SII+SII-III+SIII+SIII-IV+SIV

1.43 is the score interval with 10 as maximum.

Professional Judgements

Step-I: Based on the reference made to the scores that has previously been assigned by different authors in their respective country of origin, and the range of pollutional class represented by each taxon in the rivers of Nepal.

Step-II: Based on the distribution pattern of each taxon (family level) in response to pollutional level. The comparison of family (taxon) distribution with the distribution of observed water quality classes was carried out to see if any family with the same ecological distribution are differently scored. If so whether or not the reasons are

Table 29.3: Occurrence of Benthic Macro Invertebrate's Community

Taxanomic Group	*Family*	*Upstream*			*Downstream*		
		No. of Individual	*Per cent of Abundance*	*NEPBIOS*	*No. of Individual*	*Per cent of Abundance*	*NEPBIOS*
	Thiridae	346	46.82	4	526	74.50	4
Mollusca	*Planorbidae*	52	7.03	4	36	5.09	4
	Lymnaeidae	138	18.67	5	77	10.90	5
	Viviparidae	44	5.95	6	28	3.96	6
	Ephemerellidae	33	4.46	7			
Ephemeroptera	*Ephemeredae*	11	1.48	6	2	0.28	6
	Heptogeniidae	1	0.13	7			
	Leptophlebiidae	3	0.40	7	1	0.14	7
Odonata	*Gamphidae*	2	0.27	4	6	0.84	4
	Scirtidae				2	0.28	6
Colleptera	*Elmidae*	10	1.35	8			
	Psephenidae	3	0.40	8			
	Ecnomidae	3	0.40	–			
	Hydropsychidae				13	1.84	6
Trichoptera	*Glossosomatidae*				1	0.14	7
	Psychomyiidae	1	0.13	6			
	Goeridae	2	0.27	9			
	Hydroptillidae	7	0.94	6			
	Lapidostomatidae	2	0.27	10			
	Ceratopogonidae				4	0.56	0

Contd...

Table 29.3–*Contd...*

Taxonomic Group	*Family*	*Upstream*			*Downstream*		
		No. of Individual	*Per cent of Abundance*	*NEPBIOS*	*No. of Individual*	*Per cent of Abundance*	*NEPBIOS*
	Chironomidae				8	1.13	1
	Limoniidae	5	0.67	8			
Diptera	*Tabanidae*				1	0.14	0
	Ephydridae				1	0.14	0
	Simuliidae	9	1.21	7			
Hemiptera	*Corixidae*	25	3.38	4			
Placoptera	*Perlidae*	3	0.40	8			
Crustacea	*Palaemonidae*	33	4.46	4			
Annelida	*Oligochaetes*	6	0.81	–			

matching. Once NEPBIOS/ASPT is calculated, reference is made to the transformation (Table 29.5) for interpretation of the water quality of the particular site investigated.

Table 29.4: Nepalese Biotic Scores (NEPBIOS) Assigned to the Macroinvertebrates

Sl.No.	*Macroinvertebrates*	*Score*
1.	Capniidae, Ephemerellidae (*Drunella* sp.), Epiophlebiidae, Helicopsychidae, Helodidae (Scirtidae), Heptageniidae (*Epeorus rhithralis),* Heptageniidae (*Rhithrogena nepalensis),* Leuctridae, Peltoperlidae, Perlidae (*Acroneuria* sp.), Perlidae (*Calicneuria* sp.), Siphlonuridae, Taeniopterygidae, Uenoidae	**10**
2.	Athericidae, Chloroperlidae, Goeridae, Leptophlebiidae (*Habrophlebiodes* sp.), Limnocentropodidae, Neoephemeridae, Perlodidae, Polycentropodidae.	**9**
3.	Baetidae (*Centroptilum* sp.), Brachycentridae, Chironomidae (*Diamesinae*), Elmidae, Euphaeidae, Glossosomatidae, Heptageniidae (*Epeorus bispinosus*), Heptageniidae (*Iron psi*), Heptageniidae (*Rhithrogena* sp.), Hydrobiosidae, Lepidostomatidae, Limnephilidae, Nemouridae, Perlidae, Philopotamidae, Psephenidae, Rhyacophilidae, Stenopsychidae.	**8**
4.	Aphelocheiridae, *Baetidae* (*Cloedodes* sp.), Baetidae (*Baetiella* spp.), Baetidae (*Baetis* spp.), Baetidae (*Baetiella ausobskyi*), *Baetidae* (*Baetis sp.*), Corydalidae, Ephemerellidae, Ephemerellidae (*Cincticostella sp.*), Ephemeridae, Gammaridae, Gyrinidae, Heptageniidae, Heptageniidae (*Cinygmina* sp.), Heptageniidae (*Notacanthurus cristatus*), Hydraenidae, Leptophlebiidae, Limoniidae, Pleuroceridae, Psychomyiidae, Salifidae (*Barbronia* sp.), Simuliidae, Tipulidae.	**7**
5.	Aeshnidae, Baetidae (*Baetis* sp.5), *Baetidae* (*Baetis* sp.4), Caenidae, Ceratopogonidae, Ecnomidae, Ephemerellidae (*Torleya nepalica*), Heptageniidae (*Electrogena* sp.), Hydrometridae, Hydropsychidae, Hydroptilidae, Potamidae, Scirtidae, Viviparidae.	**6**
6.	Baetidae (*Baetis* sp.2), Baetidae (*Baetis* sp.3), Bithyniidae, Chlorocyphidae, Coenagrionidae, Corduliidae, Dryopidae, Hydrophilidae, Leptophlebiidae (*Euthraulus* spp.), Lymnaeidae, Odontoceridae, Protoneuridae, Sphaeriidae, Unionidae.	**5**
7.	Calopterygidae, Chironomidae (*Microtendipes* sp.), Chironomidae (*Polypedilum* sp.), Corbiculidae, Dytiscidae, Gerridae, Glossiphoniidae, Micronectidae, Naucoridae, Nepidae, Palaemonidae, Planorbidae, Ranatridae, Salifidae (*Barbronia weberi*), Thiaridae.	**4**
8.	Corixidae, Libellulidae, Lumbricidae, Noteridae, Notonectidae, Salifidae	**3**
9.	Culicidae, Physidae, Tubificidae	**2**
10.	Chironomidae [*Chironomus* group *riparius* (=thummi) and group plumosus]	**1**

On the basis of NEPBIOS Score System the species present in the samples of the site at Omkareshwar Dam on Narmada River U/S due to stagnant water show that the water quality of river belongs to class III and the D/S due to Holy dip and contamination of sewage water showed the same class III.

Conclusion

1. On the basis of our study Benthic macro invertebrates are best indicator of water quality.

Table 29.5: Water Quality Scores Based on NEPBIOS Score System

NEPBIOS/ASPT	Water Quality
8.00-10.00	I
7.00-7.99	I-II
5.50-6.99	II
4.00-5.49	II-III
2.50-3.99	III
1.01-2.49	III-IV
1.00	IV

2. The result shows that the physio chemical analysis of water and NSF WQI U/S and D/S of Omkareshwar Dam is of class III (moderately polluted) and the NEPBIOS Score System is also shows the range of water quality class III (moderately polluted). Both results are same that means Benthic Macro invertebrates also show the water quality as well.

References

A.P.H.A. 2002: Standard method for examination of water and wastewater American Public Health Association (22th Edition) APHA New York.

Kumar, Arvind. 1994: Role of species diversity of aquatic insects in the assessment of population in wetlands of Santhal Parganas (Bihar). J. Environment and Pollution (1, 3 and 4), 117-120.

Bath, K.S. and Kaur, 1997: Aquatic insects as bio-indicators at Harike reservoir in Punjab – India. *Indian journal of Environmental Sciences*, 2: 133-138.

Dhillon S.S., Bath, K.S. and Mander, G. 1995: Invertebrate fauna of freshwater bodies existing in and around Patiala. *Journal of Environment and Pollution* 2: 163-167.

Flotemersch, J.E., Blocksom, K., Hutchens, J.R. and Autrey, B.C. 2006: Development of a standardized large river bio-assessment protocol (LR-BP) for macroinvertebrate assemblages. River Research and Applications, 22, pp 775-790.

Golterman, H.L. 1975: River ecology (Edited by B.A. Whitton) Black-Weel Scientific Publications Oxford, London, Edinburg Melbourne PP. 39-80.

Goethals, Peter and Pauw, De Nidr 2001: Development of a concept for integrated ecological river assessment in Flassders, Belgium" *J. Limnology* 60, pp 7-16.

Hynes, H.B.B. 1972: The Ecology of Running water. University of Toronto Press. pp 555.

Hynes, H.B.N. 1994: Historical Perspective and feature direction of Biological Monitoring of aquatic ecosystem. In: S. L. Loeb and A spacic. Eds., Biological Monitoring of Aquatic systems. Lewis publishers, BOCA Paton, Fl.

Malhotra, Y.R., Gupta, K. and Khajuria, A. 1990: Seasonal variation in the population of macro-zoobenthos in relation to some physico-chemical parameters of lake Mansar. *J. Freshwater Biol.* 2: 123-128.

Merrit, R.W. and Cummins, K.W. 1984: An Introduction to the aquatic insects of North America. 2nd ed. Kendall/Hunt Publishing Company, Dubuque.

Needham, James C. and Needham, Paul R. 1988: A Guide to the study of fresh-water Biology. Reiter's Scientific and professional books, Washington, D.C.

Pennak, Robert W. 1989: Fresh-water invertebrates of the United States: Protoza to Mollusca. 3rd ed. John Wiley and Sons, New York.

Roy, S.P. and Sharma, U.P. 1983: Dipteran larval population in a freshwater wetland at bhagalpur (Bihar) in relation to physico-chemical parameters. Ind. J. Ecol. 10:129-136.

Roux, D.J., Kleymhans, C.J, Thirion, C. Hill., Engelbrecht, J.S., Deacon, A.R. and Kemper, N.P. 1999: Adoptive Assessment and management of riverine ecosystems: The crocodile/Elands river case study water SA, 25 (4), pp 501-511.

Sharma, Shailendra., Chaube, U.C., Sharma, S.P. and Kumar, P. 2008: Water quality assessment of River Satluj using benthic macroinvertebrates. Proceedings of the Scientific Conference Rivers in the Hindu Kush-Himalaya –Ecology and Environmental Assessment. Katmandu University Nepal (3rd to 5th March- 2008).

Srivastava, H.N. 1962: Aquatic fauna as indicators of Pollution. *Env. Hlth.* 106-113.

Subba Rao, N.V. 1993: Freshwater Molluscs of India in Recent Advances in freshwater Biology" (ed. K.S.Rao) 187-220, 4+61.

Tonapi, G.T. 1980: Freshwater animals of India-An ecological approach. Oxford and IBH Publishing Co. New Delhi.

Tyagi, P. 2006: Occurrence of benthic macroinvertebratres families encountered in river Hindan in Uttar Pradesh (India). J.Zool.India, Vol.9, No 1, pp 209-216.

VOS P, Wepener V and Cyrus D.P. 2002: Efficiency of the SASSY rapid bioassessment proposal in determining the health: A case study on the Mhlathinze River, Kwazula-Natal, South Africa" Water SA, 28(1), pp 13-22.

Welch, P. S. 1952: Limnological methods. Blakiston Co; Philadelphia. Pa.

Weltch. 2002: Limnology: Lake and river Ecosystem, 3rd ed. Academic Press. ISBN–12-744760-1.

Williams, D.D. and Feltmate, B.W. 1992: Aquatic insects. C.A.B. International, Wallingford, Oxon, UK.

2013, Climate Change Effects on Agriculture and Economy *Pages* ***259–262***
Editors: **D.R. Khanna, A.K. Chopra, Vikas Singh, Rakesh Bhutiani & Gagan Matta**
Published by: **BIOTECH BOOKS, NEW DELHI**

Chapter 30

Water Quality Index of Sitaput Pond at Dhar Town (M.P.)

Waskel Darasingh
Department of Zoology, Govt. P.G. College, Dhar, Madhya Pradesh

Water Quality Index (WQI): Study of physico-chemical and bacteriological parameters of Sitapat pond at Dhar town (M.P.) to ascertain the quality of water for public consumption. Physico–chemical and bacteriological parameters of this pond were monitored seasonally during the 2007 and 2008. Results obtained from the study revealed that the WQI of the pond was well within the permissible limits (WHO and BIS) and water is safe for drinking purpose.

Keywords: WQI, Physico-chemical, Bacteriological, WHO, BIS.

Introduction

WQI is one of the most effective ways to communicate information on water quality trends are with indices. Water Quality Index (WQI) is commonly used for detection and evolution of water pollution and may be defined as a rating reflecting the composite influence on the overall quality of a number of water quality parameters. Water quality refers to physical, chemical and biological characteristics of water. The physical characteristics include the temperature, turbidity, colour and similar qualities. Chemical water characteristics include the presence of organic and inorganic substances in water. Biological characteristics of water include identity and impact of organisms which are present in water. In response to the need for a uniform understandable yardstick of water quality, scientists worked out to compile all the water quality parameters in to some convenient approach to represent an index, a procedure generally described as a "Water Quality Index (WQI)" (Horton, 1965; Brown, 1972; Otta, 1978; Lee and Irving, 1971).

Materials and Method

In the present study, WQI has been calculated on the basis of ten physico-chemical characteristics. The main objective of the study area was to know the water pollution of potable water from Sitapat pond resources of Dhar town. We know that water is essential for life of various organisms including human beings who depend on good quality of water. The water quality index (WQI) was calculated using weigted arithmetic index method and the quality rating/sub index (Qi) corresponding to the Ith Parameter pI, is a number reflecting the relating value of this parameter. The W QI was calculated by using following expression:

$$WQI=\sum_{i=1}^{n} [\{ M_1 (-) I_1 / (S_1-I_1) \}] X 100$$

where, M_1= Estimated values of the parameter in the laboratory, I,-Ideal values of the I[st] parameter and SF standard values of the I[st] parameter. The sign [-] indicates the numerical difference of the two values ignoring the algebraic sign. All the ideal values (Ii) are taken as zero except for pH= 7, D= 14.6 and fluorides= 1.0 in the present study unit weight (Wi) was calculated by a value inversely proportional to the recommended standard (Si) of the corresp-onding parameter. The overall water Quality index (WQI) was calculated by aggregating the qulity rating (QI) with unit weight (WJ) linearly.

$$WQI = [(\sum_{i=1}^{n} Q_1 W_1) / (\sum_{i=1}^{n} W_1)]$$

Sitapat pond from 6 selected stations is studied carefully and the calculations are made below 100, it means the water is permissible for dinking. Over all WQI of 67 per cent sampling stations, were found of "very good" quality and 33 per cent sampling stations were found "good" quality of water during 2007. In year 2008, water quality rating was "very good" at two stations and four stations are of "very good" status. Disinfection of community surface water, proper supply system, periodical quality monitoring of drinking water sources, simple and economical water treatments like filtration, boiling, reverse osmosis etc. would prove beneficial to avoid water born diseases in the study area.

Obtained results indicate that bacteriological contamination of water is seen, sometimes suggesting that the need of source protection and regular treatment of water at local filter plant as any lacuna may result in serious health problems. An action plan for drinking water safety should be applied; the plan should be based on the principles of multi barrier approach where by protection barriers were used to ensure the cleanliness, safety and reliability of drinking water.

Results and Discussion

The result obtained from the study reveals that WQI of Sitapat pond are well within the permissible limit and the overall status of the pond water remained good during the study period.

During the study period minimum WQI of pond water remained 33.780 at Station III during 2008 while maximum WQI of the same was 56.268 at Station V during 2007.

Overall WQI of all the sampling stations during 2007 was 45.886 while the overall WQI during 2008 was 44.333 which indicate 'good' to 'very good' status of water quality.

Table 30.1: Rating Scale for WQI

Sl.No.	WQI Value	Water Quality Rank
1	0-25	Excellent
2	26-50	Very Good
3	51-75	Good
4	76-100	Poor
5	100 <	Unfit For Drinking

Table 30.2: Station-wise Water Quality Index of Sitapat Pond during 2007-2008

Sl.No.	Stations	2007			2008		
		Qiwi	ΣWQI	Status	Qiwi	ΣWQI	Status
1.	I	51.17	41.6	Very Good	52.42	42.62	Very Good
2.	II	46.04	37.43	Very Good	45.79	37.23	Very Good
3.	III	46.42	37.74	Very Good	41.55	33.78	Very Good
4.	IV	56.6	46.02	Very Good	57.51	46.76	Very Good
5.	V	69.21	56.27	Good	68.29	55.52	Good
6.	VI	65.25	53.05	Good	64.55	52.48	Good
Overall		56.44	45.89	Very Good	54.53	44.33	Very Good

Table 30.3: Season-wise WQI of Sitapat Pond during 2007-2008

Sl.No.	Stations	2007			2008		
		Qiwi	ΣWQI	Status	Qiwi	ΣWQI	Status
1.	Rainy	52.58	42.75	Very Good	58.21	47.33	Very Good
2.	Winter	55.67	45.26	Very Good	54.08	43.97	Very Good
3.	Summer	51.23	41.65	Very Good	52.9	43.01	Very Good

During 2007, WQI of 67.0 per cent sampling stations indicate 'very good' status while WQI of remaining 33 per cent stations indicate 'good' water quality status. On the other hand during 2008, WQI of 77.0 per cent sampling stations indicate 'very good' status while WQI of 33.0 per cent represent 'good' status of water quality. During the study period WQI in all seasons indicate 'very good' status.

The careful study by the analysis of WQI of different sampling stations during the study period reveals that the drinking water quality of Sitapat pond is good and water is suitable for drinking and domestic purpose.

The present study also reveals that out of six sampling stations, four stations were having WQI below 50 indicating that the water quality is 'very good'.

Remaining two sampling station with higher WQI ranged between 50 to 60 during the study period indicating that the water quality is inferior to other stations. Still the status of water quality is good and fit for human consumption.

The results obtained from this study clearly indicate usability of water but the agriculture land around the pond and uses of pesticides, improper disposal wastes, creating man made pollution inside the water body and it can be harmful for public health. Overall water quality index application in these studies is useful in assessing the quality of water and gives right judgment on quality of sampling stations. This method appears to be more systematic and gives comparative evalution of the water quality of different sampling stations.

Results obtained also indicate the bacterial contamination of water suggesting the need of source protection and regular treatment of water before supply for drinking purpose as any lacuna may result in serious health problems.

An action plan for drinking water safety should be applied; the plan should be based on the principles of multi barrier approach where by protection barriers were used to ensure the cleanliness, safety and reliability of drinking water. Such protection barriers include source protection, proper and regular treatment, water system operation, maintenance, the monitoring and reporting of drinking water quality.

References

Brow, R.M., McCleiland, N.J., Deiminger, R.A. and O' Conner, M.F. 1972: A water quality index erasing the psychological barrier (Edt.bys.H.Jenkis) proc lnt Conf. on water poll Res. Jerusalem 6:787-797, 1972.

Horton, R.K. 1965: An index number system for rating water quality. *J. Water Poll Cont.* Fed 3:300-305.

Otta, W.R. 1978: Environmental indices: theory and practice Anna Arbor Science Publishing, Michigan (USA).

2013, Climate Change Effects on Agriculture and Economy *Pages* ***263–273***
Editors: **D.R. Khanna, A.K. Chopra, Vikas Singh, Rakesh Bhutiani & Gagan Matta**
Published by: **BIOTECH BOOKS, NEW DELHI**

Chapter 31

Seasonal Variations in the Physico-chemical Parameters of Kanhala Pond of Central India

N.V. Harney[1], A.A. Dhamani[2] and R.J. Andrew[3]

[1]*Department of Zoology, Nilkanthrao Shinde Science and Arts College, Bhadrawati Dist. Chandrapur, Maharashtra*

[2]*P.G.Department of Zoology, N.H.College, Bramhapuri, Dist. Chandrapur, Maharashtra*

[3]*P.G. Department of Zoology, Hislop College, Nagpur, Maharashtra*

The seasonal variations in the physico-chemical parameters of Kanhala pond of central India, located near Bhadrawati town of Chandrapur district of Maharashtra was studied from October, 2005 to September, 2007. The parameters that were determined included temperature, pH, transparency, conductivity, dissolved oxygen (DO), free carbon dioxide, alkalinity, hardness (calcium, magnesium), chloride, biological oxygen demand (BOD), chemical oxygen demand (COD), phosphate, sulphate and nitrate concentrations.

In the present investigation, the atmospheric temperature was found to be higher than the water temperature and as expected, the maximum water temperature was recorded during the summer and minimum during the winter season. Maximum transparency was recorded during winter and minimum during the monsoon season, while maximum conductivity, pH and BOD were recorded during summer and minimum during the monsoon season. Free carbon dioxide, total hardness, calcium and magnesium hardness, chemical oxygen demand and phosphate and sulphate concentration were maximum during summers and minimum during the winter season while DO and chloride showed the maximum levels during the summer and minimum during the monsoon season. Alkalinity and nitrate showed maximum level during the monsoon and minimum during the winter season. The comparative higher value of conductivity, total alkalinity, chloride, phosphate and nitrate in Kanhala pond indicates its contaminated status.

Keywords: *Water, Physico-chemical parameters, Kanhala pond, Bhadrawati, Pollution.*

Introduction

The physico-chemical characteristics of the aquatic environment directly influence the life inhabiting it. Fluctuation in these constituents often create an adverse environment to organisms, limiting their growth and interfering in the physiological processes, which reduce their ability to compete with other populations within the environment, ultimately changing the community structure (Kedar and Patil, 2011). The central region of the country is dotted with many small and large water bodies which form a very important component of its physical environment. Due to heavy anthropological pressure, many have shrunk or become polluted with the discharge of untreated domestic and industrial effluent.

The Kanhala pond lies at the central region of the country and is situated at 7905'46" E longitude and 2005'57" N latitude. This pond is located on the west of Bhadrawati town at about 196 m above mean sea level. It receives water from the surrounding catchment area of about 10 acres. The pond is presently surrounded and encroached by residential complexes and has become a perfect site for drainage of municipal waste. The area of Kanhala pond spreads over 66 acres. The depth of water is 25 feet during the monsoon and 5 to 6 feet during the summer season. The water of this pond is primarily used for washing, bathing and fishing. The present study deals with the seasonal variation in the physico-chemical parameters of this pond from October, 2005 to September, 2007.

Materials and Method

The water samples were collected from the marginal areas at 1.0 to 1.5 m depth from three sides of the Kanhala pond with the help of Ruffner's sampler. The water samples were collected in dried plastic cans of 2 litre capacity and brought to the laboratory and were immediately analyzed. The sample collection was made during the morning hours between 8.30 to 10.30 am. The samples were collected every month for the period of 2 years *i.e.* October, 2005 to September, 2007. The recorded data was yearly segregated in three seasons, winter (October to January), summer (February to May) and monsoon (June to September). During the monitoring, physico-chemical parameter like atmospheric and water temperature was measured using Mercury thermometer; pH was determined using pH meter and transparency by using Secchi disc at the sampling spot. The conductivity of water was measured using conductivity meter (Electronic India, model no. EQ-660). The dissolved oxygen was measured using modified Winkler's method. Free CO_2, alkalinity, chloride and hardness were measured by titrimetric methods described in APHA (2005). BOD was estimated using APHA (2005) while the COD distillery assembly was used to measure the

COD. Phosphate, sulphate and nitrate were measured by using Spectrophotometer (model no.UV-108) (APHA, 2005; Trivedi and Goel, 1984).

Results and Discussion

The physico-chemical parameters are essential and fundamental to distinguish the tropical status of an aquatic ecosystem. Tables 31.1 and 31.2 represents the monthly variation of the physico-chemical parameters of the Kanhala pond while Tables 31.3 and 31.4 represents the seasonal variation in the physico-chemical parameters of the same pond.

In the present investigation atmospheric temperature varied from 23.70 °C to 44.12 °C. Rise in temperature speed up the biochemical reactions and reduce the solubility of gases. The atmospheric temperature was always found higher than the water temperature.The large lakes tend to moderate local climate, provide longer growing seasons for aquatic life, and serve as integrated recorders of recent climatic phenomena (Bartholow *et al.*, 2001). The water temperature fluctuated between 20.12 °C to 34.99 °C. All the metabolic and physiological activities of life processes are greatly influenced by water temperature. Water temperature is important for calculating the solubility of oxygen and carbon dioxide and bicarbonate and carbonate equilibrium (Hutchinson, 1957). In the present investigation there is a gradual decline in the temperature of water from August to December and it increases till May. The temperature of water thus was clearly influenced by seasonal changes in atmospheric temperature.

In the present investigation, Secchi disc transparency varied from 15.80 cm to 35.00 cm and the maximum (32.250 cm ± 1.314) during the summer season in 2005-06while minimum (20.475 cm ± 3.369) during the monsoon season in 2006-07. The minimum transparency in rainy seasons is attributed to the turbid conditions of water, influenced of suspended silt with colloidal particles, high waves and cloudy weather. In the present investigation minimum, transparency was recorded during the monsoon seasons as also reported by Singhai *et al.* (1986), Meshram (1996), Bhatt *et al.* (1999) and Ade (2001). However, high transparency was recorded during the winter season is mostly due to clear atmosphere, high light penetration and less phytoplankton productivity. The values of conductivity recorded were 150 µmhos/cm to 242 µmhos/cm and the maximum (212.250 µmhos/cm ± 20.204) during the summer season in 2006-07 while minimum (168.750 µmhos/cm ± 11.189) during the monsoon season in 2005-06. Similar findings were recorded by Narayana *et al.* (2005) and Kedar *et al.* (2007). Seasonal fluctuations in these factors are caused mainly by variation in the inorganic preparation and the diluting effect of rains (Welch, 1952). According to Sreenivasan (1966), electric conductivity above 400 µmhos/cm does not limit productivity. Higher values of conductivity were observed during the summer season due to increase in the concentration of minerals and organic matter mostly due to evaporation.

The pH value varied between 6.12 to 8.20 and season wise, the maximum (8.058 ± 0.083) was recorded during the summer season in 2005-06 while minimum (6.985 ± 0.250) during the monsoon season in the same study period. According to Wetzel (2001) the carbon dioxide in turn is obtained from conversion of carbonates to

Table 31.1: Monthly Variation in Physico-chemical Parameters of Kanhala Pond Water in 2005-06

Parameters	*Oct.*	*Nov.*	*Dec.*	*Jan.*	*Feb.*	*March*	*April*	*May*	*June*	*July*	*August*	*Sep.*
Temperature atm. (° C)	32.12	27.40	23.70	26.36	32.37	36.62	40.47	44.03	42.37	35.50	34.02	34.08
Temperature water (° C)	26.30	25.70	20.12	22.40	23.80	27.87	30.87	34.14	34.07	31.12	30.12	28.12
pH	6.90	7.10	7.30	8.10	8.2	8.01	8.02	8.00	6.62	6.92	7.1	7.30
Transparency (cm)	25.4	28.6	30.9	35.0	33.8	32.8	32.2	30.2	27.69	15.80	20.9	18.40
Conductivity (μmhos/cm)	205.00	191.00	187.00	181.00	165.00	187.00	205.00	230.00	179.00	175.00	171.00	150.00
D.O. (mg/l)	2.86	2.56	2.83	4.96	4.76	3.36	2.03	1.06	1.53	2.2	3.06	3.13
CO_2 (mg/l)	5.10	3.08	1.90	4.27	6.17	8.31	9.02	8.31	4.30	3.56	3.08	3.50
Total Alkalinity -M.O. (mg/l)	60.00	40.00	26.10	35.00	43.00	50.00	58.00	65.00	70.00	73.00	84.60	80.00
Hardness – Total (mg/l)	108.72	86.50	90.12	98.30	102.33	110.03	137.33	168.00	157.13	148.48	142.32	130.12
Hardness - Ca. (mg/l)	60.88	52.32	51.55	50.05	59.33	73.12	86.88	106.11	96.70	94.22	91.12	81.33
Hardness - Mg. (mg/l)	47.84	34.18	10.95	8.45	20.15	36.19	50.45	61.89	60.43	54.26	51.20	48.79
Chloride (mg/l)	23.30	28.62	34.00	34.02	37.00	38.52	39.39	38.77	30.12	23.33	18.04	19.87
B.O.D. (mg/l)	8.00	9.00	8.00	10.00	12.00	13.00	15.00	17.00	15.00	14.00	9.00	7.00
C.O.D. (mg/l)	8.00	6.00	8.00	10.00	16.00	17.00	19.00	20.00	14.00	12.00	10.00	9.00
Phosphate (mg/l)	0.087	0.064	0.061	0.059	0.6	0.7	0.8	0.092	0.080	0.072	0.070	0.070
Sulphate (mg/l)	1.2	1.1	1.4	1.35	3.2	3.0	3.1	3.3	1.3	1.8	2.1	2.6
Nitrate (mg/l)	0.11	0.28	0.30	0.35	0.30	0.24	0.30	0.20	0.32	0.38	0.38	0.40

Table 31.2: Monthly Variation in Physico-chemical Parameters of Kanhala Pond Water in 2006-07

Parameters	*Oct.*	*Nov.*	*Dec.*	*Jan.*	*Feb.*	*March*	*April*	*May*	*June*	*July*	*August*	*Sep.*
Temperature atm. (° C)	32.22	27.43	24.00	27.00	33.00	37.00	41.02	44.12	43.02	34.00	35.00	33.00
Temperature water (° C)	26.38	25.80	20.40	23.00	24.02	28.12	31.12	34.50	34.99	30.87	31.42	27.17
pH	6.36	7.02	7.20	7.9	7.98	8.00	8.00	8.02	6.98	6.12	7.2	8.0
Transparency (cm)	24.8	28.00	29.2	34.20	33.00	33.6	31.8	29.9	25.4	16.20	19.00	21.30
Conductivity (µmhos/cm)	232.00	210.00	202.00	165.00	185.00	210.00	212.00	242.00	180.00	170.00	191.00	160.00
D.O. (mg/l)	2.91	2.30	3.87	1.92	1.47	1.49	1.40	2.09	2.02	2.1	2.87	2.80
CO_2 (mg/l)	3.00	2.00	3.15	4.30	6.65	8.80	9.10	8.87	6.87	5.62	5.42	5.40
Total Alkalinity -M.O. (mg/l)	40.00	32.8	35.00	47.00	60.00	72.00	87.00	90.00	96.00	102.03	100.00	65.00
Hardness – Total (mg/l)	110.11	89.00	96.42	112.12	127.33	132.12	150.77	149.22	146.88	142.05	140.00	133.72
Hardness - Ca. (mg/l)	72.51	58.17	52.63	50.72	69.88	87.88	99.12	95.17	93.22	92.30	87.37	85.42
Hardness - Mg. (mg/l)	37.60	30.83	20.19	18.12	18.70	44.24	51.65	54.05	53.66	49.75	52.63	48.30
Chloride (mg/l)	21.77	24.00	24.80	29.12	34.77	36.00	37.00	39.42	30.62	27.33	22.00	17.20
B.O.D. (mg/l)	9.00	10.00	9.00	9.00	13.00	14.00	16.00	18.00	16.00	15.00	8.00	8.00
C.O.D. (mg/l)	10.00	5.00	9.00	13.00	17.00	19.00	20.00	21.00	18.00	15.00	14.00	13.00
Phosphate (mg/l)	0.092	0.069	0.073	0.045	0.091	0.098	0.099	0.085	0.051	0.079	0.096	0.081
Sulphate (mg/l)	3.7	1.00	1.6	1.20	3.8	3.47	3.80	3.7	1.2	3.10	2.12	2.1
Nitrate (mg/l)	0.13	0.20	0.26	0.27	0.51	0.46	0.29	0.53	0.47	0.59	0.57	0.41

bicarbonates with rise in pH, such conditions have also been reported by earlier studies of Adoni (1985), Jakher (2002) and Jakher and Rawat (2003). In the present investigation maximum values of pH was recorded during the summer which may be due to rise of temperature with increase in rate of photosynthetic activity and resultant higher consumption of carbon dioxide as postulated by Jakher and Rawat (2003).The values of dissolve oxygen (DO) recorded ranges between 1.06 mg/l to 4.96 mg/l and the maximum (303.00 mg/l ± 0.964) during the winter season in 2005-06 while minimum (1.613 mg/l ± 0.278) during the summer season in 2006-07. Hazealwood and Parker (1961) stated that maximum dissolve oxygen in winter may be due to low atmospheric temperature and minimum dissolve oxygen in summer may be due to high metabolic rate of organisms. Seasonal changes affect dissolved oxygen concentrations. Warmer temperatures during summer speed up the rates of photosynthesis and decomposition. The range of free carbon dioxide recorded is 1.90 mg/l to 9.10 mg/l and the maximum (8.355 mg/l ± 0.991) during the summer season in 2006-07 while minimum (3.113 mg/l ± 0.816) during the winter season,in same study period. Similar findings were recorded by Jain and Seethapati (1996). In the present investigation, the maximum free carbon dioxide in summer is due to the decomposition of organic matter and the respiration of aquatic flora and fauna. The total alkalinity ranged between 26.10 mg/l to 102.03 mg/l and maximum (90.758 mg/l ± 15.029) during the monsoon season in 2006-07while minimum (38.700 mg/

Table 31.3: Seasonal Variation in Physico-chemical Parameters of Kanhala Pond Water during 2005- 06

Parameters	*Winter*	*Summer*	*Monsoon*
	Mean±S.D.	*Mean±S.D.*	*Mean±S.D.*
Temperature atm. (° C)	27.395±3.043	38.373±4.345	36.493±3.445
Temperature water (° C)	23.630±2.512	29.170±3.812	30.858±2.146
pH	7.350±0.456	8.058±0.083	6.985±0.250
Transparency (cm)	29.975±3.497	32.250±1.314	20.698±4.422
Conductivity (μmhos/cm)	191.000±8.832	196.750±23.858	168.750±11.189
D.O. (mg/l)	3.303±0.964	2.803±1.394	2.480±0.660
CO_2 (mg/l)	3.588±1.210	7.953±1.069	3.610±0.439
Total Alkalinity -M.O. (mg/l)	40.275±12.429	54.000±8.276	76.900±5.738
Hardness – Total (mg/l)	95.910±8.542	129.423±25.792	144.513±9.835
Hardness - Ca. (mg/l)	53.700±4.225	81.360±17.293	90.843±5.837
Hardness - Mg. (mg/l)	25.355±16.407	42.170±15.637	53.670±4.358
Chloride (mg/l)	29.985±4.443	38.420±0.879	22.840±4.612
B.O.D. (mg/l)	8.750±0.829	14.250±1.920	11.250±3.345
C.O.D. (mg/l)	8.000±1.414	18.000±1.581	11.250±1.920
Phosphate (mg/l)	0.068±0.011	0.548±0.273	0.073±0.004
Sulphate (mg/l)	1.263±0.119	3.150±0.112	1.950±0.472
Nitrate (mg/l)	0.260±0.090	0.260±0.042	0.370±0.030

l ± 5.456) during the winter season in same study period. In the present investigation, maximum values of alkalinity during monsoon maybe due to household detergent products added in the pond only in monsoon season due to flooding of city channels after their saturations in summer and minimum values during winter and summer is due to the use of bicarbonates by the pond biota directly for their load which leads to the depletion of bicarbonates in water resulting in low values of total alkalinity as reported by Guaraiswamy (1988). Rai (1974) reported that pollution by sewages and its consequent decomposition have been the major causes of increase in bicarbonates alkalinity. The total hardness was recorded in the range of 86.50 mg/l to 168.00 mg/l and the maximum (144.513 mg/l ± 9.835) during the monsoon season in 2005-06 while minimum (95.910 mg/l ± 8.542) during the winter seasons. Kataria *et al.* (1996) also recorded maximum value of total hardness in summer, moderate in monsoon and minimum in winter at Kolar reservoir in Bhopal, Madhya Pradesh. The contents release from dead molluscans shell may also increase the concentrations of total hardness (Khan and Chaudhary, 1994). The calcium hardness was recorded between 50.05 mg/l to 106.11 mg/l and the maximum (90.843 mg/l ± 5.837) was recorded during the monsoon in 2005-06 season while minimum (53.700 mg/l ± 4.225) during the winter season in same study period.

Table 31.4: Seasonal Variation in Physico-chemical Parameters of Kanhala Pond Water during 2006-07

Parameters	*Winter*	*Summer*	*Monsoon*
	Mean±S.D.	*Mean±S.D.*	*Mean±S.D.*
Temperature atm. (° C)	27.663±2.944	38.785±4.187	36.255±3.969
Temperature water (° C)	23.895±2.389	29.440±3.858	31.113±2.772
pH	7.120±0.548	8.000±0.014	7.075±0.669
Transparency (cm)	29.050±3.380	32.075±1.413	20.475±3.369
Conductivity (µmhos/cm)	202.250±24.149	212.250±20.204	175.250±11.519
D.O. (mg/l)	2.750±0.737	1.613±0.278	2.448±0.389
CO_2 (mg/l)	3.113±0.816	8.355±0.991	5.828±0.608
Total Alkalinity -M.O. (mg/l)	38.700±5.456	77.250±12.070	90.758±15.029
Hardness – Total (mg/l)	101.913±9.595	139.860±10.290	140.663±4.723
Hardness - Ca. (mg/l)	58.508±8.535	88.013±11.218	89.578±3.273
Hardness - Mg. (mg/l)	26.685±7.935	42.160±14.019	51.085±2.154
Chloride (mg/l)	24.923±2.666	36.798±1.708	24.288±5.119
B.O.D. (mg/l)	9.250±0.433	15.250±1.920	11.750±3.767
C.O.D. (mg/l)	9.250±2.861	19.250±1.479	15.000±1.871
Phosphate (mg/l)	0.070±0.017	0.093±0.006	0.077±0.016
Sulphate (mg/l)	1.875±1.076	3.693±0.135	2.130±0.672
Nitrate (mg/l)	0.215±0.056	0.448±0.094	0.510±0.073

In the present investigation the magnesium hardness was recorded between 8.45 mg/l to 61.89 mg/l and maximum (53.670 mg/l ± 4.358) was recorded during the monsoon season in 2005-06 while minimum (25.355 mg/l ± 16.407) during the winter season in same study period. In the present investigation, magnesium hardness was minimum during winter and maximum during the summer season. Ravikumar *et al.* (2005) reported maximum calcium hardness in the months of April in Ayyanakere tank in Harapanahalli town in Davangere district of Karnataka. The maximum values of magnesium hardness were due to low water levels and bacterial decompositions of the organic matter in the lake.

The values of chloride recorded were 17.20 mg/l to 39.42 mg/l and the season wise maximum (38.420 mg/l ± 0.879) level was recorded during the summer in 2005-06 while minimum (22.840 mg/l ± 4.612) during the monsoon season in the same study period. According to Ganpati (1960) high concentration of chloride in freshwater is an indicator of organic pollutions since the most important sources of chloride in freshwater is the discharge of domestic and industrials sewage. Pejawar *et al.* (2004) reported low chloride range during rainy seasons and high in summer and further suggested that low chloride specifically observe during rainy season indicates the dilution effect and high in summer due to higher temperature and rapid reduction in water mass.

The biochemical oxygen demand (BOD) values recorded were 7.00 mg/l to 18 mg/l and maximum values were recorded (15.250 mg/l ± 1.920) during the summer season in 2006-07 while minimum (8.750 mg/l ± 0.829) during the winter seasons. Bhatt *et al.* (1999) stated that high values of biochemical oxygen demand was may be due to higher rate of organic decompositions and gradual decline of biochemical oxygen demand from monsoon followed by winter, which could be due to decrease in temperature which in turn retards the microbial activities. The chemical oxygen demand values recorded were 5 mg/l to 21 mg/l and the maximum (19.250 mg/l ± 1.479) was during the summer season in 2006-07 while minimum (8.000 mg/l ± 1.414) during the winter season in 2005-06. However Salve and Hiware (2006) noted maximum values in summer and monsoon and minimum in winter due to fewer amounts of water, death and decay of aquatic flora and fauna and minimum chemical oxygen demand in winter due to settlement and dilution effect. In the present investigations, higher value during summer maybe due to higher decomposition activities and low levels of water. However minimum chemical oxygen demand in winter is due to low temperature, low decomposition activities and dilution effect.

The phosphate level was recorded between 0.045 mg/l to 0.8 mg/l and the maximum (0.548 mg/l ± 0.273) during the summer season while minimum (0.068 mg/l ± 0.011) during the winter seasons. The high values of phosphate during summer seasons are probably due to rapid evaporation and mineralization of decomposed materials in pond water. The main source of phosphorous in natural water comes from weathering of phosphate bearing rocks and leaching of soil in the catchment area (Jhingran, 1982). The concentration of phosphate however may increase in the water by contaminations or by receiving drainage from agriculture surface runoff (Devi, 1985). The values of sulphate recorded were 1.0 mg/l to 3.80 mg/l and the maximum (3.693 mg/l ± 0.135) during the summer season in 2006-07 while minimum

(1.263 mg/l ± 0.119) during the winter season in 2005-06. The higher concentration of sulphate recorded in summer is probably due to the activity of biodegradation (Munawar, 1970) whereas dilution and utilization of sulphate by the aquatic plants gradually decreased the sulphate concentration in monsoon followed by winter. The values of nitrate range between 0.11 mg/l to 0.59 mg/l and the maximum (0.510 mg/l ± 0.073) was recorded during the monsoon season in 2006-07 while minimum (0.215 mg/l ± 0.056) during the winter season in same study period. Shrishail and Mathad (2008) recorded maximum nitrate during the monsoon and minimum during the winter in Khaji Kotnoor reservoir in Gulbarga. Most of the unpolluted sources of water are deficient of nitrates because it exists only in few natural sources (Trivedi and Goel, 1984). Domestic sewage may be the principal contributor of nitrogenous substance in a waterbody. Small quantity of nitrogen is enough for rapid growth of blue green algae in a water body (Singh and Swarup, 1979).

The seasonal physico-chemical parameters of Kanhala pond exhibit cyclical variations. The higher value of conductivity, total alkalinity, chloride, phosphate and nitrate in Kanhala pond indicates its polluted nature and it exhibits a slightly polluted status.

References

Ade, P.P. 2001: *Limnological studies of Amravati University reservoir with reference to trophic status and conservation*. Ph.D. Thesis, Amravati University, Amravati.

Adoni, A.D. 1985: *Work book on Limnology*, Dept. of Environment, Govt. of India, Bandana printing service, New Delhi.

APHA. 2005: *Standard Methods for Examination of Water and Wastewater*. 21st Edn. Washington D.C.

Bartholow, J.M., Hanna, R.B., Saito, L.D. Lieberman and Horn, M. 2001: Simulated limnological effects of the Shasta Lake temperature control device. *Environmental Management* 27(4):609-627.

Bhatt, L.R., Lacoul, P., Lekhak, H.D. and Jha, P.K. 1999: Physico-chemical characteristics of Taudaha lake, Kathmandu. *Poll. Res.* 18(4): 353-358.

Devi, M.J. 1985: Ecological studies of the Limnological plankton of three water bodies of Hyderabad. Ph.D. Thesis Osmania University.

Ganpati, S.V. 1960: Ecology of tropical waters. *Proc. Symp. Alga*. 200-218.

Guraiswamy, K. 1988: *Limnological studies on certain freshwater bodies of Salem*. M.Phil. Disser., Madras Univ., India.

Hazelwood, D.H. and Parker, R.A. 1961: Population dynamics of some freshwater zooplankton. *J. Ecology*, 42:266-274.

Hutchinson, G.E. 1957: *A Treatise on Limnology* Vol. I and II Geography, Physics and Chemistry. John Willey Inc. New York 1015.

Jain, C.K. and Seethapati, P.V. 1996: Limnological studies of kay Amkula Lake. *Indian J. Environ. Protection*. 16: 61-568.

Jakher, G.R. 2002: Variation in different bacterial count in a freshwater pond of Jodhpur, Rajasthan, India. *Oikoassay.* 15 (1 and 2): 1-5.

Jakher, G.R. and Rawat, M. 2003: Studies on physico-chemical parameters of a tropical lake, Jodhpur, Rajasthan, India. *J. Aqua. Biol.* 18(2): 79-83.

Jingran, V.G. 1982: *Fish and fisheries of India*, Hindustan Pub. Corp. New Delhi, India. 268-269 pp.

Kataria, H.C., Iqbal, S.A. and Chandilya, C.B. 1996: Limnochemical studies of Tawa reservoir. *Indian J. Enviro. Prot.* 16(11): 841-846.

Kedar, G.T., Patil, G.P. and Yeole, S.M. 2007: Rotifer biodiversity of Yedshi lake, Maharashtra. *J. Aqua. Biol.* 22(1): 8-12.

Kedar, G.T. and Patil, G.P. 2011: Study on seasonal fluctuation in physico-chemical parameters of Rishi lake, Karanja (Lad), Dist.Washim, Maharashtra. Proc. Nat. Conf. Environ.l Biology and Biodiversity, Innani College, Karanja lad, (MS), pp.120-122.

Khan, M.A.G. and Chaudhary, S.H. 1994: Physical and chemical limnology of lake Kapta, Bangladesh. *Trop. Ecol.* 35(1): 35-51.

Meshram, C.B. 1996: *Limnological studies of Wadali lake, Amravati.* Ph.D. Thesis, Amravati University, Amravati.

Munawar, M.C. 1970: Limnological studies on freshwater ponds of Hydrabad-1. The Biotope. *Hydrobiologia.* 25 (1): 127-162.

Narayana, J., Purushothama, R., Kiran, B.R., Ravindrakumar, K.P. and Puttah, E.T. 2005: Investigation of drinking water quality of Basavanahole Tank with reference to physical chemical characteristics. *Fundamental of limnology*, 201-206.

Pejawar, M. Somani, V. and Goldin Quadros. 2004: Physico chemical parameters of two quarry lakes near thane city, Maharashtra. *J. Aqua. Biol.* 19(1): 107-110.

Rai, H. 1974: Limnological observation on the different rivers and lakes in the Ivory coast, *Hydrobiologia.* 44(2 and 3): 301-317.

Ravi Kumar, M., Manijappa, S., Kiran, B.R. and Puttaiah, E.T. 2005: Hydrochemistry of Ayyanakere Tank in Harpanhalli Town, Davanagere, Karnataka. *J. Aqua. Biol.* 20(2): 117-120.

Salve, B.S. and Hiware, C.J. 2006: Studies on water quality of Wanaprakalpa reservoir, Nagapur near Parali Vaijnath, Dist. Beed, Marathwada region. *J. Aqua. Biol.* 21(2): 113-117.

Shrishail, V.G. and Mathad, P. 2008: Distribution and periodicity of phytoplankton in Khaji Kotnoor reservoir of Gulbarga region, India. *Ecol. Environ. and Cons.* 14(2 and 3): 429-433.

Singh, S.R. and Swarup, K. 1979: Limnological studies of Surha Lake (Ballia)II. The periodicity of plankton. *Indian Bot. Soc.* 58: 319-329.

Singhai, R.N., Swarnjeet, A. and Davies, R.W. 1986: The physico-chemical environment and the plankton of managed pond in Haryana *India. Proc Indian Acad. Sci. Anim. Sci.* 9(3): 353-363.

Sreenivasan, A. 1966: Limnology of tropical impoundments. Hydroriological and fish production in Stanley reservoir, Mettur Dam. *Hydrobiology* 51(2): 295-306.

Trivedi, R.K. and Goel, P.K. 1984: *Chemical and Biological methods for water pollution studies.* Karad Environmental publication, pp. 1-251.

Welch, M.N. 1952: *Limnology,* McGraw Hill Book Co. Inc. NY.pp.538.

Wetzel, R.G. 2001: *Limnology,* W.B. Saunders Co. Philadelphia, pp.743.

2013, Climate Change Effects on Agriculture and Economy *Pages* **275–283**
Editors: **D.R. Khanna, A.K. Chopra, Vikas Singh, Rakesh Bhutiani & Gagan Matta**
Published by: **BIOTECH BOOKS, NEW DELHI**

Chapter 32
Fluoride Toxicity in Populace of Deoli Tehsil of Tonk District (Rajasthan) and its Effect on Health

***K.S. Meena*[1], *R.K. Gunsaria*[2] *and Kanta Meena*[3]**
[1]*Department of Chemistry,* [3]*Department of Botany, M.L.V. Govt. College, Bhilwara, Rajasthan*
[2]*P.G. Department of Chemistry, Govt. P.G. College, Tonk, Rajasthan*

Fluoride is universally present in the earth's surface and in water in trace concentration. If fluoride level is higher than 1.5 ppm than it is toxic and causes dental and skeletal fluorosis, decalcification, digestive and nervous disorders. This study aims to identify the toxic effect of fluoride on human health. For this purpose 130 samples were collected from different villages of Deoli Tehsil (Tonk), Rajasthan during the month of May-June, 2009 in clean polyethylene bottles. The fluoride concentration along with physico-chemical parameters in drinking water samples was determined by using standard techniques.

The result revealed that populace of 80 per cent villages are affected with high concentration of F^- lower in 14 per cent villages, However 6 per cent villages contained optimum limit of F^- concentration. Permissible limit for F^- concentration is 1-1.5 ppm according to WHO (1996). People living in these villages are dangerously affected with fluorosis. They become old in short age. Their joints are almost finished after 30-40 age. Removal of fluoride from drinking water is suggested through various defluoridation techniques available including quick reverse osmosis, electrodialysis and hit and trial method.

Keywords: *Fluoride toxicity, Fluorosis, Defluoridation, Groundwater.*

Introduction

Fluoride concentration is an important aspect of hydrogeochemistry because of its impact on human health. India has 14.1 per cent of total fluoride deposits on the earth's crust. It is not surprising, therefore, that the fluorosis is endemic in 17 states of India (UNICEF, 2002). Research has shown that fluoride concentrations between 0 and 0.5 mg/l favor dental caries development, whereas concentrations between 1.5 and 5 mg/l can result in dental fluorosis. Ingestion of 5– 40 mg/day of fluoride *via* drinking water can produce skeletal deformities and knock knees have been reported in adolescents receiving >10 mg/day in water, accumulated from birth. However, fluoride contents between 0.5 and 1.5 mg/l have a beneficial effect, reducing caries development (WHO, 1996). Thus, fluoride concentration in drinking water is a like two-edged knife; both its higher and lower concentration is not safe. The Bureau of Indian Standards (BIS) prescribed a limit between 1.0 and 1.5 mg/l whereas USPHS has set a range of allowable concentrations for fluoride in drinking water for a region depending on its climatic conditions because the amount of water consumed and consequently the amount of fluoride ingested being influenced primarily by the air temperature. Accordingly, the maximum allowable concentration for fluoride in drinking water in Indian conditions comes to 1.4 mg/l, while as per Indian standard it is 1.5 mg/l. There is also evidence that the adverse health effects of fluoride are enhanced by a lack of Ca, vitamins and protein in the diet. In response to the potentially

Table 32.1: Drinking Water Standards Prescribed by different Agencies

Parameter		BIS	WHO	MUD	MWH	ICMR
pH	Desirable Limit	6.8–8.5	7.0–8.5	7.0–8.5	7.0–8.5	7.0–8.5
	Max. per limit	No Relaxation	6.5–9.2	6.5–9.2	6.5–9.2	6.5–9.2
Fluoride	Desirable Limit	1	0.7	1	–	–
	Max. per limit	1.5	1.5	1.5	–	–
Total Dissolved	Desirable Limit	500	500	500	–	500
Solids	Max. per limit	2000	1500	1500	–	1500
Chloride	Desirable Limit	250	200	200	200	200
	Max. per limit	1000	600	1000	1000	1000
Nitrate	Desirable Limit	45	45	45	45	20
	Max. per limit	100	–	100	45	50
Ca Hardness	Desirable Limit	75	75	75	–	75
	Max. per limit	200	200	200	–	200
Mg Hardness	Desirable Limit	30	30	30	–	50
	Max. per limit	150	150	150	–	150
Alkalinity	Desirable Limit	200	–	–	–	–
	Max. per limit	600	–	–	–	–

All value are in mg/L except pH.

BIS: Bureau of Indian Standard; ICMR: Indian Council of Medical Research; WHO: World Health Organization; MWH: Ministry of Works and Housing; MUD: Ministry of Urban Development.

harmful effects of high-fluoride waters, the World Health Organization (WHO, 1996) has set an upper drinking water quality guideline of 1.5 mg/l. Conversely, the WHO also recommends intakes of water containing 0.5–1.0 mg/l for the prevention of dental caries (Table 32.1).

Fluoride is available in nature mainly in the minerals such as fluorite and cryolite and in separate combination with sodium, magnesium and aluminum. Fluoride frequently occurs in igneous as well as in metamorphic rocks. Especially in alkali rocks, Granite, Basalt, Shales, Clays and Calcium Phosphate rocks are the main sources of fluoride. The dead remains of invertebrates in the form of bones and teeth settle to the bottom of lakes and rivers form where it gradually dissolved and recycled in nature. Thus, fluoride concentration of surface sources on account of natural forces rarely exceeds 1.0 mg/l. The availability of fluoride in groundwater is primarily connected with the presence of minerals in the groundwater.

The fluoride level studies of groundwater have been carried out in various countries. Ncube and Schutte (2005) identified areas of high fluoride concentration in ground of South Africa and reported dental fluorosis. Webber (2009) published pollution of drinking water with fluoride by artical fluoridation. Alvarado *et al.* (2010) have studied fluoride and dental fluorosis in student of Tula de Allende Hidalgo of Mexico. Farfan *et al.* (2011) discussed fluoride consumption and its impact on oral health. Susheela (1999) have studied fluorosis management programmed in India. In this study some point involves such as present status of fluorosis in India, types of fluorosis, remedies and recommendations. Jadeja *et al.* (2005) studied the physico-chemical and bacteriological characteristics of groundwater at Dharmpur Industrial Area in Porbandar city were rated as unacceptable for their taste on the basis of total hardness. Results show that the groundwater of Dharmpur Industrial area, Porbandar City is suitable for drinking purpose, subject to proper disinfection to ensure health of population. Dutta *et al.* (2010) studied monitoring of fluoride concentration in groundwater of small tea gardens in Sohitpur district of Assam. Rajasthan state is thought to be most seriously affected by high fluoride. All 33 districts are endemic for fluoride. Choubisa (2001) reported endemic fluorosis in Southern Rajasthan. Sharma *et al.* (2005) studied fluoride contamination in groundwater in Agra district (U.P.), fluoride analyzed by fluoride ion-selective electrode method. Sharma and ojha has studied (2001) groundwater quality of an industrial town Bhilwara. Seth *et al.* (2005) studied Geochemical of fluoride in groundwater of Rajasthan. Sharma *et al.* (2005) studied quality status of groundwater of Sanganer tehsil in Jaipur district. Tailor and Chandel (2010) studied quality of groundwater of Malpura teshil, Tonk district, Rajasthan. Yadav and Khan (2010) reported fluoride and fluorosis status in groundwater of Todaraisingh of Tonk district, Rajasthan. Most people of Todaraisingh area are suffering from dental and skeletal fluorosis such as mottling of teeth.

Materials and Method

Deoli is a city and a municipality in Tonk district in the state of Rajasthan. Deoli is situated 971 ft above sea level on the National Highway No. 12 and 111 kms south to Jaipur. It is located between co-ordinates 25° 58' N 73° 62' E. It has an average elevation of 296 meters.

Deoli is bounded on the north by Jaipur, south by Bhilwara and Bundi, east by Sawaimadhopur and west by Ajmer. Among the famous temples, there are temple of Muchkundeshwar, Mahadev temple nearby Mandakala, 'Doonija Mata' temple is situated in Dooni Lord, Gokarneswar temple, Chandali Mataji and Bijashan Mataji temple.

Sampling

130 samples of groundwater were collected from fluoride affected villages of Deoli Tehsil in clean polyethylene bottles from different sources *viz.* hand pumps, open wells, tube wells and PHED water supply. Water quality parameters pH, nitrate, TDS, chloride, EC, CO_3^{-2}, HCO_3^- were analysed by conventional techniques. Na^+ and K^+ were analyzed by Flame photometer. Ca-H, Mg-H, TH were analyzed by EDTA-Titrimetric methods.

Fluoride Ion-Selective Electrode Method

Apparatus

Ion-Selective Meter, Fluoride Electrode, Magnetic Stirrer

Reagent

Fluoride Standards of various ranges (0.2-20ppm), Fluoride Buffer (TISAB-Total Ionic Strength Adjustment Buffer).

Procedure

Firstly, instrument was calibrated and 10 ml of sample and 10 ml of buffer solution was taken in a beaker. Stirring bar and immerse electrode was put and solution was stirred until reading became constant and then electrode was withdrawn.

Results

Groundwater samples collected in the Deoli Tehsil were clear without any visible color, order and turbidity. The fluoride range in different samples are shown in the (Table 32.2) Fluoride is a geochemical contaminant and natural sources account for much of the concentration of fluoride in these waters depends principally on the solubility of fluoride containing rocks with the water is in contact. Fluoride containing in the present study ranged from 0.26 to 9.60 ppm. This fluoride concentration is very high and harmful for human health which causes fluorosis diseases. Lowest value of 0.26 ppm was found in Durga mata Temple, Duni village and highest value 9.60 ppm was found in Akodia village (Figure 32.1).

The data revealed that 80 per cent villages are affected with high concentration of F^- Lower in 14 per cent villages, However 6 per cent villages containe optimum limit of F^- concentration (Table 32.2 and Figure 32.2). Permissible limit for F^- concentration is 1-1.5 ppm according to WHO (1996). In general, relatively high pH conditions have a tendency to displace fluoride ions from the mineral surface. From the above it is clear that relatively high alkalinity favors high fluoride concentration in groundwater of Deoli Tehsil.

Table 32.2: Showing Permissible Limit and Percentage of Villages of Deoli Tehsil

Parameters	*Permissible Limit*	*Villages per cent*		
		Below	*Optimum*	*Higher*
pH	6.9 - 9.2	–	100 per cent	–
F-	1 - 1.5 ppm	14 per cent	6 per cent	80 per cent
EC	300 µmhos/cm	–	–	100 per cent
TDS	500-1500 mg/l	5.38 per cent	70 per cent	24.62 per cent
Ca-H	75-200 mg/l	10 per cent	48 per cent	42 per cent
Mg-H	30-150 mg/l	–	92 per cent	8 per cent
TH	100-500 mg/l	0.76 per cent	95.38 per cent	3.84 per cent
Cl^-	200-600 mg/l	48.47 per cent	46.15 per cent	5.38 per cent
Alkalinity	200 mg/l	3.85 per cent	–	96.15 per cent
Na^+	50 - 60 mg/l	26.15 per cent	13.07 per cent	60.76 per cent
K^+	20 mg/l	100 per cent	–	–
No_3^-	40 - 50 mg/l	57.70 per cent	10 per cent	32.30 per cent

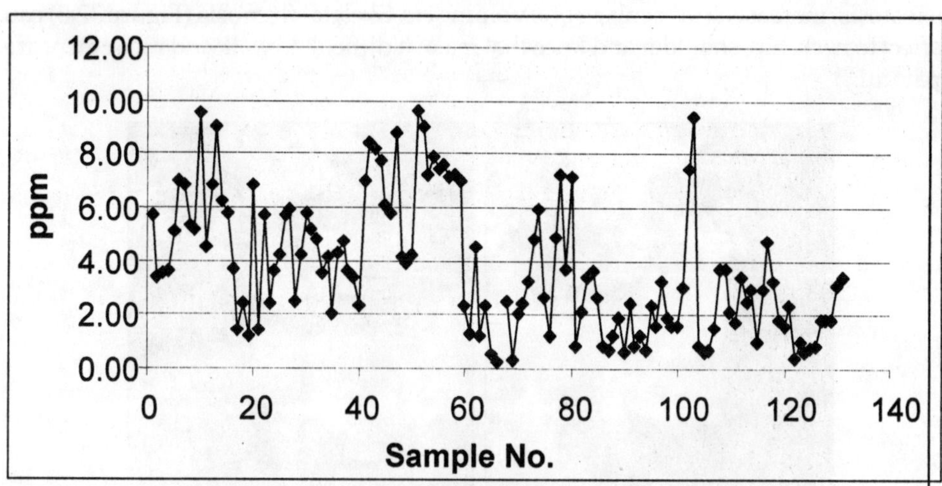

Figure 32.1: Fluoride in Villages of Deoli Tehsil

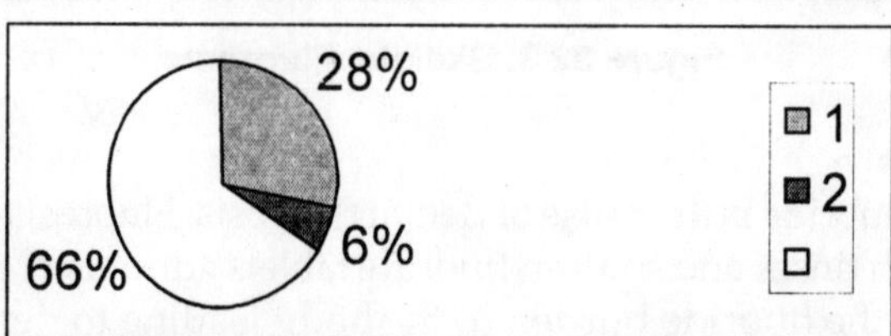

Figure 32.2: F per cent of Villages of Deoli Tehsil

According to fluoride concentration 20 per cent villages have quality fit for drinking purpose. In study area villages are dangerously affected with fluoride concentration having fluoride more than 1.5 mg/l. These villages are Akodia, Gadigram, Poliyada, Poliyadi, Mundia Khurd, Sawatghard, Sarakawas, Digariya, Kawarpura, Alegendarpura, Benediya, Khurd, Thawala, Dalawasa, Dahaber Khurd, Ramthala, Dabarkala, Bijawad, Baida, Panwer, Gopipura, BasLaxamana, Ambapura, Ruparail, Rathanpura, Dhakadiyawas, Niwariya, Kawarpura, Balapura, Shrinagar Bhimpura, Sagarapura, Syawata *etc*.

Fluoride Toxicity and Human Health

Fluoride element is the most exclusive bone seeking anion to the affinity for calcium phosphate. In human body 99 per cent of fluoride is present in bone. Fluoride is beneficial and detrimental for the human health. High amount of fluoride effected bone formation, cripping, fluorosis mainly manifests in three forms – skeletal, dental and non- skeletal.

Skeletal Fluorosis

The effect of fluoride on bones and skeletal system depends on the type of bone and skeletal on the bone and its inorganic and organic constituents. A fluoride bone shows characteristic structural changes like- increased bone mass and density, exostosis, increased trabecular bone volume *etc*. Skeletal fluorosis (Figure 32.3) can affect to neck, hip, shoulder and knee that make it difficult to walk and movement are painful.

Figure 32.3: Skeletal Fluorosis

Dental Fluorosis

Ingestion of fluoride is the cause of dental fluorosis. Fluorosis in drinking water, tooth paste, mouth rinses and sodium fluoride tablets administered on prescription, can contribute to the fluoride burden of the body leading to dental fluorosis. The discoloration of the teeth may progress from white, yellow, brown and black. The discoloration might is in spots or as streaks invariably horizontally during tooth

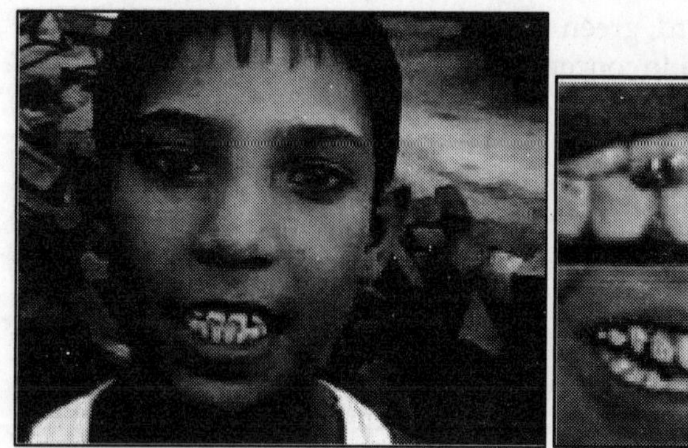

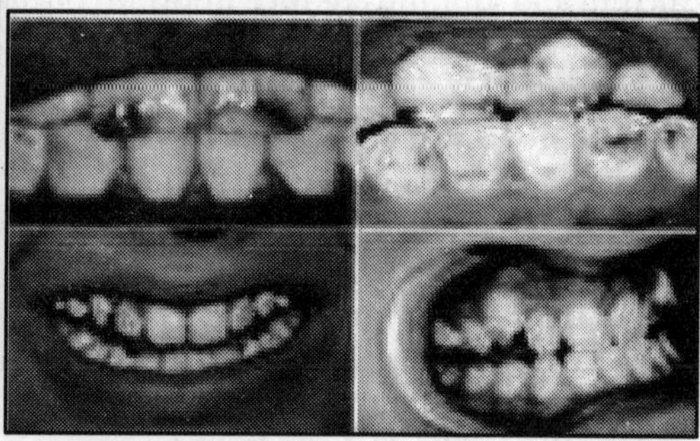

Figure 32.4: Dental Fluorosis

development. The natural shine or luster of the teeth disappears. Dental fluorosis affects both the inner and outer surface of the teeth (Figure 32.4). The disease has most cosmetic implications and has no treatment.

Non-Skeletal Fluorosis

The soft tissues of the body may be affected by excessive consumption of fluoride. The symptoms are pain in joints without visible signs of fluid accumulation, pain in abdomen, muscle weakness, fatigue, anemia with very low hemoglobin in level, neurological manifestations. Cardiac problems may arise due to cholesterol production. Repeated abortions or still birth, male infertility due to sperm abnormalities are also some of the complications.

Discussion

The above cited results conclude that the drinking water quality of Deoli Tehsil is very poor because there are few villages where the water is fit for drinking purpose with all of the parameters under permissible limit, rest villages have very poor drinking water where one or more parameters are more than permissible limit. People living in these villages are dangerously affected with fluorosis. They become old in short age. Their joints are almost finished after 30-40 age. They are using belts and bandages for their joints. Most of the persons above 50 years are knelt down on their knees. There teethes are in sympathetic conditions. In Deoli Tehsil water quality is very poor because of dry region, less rainfall, low groundwater level, less awareness about pure water in people, scarcity of alternative water sources and ignorance of government machinery and high concentration of minerals in soil.

This is the main problem of Deoli Tehsil many people are affected and suffering with the problem of pure drinking water. 20 to 30 persons are affected with fluorosis in each village. Back bone, joint and teeth are highly affected with fluorosis. Many old persons are knelt double on their knees. As per medical science do not have proper treatment of fluorosis. They recommend pain killers to the fluorosis patient, but that is not a perfect and permanent solution. The education in populace of study area is low so that fluorosis awareness is discussed. It is also recorded that poor populace are affected with fluorosis because of their low nutrient food. The person who used calcium rich food such as milk, curd, green vegetable *etc.* are less affected because calcium and vitamins reduce fluoride concentration in body.

Conclusion

The populace of study area was found affected with skeletal fluorosis, bone fluorosis and dental fluorosis. The presence of excessive quantity of fluoride in drinking water is accompanied by a characteristic sequence of changes in teeth, bone and periarticular tissues. Therefore drinking water of Deoli Tehsil is not potable. Proper treatment of groundwater is suggested prior to its use for drinking purpose. Removal of fluoride from drinking water is suggested through various defluoridation techniques available including quick reverse osmosis, electrodialysis and hit and trial method. The Nalgonda technique is an economical way for defluoridation. Activated alumina technology may be used which is based on ion exchange resin. Public awareness and health education are most important measure to be widely adopted. This can be done by using audio-visual aids, seminars, conferences, symposium and training programme, N.G.O. must be encouraged in such programmed for public welfare. Since these people are dependent on groundwater for domestic use, remedial measures such as importing of drinking water and rain water harvesting are needed. Nutrional diet such as calcium and phosphorus rich food should be recommended to those affected with fluorosis as it decreases rate of accumulation of fluoride in the human body.

Acknowledgement

Authors are thankful to Dr. Surendra Pal Verma, Jr. Chemist PHED Deptt. Rajsamand, Rajasthan and Mr. S.L. Meena Principal Govt. P.G. College Tonk, Rajasthan.

References

Alvarado, P.V., Gareia, F.P., Oilvares, C.C., Martinez, A.J.G., Espinosa, R.M.O. 2010: *J. Toxi and Environ. Helth.*, 2(3), 24.

APHA, 1989: Standards Methods in the Examination of Water and Wastewater, 17th ed. Washington, D.C.10.

WHO. 1996: *Anonymous, Guidelines for drinking water quality*, (2), 231.

BIS. 1991: Bureau of Indian Standards, Drinking Water Specification, IS: 10500.

Choubisa, S.L. 2001: *Research Report fluoride*, 34(1) 61.

Dutta, J., Nath, M., Chetai, M. and Misra, A.K. 2010: *Correlation Physico-chemical parameters"* 2(2), 1199.

Farfan, M.D.J., Guerrero, J.C.H., Lopez, L.A.J., Aleman, F.J. and Hernandez, J.D.F. 2011: *Int. J. Environ. Res. Public Health*, (8), 148.

Jadeja, B.A., Odedra, N.K. and Thakur, M.R. 2005: *Plant Archives*, 6(1), 341.

Ncube, E.J. and Schutte, C.F. 2005: *Water SA*, 31 (1), 35.

Seth, G., Kumar, A. and Samota, M.K. 2005: *CAIJ*, 2(6), 191.

Sharma, B.S., Agrawal, J. and Gupta, A.K. 2011: *Asian J. Exp. Biol. Sci.*, 2(1),131.

Sharma, J.D., Jain, P. and Sahu, D. 2005: *Nature Environ. and Pollution Technology*, 4(2), 203.

Sharma, K.C. and Ojha, K.G. 2001: *Journal of Environmental and Pollution*, 8(1), 109.

Susheela, A.K. 1999: *Current Science*, 77(10), 1050.

Tailor, G.S. and Chandel, C.P.S. 2010: *Nature and Science*, 8(11), 20.

UNICEF Report, 2002: Published by Rajasthan Patrika Press.5.

Webber, J.T. 2009: *"Fluoride compounds" Australian Fluoridation news*, 45(1).

Yadav, A.K. and Khan, P. 2010: *Int. J. Chem. Environ. Pharm. Res.*, 1(1), 6.

2013, Climate Change Effects on Agriculture and Economy Pages 285–289
Editors: D.R. Khanna, A.K. Chopra, Vikas Singh, Rakesh Bhutiani & Gagan Matta
Published by: BIOTECH BOOKS, NEW DELHI

Chapter 33

Textile Industry: The Major Cause of Environmental Pollution

Sapna Pathak and Anita Rani
College of Home Science, G.B. Pant University of Agriculture & Technology, Pantnagar, Uttarakhand

Pollution is defined as an unfavourable alteration of the environment with the effect of changes in energy pattern, radiation level, chemical and physical constitution or the abundance of organism. Pollution dealing with atmospheric changes is called air pollution; one which contaminates water, irrespective of it being used for drinking or industrial purposes is called water pollution. The textile industry continuously involve in increasing air pollution, water pollution and soil pollution. Above these three types of environmental pollution, water pollution is major concern in textile industry owing to the consumption of large amount of water in wet processing.

Keywords: Pollution, Textile industry, Industrialization, Environment, BOD, COD.

Introduction

The extensive use of natural resources to support industrialization is also putting tremendous pressure on ecology and ecological balance. The interaction of the atmosphere, lithosphere, hydrosphere and biosphere has continued for years together. Unfortunately, on account of the various activities of man, the composition and complex nature of environment get changed. Such activities include industrialization in different sectors construction and transportation etc. These activities although desirable for human development and welfare, lead to generation and release of objectionable materials in to environment resulting in ecological imbalance. Since World War-II and the birth of 'chemical age' water quality has been heavily impacted widely by industrial and agricultural chemicals. Between these two major areas of development the textile industry is one which exerts the pollution load on the environment.

Pollution is defined as an unfavourable alteration of the environment with the effect of changes in energy pattern, radiation level, chemical and physical constitution or the abundance of organism. Pollution dealing with atmospheric changes is called air pollution; one which contaminates water, irrespective of it being used for drinking or industrial purposes is called water pollution. The pollution caused by water after use is called wastewater or effluent also causes pollution. Other causes of pollution were due to solid waste, hazardous waste or biomedical waste which is rising in alarming proportion. The textiles industry contributes to pollution in various ways starting from production till its disposal.

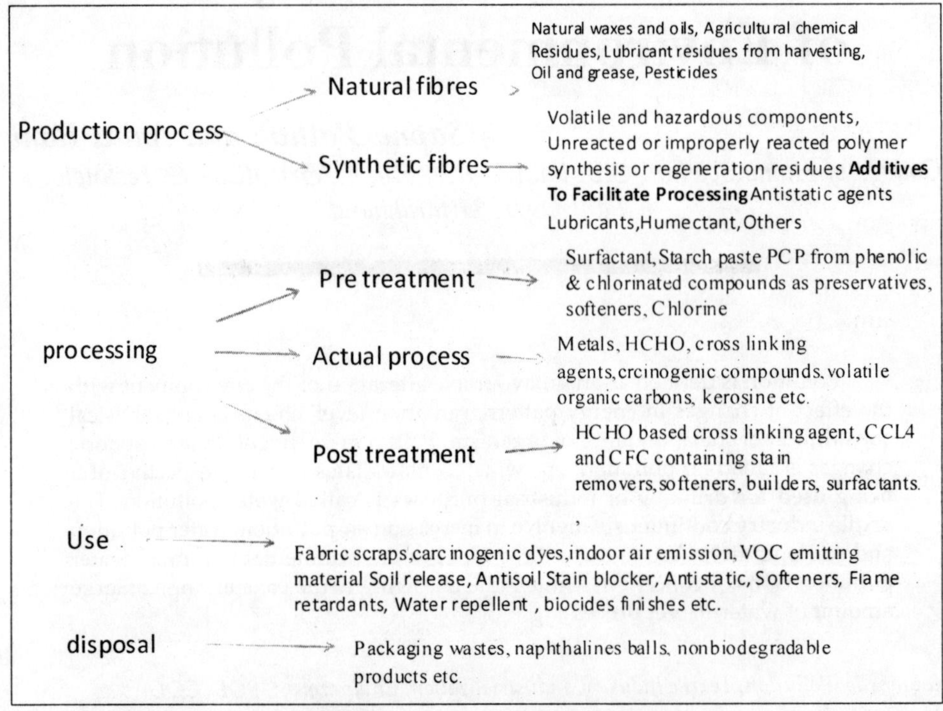

Figure 33.1: Pollution from Textiles Industry

The contaminants found in natural fibers along with a brief description of the resulting pollution problems. For example, waxes and oils from animal-derived fiber can contribute to BOD, COD as well as fats, oil and grease (FOG), and pesticide residues can contribute to aquatic toxicity. Metals can accumulate in sludges or in the treatment system itself, causing potential longterm problems. Many fiber contaminants are metal and organics, which are undesirable in wastewater and can cause air pollution. Some of these impurities are produced during polymerization of synthetic fibers, while others are intentionally added to control the surface and electrical properties of fibres can interfere with downstream processing and quality, and create pollution problems. Effluents pose disastrous effect on environment on the environment. Therefore the major industries which involve in producing these effluents. These are as follow:

- ☆ Soap and detergent industry
- ☆ Synthetic resin industry
- ☆ Textiles processing industry
- ☆ Viscose rayon manufacturing industry.

The textile industry continuously involve in increasing air pollution, water pollution and soil pollution. Above these three types of environmental pollution, water pollution is major concern in textile industry owing to the consumption of large amount of water in wet processing.

Table 3.1: Classification of Textiles Effluent-According to EPA (1996)

Sl.No.	*Waste Type*
1.	Hard to treat waste
2.	Hazardous or toxic waste
3.	High volume waste
4.	Dispersible waste

The waste stream released from the industry is called effluent.The wastewater stream which emerges out after use from industries is called as a industrial effluent, such effluent have no definite composition, as anything which is not required is carelessly dumped in to its stream. The quality of water depends upon its source and history. The history of water signifies the terrain through which water is flowing and to which extent it is contaminated by various ways impurities, sediments and industrial waste.

Table 3.2: Effluent Quality Ascertain by

Sl.No.	*Physical Examination*	*Chemical Examination*	*Biological Examination*
1.	Colour and odour	Acidity and alkalinity	Presence of pathogenic bacteria
2.	Turbidity	DO	*E. coli*
3.	Temperature	BOD	–
4.	Hardness	COD	–
5.	Solid	Nitrogen	–
6.	–	Phosphorous	–
7.	–	Hg	–
8.	–	Cu	–

Liquor or wastewater released from pre treatment processes are heavily loaded with sizes and fibre impurities. Waste from continuous finishing are low in volume, acidic in nature and pH. The effluent from dyeing and printing are strongly coloured and COD was exceptionally high for the carrier dyeing of polyester with intermediate value of vat, sulphur or reactive dyeing on cotton and conventional disperse dyeing of polyester or nylon. These industries utilize pure water for there processing and then out let the water with wetting agent, soda ash, peroxides, sodium hypochlorides, bleaching powder, common salt, dye stuff, soap oil, fixing and finishing agents *etc*.

This wastewater is very harmful to the environment as well as flora and fauna in both ways either by direct discharge of effluent or by disposing the sludge which is obtained after the effluent treatment. The purpose of wastewater treatment is to remove the contaminants from the effluent so that the treated water can meet the acceptable quality standards. The quality standards usually depend upon whether water will be reused or discharged into the receiving stream.

The general effluent treatment approach consists of following three stages of treatment:

1. Primary Screening

   Equalization → Neutralization → Coagulation → Floatation
2. Secondary Treatment
   - ✰ Aerated Lagoons
   - ✰ Trickling Filtration
   - ✰ Activated sludge
   - ✰ Multimedia filtration
3. Tertiary Treatment
   - ✰ Chemical coagulation
   - ✰ Chemical precipitation
   - ✰ Dialysis
   - ✰ Activated carbon adsorption

   The hazardous and toxic waste are non-biodegradable so that some efforts are being made to reduce and minimize the harmful effects that are:

   - ✰ Recycling - Swift Textiles, Inc. is using recycled soda bottles to make denim. The fabric contains 80 percent cotton and 20 percent polyethylene terephthalate, which comes from recycled soda bottles.The empty soda bottles are sorted by color, washed, dried and melted into pellets.
   - ✰ Reuse- The biggest challenge Mount Vernon Mills, Inc. faced was to help improve the operation of a 30-year-old local wastewater treatment facility. Because Mount Vernon Mills supplies almost all of the wastewater to the facility, the company funded the vast majority of an $11 million project to upgrade the facility. The company is trying to improve the quality of wastewater is sends to the treatment facility by using the most environmentally friendly dyes and chemicals and by working with its vendors to see whether some of those dyes can be recycled. In addition, by recycling packaging materials collected from its customers and reusing solid waste that is generated internally, Mount Vernon Mills has reduced the amount of waste it has sent to the landfill

during the last two years by 32 percent. **ATMI's E3 Program: Encouraging Environmental Excellence Report 1995)**

- ☆ Conservation of water and energy

Greatest Liability

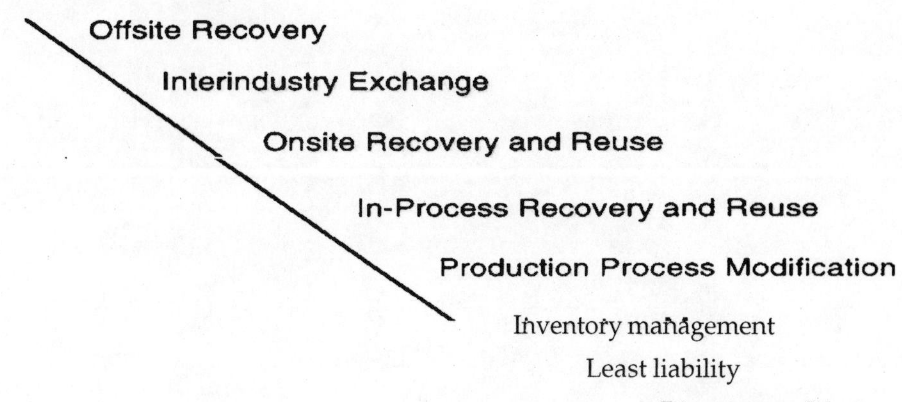

Inventory management

Least liability

Diagramme: Waste reduction hierarchy

References

Sasmira Development for Dye Bath Reuse for Reduced Dyeing Costs. *The Synthetic and Art Silk Mills' Research Association Sasmira Marg, Worli Mumbai – 400 025page-* 4,5,6

Best Management Practices for Pollution Prevention in the Textile Industry, 1996: U.S. Environmental Protection Agency, Office of Research and Development, National Risk Management Research Laboratory, Center for Environmental Research Information,Cincinnati, Ohio September. page:242-243, 299.

Menezes, E. 2003: Textile Trends page: 27-32.

Khopakar, S.M. environmental pollution monitoring and control, page-3, 19, 269, 271.

Rao, C.S. Environmental Pollution Control Engineering, page.295- 305.

2013, Climate Change Effects on Agriculture and Economy *Pages* ***291–301***
Editors: **D.R. Khanna, A.K. Chopra, Vikas Singh, Rakesh Bhutiani & Gagan Matta**
Published by: **BIOTECH BOOKS, NEW DELHI**

Chapter 34
Microbial Water Quality Assessment of Yamuna River

***Manish Dubey*[1]*, Ankur Gupta*[2]*, Surabhi Mahajan*[2] *and R.K. Gautam*[3]**

[1]*Microbiology Research Laboratory, Department of Botany, R.B.S. College, Agra, Uttar Pradesh*
[2]*Department of Microbiology,* [3]*Department of Zoology, School of Life Sciences, Dr. B.R.A. University, Khandhari Campus, Agra, Uttar Pradesh*

The present study aims to evaluate of the microbial quality of Yamuna water at Agra region. Water samples were collected from different location of Agra city, during the period from March, 2010 to June, 2010. The water samples were analyzed for physiochemical parameter such as temperature, TDS, resistance, which are commonly encountered in the water of Yamuna River. The results of physicochemical parameters showed that the temperature values varied from 18.9 – 37.4 °C and TDS from 200–700 mg/ml, whereas the turbidity varied from 4 – 52 NTU and pH values from 6.1 to 9.2. The bacteriological analysis showed that TVBCs ranged from 8.8 x 10^7 to 152 x 10^7 cfu/ml at 37 °C.The cumulative resistance of the *E. coli* isolates to the antibiotics ranged from 6.6 – 33.3 per cent. Among the 8 antibiotics tested, three patterns of antibiotics ranged from one to three. The results indicate that persistent use of antibiotics against human diseases and other life forms may pollute the Yamuna water and their impact on developing antibiotic resistant *E. coli* may become a serious threat to both health and environment.

Keywords: Microbial water quality, Antibiotic resistance, Escherichia coli, Yamuna river.

Introduction

In all known forms of life continuation, water is an essential substance of the natural sources. The spoilages of the water quality and water's natural balance in its environment are known as water pollution (Akman *et al.*, 2000). The aquatic ecosystem consists of several components which are directly or indirectly affected by pollution (Kosygin *et al.*, 2007). The pollution of a particular water body can always be linked to an industry, sewage or agricultural runoff (Subramanyam and Sambamurty, 2006; Sathware *et al.*, 2007). Before operating the water refinement facilities in some developed or developing countries, many rivers are being used as a place to discharge urban wastewater. Wastewater is a source of serious public health problems because it contains pathogenic bacteria and toxic substances (Toroglu *et al.*, 2006).

The microbiological criterion for water quality has been directed towards protections of consumers from possible microbial pollution which may cause public health hazards. Therefore, the bacteriological standards for drinking water are based mainly on total coliform, fecal coliform, fecal streptococci and heterotrophic densities (Geldriech, 1982).

Another problem of prime concern to public health is the presence of antibiotic resistant bacteria in drinking water. Bacterial antibiotic resistance may arise by spontaneous mutation or extra chromosomal inheritance in man and other animals brought about by selective pressure of antibiotics used in therapy and prophylaxis or for growth promotion (Linton, 1977).

Antibiotic resistance among bacteria is a worldwide problem. The emergence of bacteria resistant to most of the commonly used antibiotics/drugs is of considerable medical significance (Khan and Malik, 2001, Leeja and Thoppil, 2007) because of the public health implications (Lateef *et al.*, 2005). Industrialization and man's activities have partially or totally turned our environment to dumping sites for waste materials. As a result, many water resources have been rendered unwholesome and hazardous for man and other living systems (Bakare *et al.*, 2003; Achudume and Olawale, 2007).

River water and natural sources are under considerable threat due to rapid ascending population, developing technology and domestic and agricultural pollutions. The aim of present study is to investigate the microbiological levels due to the domestic and agricultural pollution of Yamuna water. To investigate the relative frequency of different types of bacteria isolated from Yamuna water. To identify different bacteria isolated from Yamuna water at Agra region. To examine the occurrence of multiple antibiotic resistance in different bacteria isolated from Yamuna water.

Materials and Method

Water samples were collected during lean season on monthly basis from seven stations at Agra region, for a period of four months from March, 2010 to June, 2010.

Sampling Sites

- ☆ Station: 1 – Moksh dwar – behind Tajmahal Tajganj, Agra
- ☆ Station: 2 – Balkeshwar ghat – Balkeshwar, Agra

- ☆ Station: 3 – Dashara ghat – Tajganj, Agra
- ☆ Station: 4 – Poiya ghat – Dayalbagh, Agra
- ☆ Station: 5 – Moti Mahal Dhobi ghat, Agra
- ☆ Station: 6 – Mehtab bag, Agra
- ☆ Station: 7 – Near Aram bag Rambagh, Agra

Water Sample Collection

Water samples were collected using sterile screw-capped glass bottles and stored at 4°C until analysis in the laboratory within 2hr after collection (Toroglu *et al.*, 2005).

Physico-chemical Characters

Some environmental parameters *e.g.* temperature, TDS, turbidity and pH were measured in the field.

Microbiological Analysis

Microbiological analyses of water samples were made by standard methods outlined by (APHA, 1985 and Khanna, 1993). The total viable bacterial counts (TVBCs) at 37.0 °C was determined using the standard plate count.

Identification of Some Bacteria

Preliminary identification of strains obtained in pure culture was based on gram staining, respiration, fermentation tests and biochemical tests (IMViC test) (Collins and Lyne, 1976; Jones *et al.*, 1986; Coyne and Howell, 1994). The identification was done by KB001 HiIMViC Biochemical test kit. Each HiIMViC kit is a standardized colorimetric identification system utilizing four conventional biochemical tests and eight carbohydrate utilization tests.

Antibiotic Susceptibility Test

All the bacterial isolates were tested for their sensitivity to antibiotics by means of a disc diffusion method (Bauer *et al.*, 1966) according to NCCLS (1997) recommendations.

The isolates were tested for sensitivity to 8 antibiotics and applied as discs (Octodiscs). The 8 antibiotics used were Chloramphenicol (C) 25 mg, Ampicillin (A) 25 mg, Tetracycline (T) 25 mg, Gentamicin (G) 30 mg, Co-trimoxazole (Co) 25 mg, Cefriaxone (Ci) 30 mg, Cefuroxime (Cu) 30 mg, Ciprofloxacin (Cf) 30 mg.

Muller-Hinton agar medium was used for drug sensitivity test. This medium was originally formulated for isolation of pathogenic bacteria (Mueller and Hindon, 1941). Now-a-day it is more commonly used in conjunction with high potency antibiotic discs for determination of antibiotic sensitivity patterns of bacteria.

Results and Discussion

The results of different physiochemical parameters of Yamuna water showed that the highest temperature value was recorded 37.4 °C from the samples taken during the June, while the lowest one was 18.9 °C observed in the month of March. The results showed a noticeable seasonal trend of the temperature.

Table 34.1: Physio-chemical Characteristics of Yamuna Water at Agra Region

Stations	March 2010				April 2010				May 2010				June 2010			
	Water Temp.	*TDS*	*Turbidity*	*pH*	*Water Temp.*	*TDS*	*Turbidity*	*pH*	*Water Temp.*	*TDS*	*Turbidity*	*pH*	*Water Temp.*	*TDS*	*Turbidity*	*pH*
1.	20	290	15	6.7	25	510	30	7.1	30	500	33	8	37	550	50	8.5
2.	19.5	260	10.3	6.5	25.5	390	38.7	7.3	31.4	520	43	7.9	35.6	680	36	8.6
3.	21	300	7	7.0	24.8	410	28	7.8	30.5	510	49	8.1	35.7	680	52	7.4
4.	22.3	200	4	6.8	26	480	34	7.4	29.9	530	41	7.2	33.8	640	49	8.9
5.	20.6	230	11	6.1	23.9	350	32	7.2	30.8	570	37	7.8	36.7	700	48	9.2
6.	13.9	400	12.5	6.6	26.1	380	35	7.2	29.6	470	38	7.5	37.4	600	42.5	9
7.	21.2	330	14	6.5	22.6	450	29	7.1	31.3	490	37	7.7	36.9	580	48	9

Station: 1: Moksh Dwar; Station 2: Balkeshwar Ghat; Station 3: Dashara Ghat; Station 4: Poiya Ghat; Station 5: Moti Mahal Dhobi Ghat; Station 6: Mehtab Bagh; Station 7: Near Aaram Bagh Rambagh.

The total dissolve solids (TDS) in the Yamuna water at Agra region was found highest as 700 mg/ml in the month of June at Moti Mahal Dhobi Ghat Agra and lowest 200 mg/ml at Poiya Ghat, Agra in the month of March.

The variation in water turbidity was influenced by the local activities, without seasonal trends. The highest value of turbidity - 52 NTU was observed at Dashara Ghat, Agra in the month of June and the lowest 4 NTU were recorded at Poiya Ghat, Agra in the month of March.

The pH values of the Yamuna water were found on the alkaline side. The highest pH value 9.2 was recorded at Moti Mahal Dhobighat, Agra in the month of June and the lowest was observed 6.1 in front of Moti Mahal Dhobi Ghat, Agra in the month of March. The recorded pH values were favorable for bacterial multiplication shown in Table 34.1 and from Figures 34.1 to 34.5.

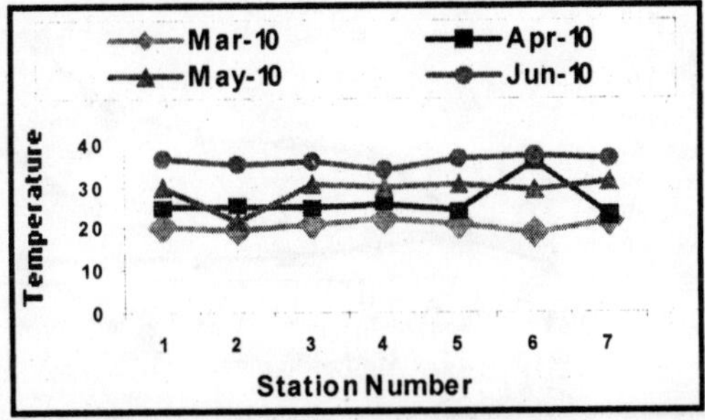

Figure 34.1: Variation of Temperature in the Samples of different Stations

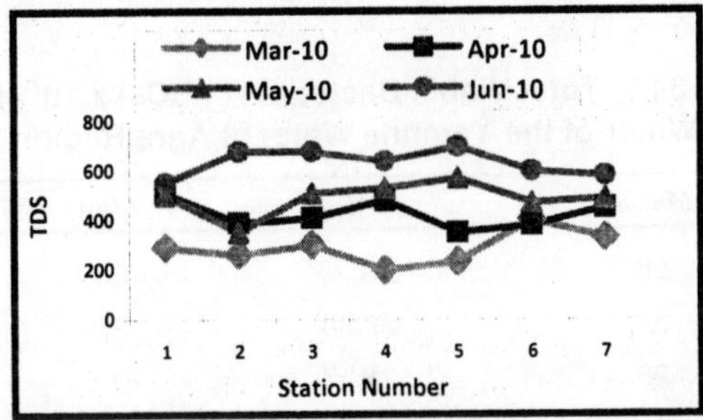

Figure 34.2: Variation of TDS in the Samples of different Stations

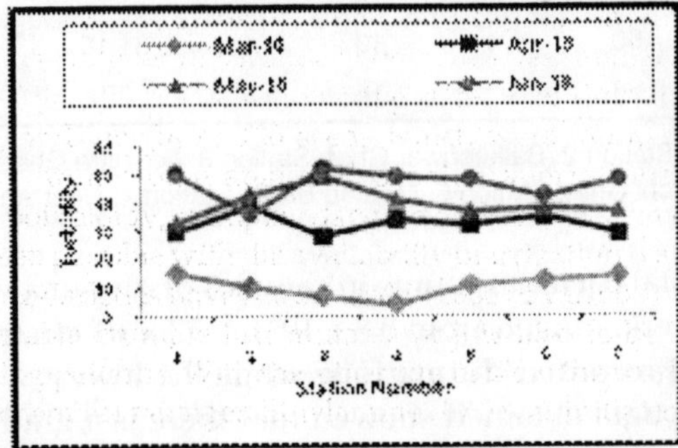

Figure 34.3: Variation of Turbidity in the Samples of different Stations

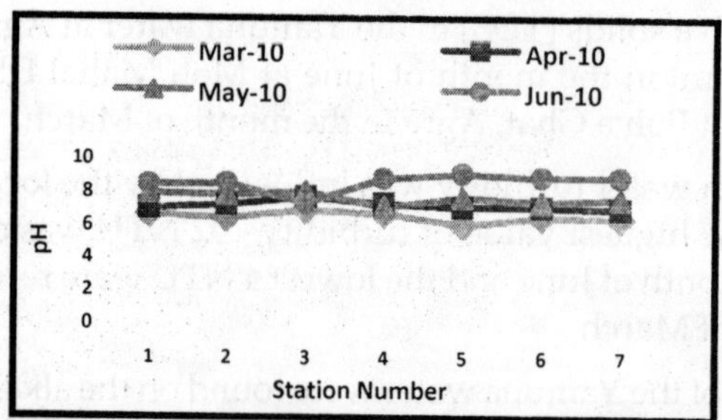

Figure 34.4: Variation of pH in the Samples of different Stations

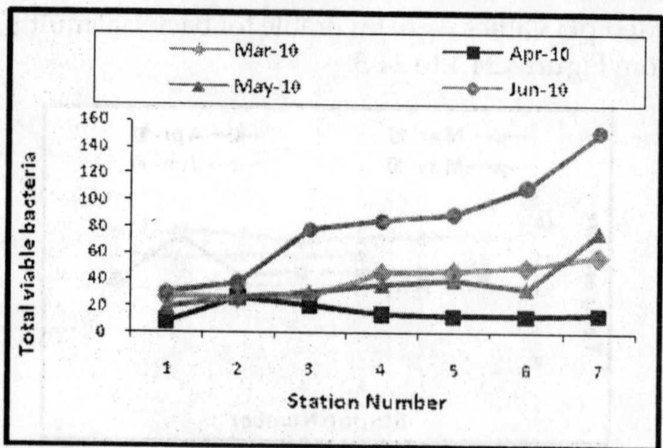

Figure 34.5: Variation of Total Viable Bacteria in the Samples of different Stations

Table 34.2: Total Viable Bacterial (TVBCs) x 10^7 cfu/ml Water of the Yamuna Water at Agra Region

Stations	*March*	*April*	*May*	*June*
1.	28	8.S	20	31
2.	37	27	27	38
3.	28	21	31.6	78
4.	46	14.6	37.2	85
5.	47	13	41	90
6.	50	13	34.18	110
7.	58	14	77	152

Station: 1: Moksh Dwar; Station 2: Balkeshwar Ghat; Station 3: Dashara Ghat; Station 4: Poiya Ghat; Station 5: Moti Mahal Dhobi Ghat; Station 6: Mehtab Bagh; Station 7: Near Aaram Bagh Rambagh.

Estimation of total bacterial counts in samples of Yamuna water revealed that total number of bacteria ranged from 8.8 x 10^7 to 152 x 10^7 cfu/ml at 37.0 °C. Their highest number was recorded during June while the lowest was observed in the month of April. The data obtained showed that there is a gradual increase in the bacterial counts from upstream to downstream *i.e.* from Moksh dwar to Rambagh,

Agra, which might be attributed to domestic, sewage and agricultural effluents discharged into Yamuna River. Thus the highest bacterial counts recorded for one group (at 37.0 °C) during March to June might be due to the high temperature prevailing during the season. These findings were in accordance with El-Fadaly *et al.* (2001) and Sabae (2004) given in Table 34.2 and Figures 34.6 and 34.7.

In the present study, the isolates of *E. coli* were differentiated and confirmed by IMVIC test as *E.coli* is a gram-negative bacteria in the water of Yamuna subjected to sewage pollution.

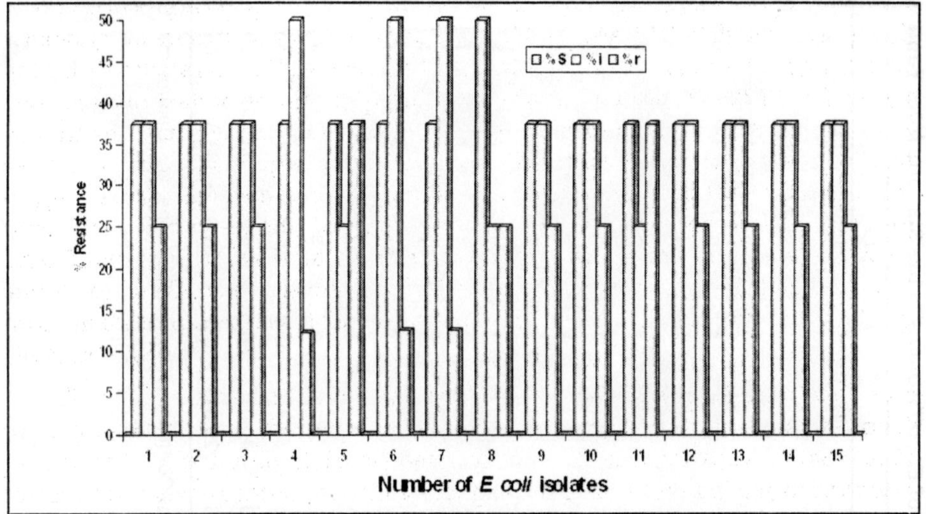

Figure 34.6: Percentage Resistance of different Antibiotics against *E. coli* Isolates

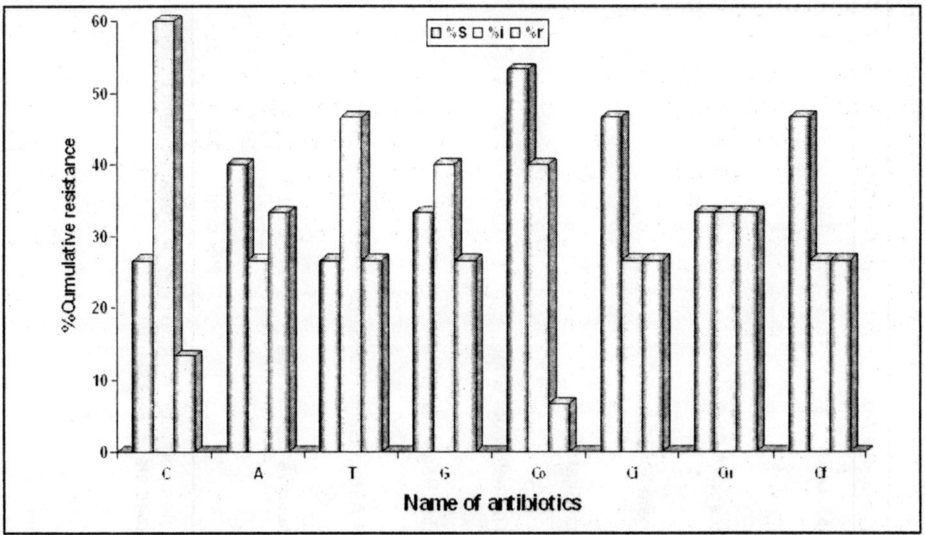

Figure 34.7: Cumulative Resistance of different Antibiotics against *E. coli* Isolates

Table 34.3: Antibiotic Resistance Level of *E. coli* Isolates in the Stations from Yamuna River

Stations No.	*Isolates No.*	*C*	*A*	*T*	*G*	*Co*	*Ci*	*Cu*	*Cf*	*% S*	*% i*	*% r*
Moksh Dwar	1	i	s	i	r	i	r	s	s	37.5	37.5	25.0
	2	s	i	r	r	i	s	s	i	37.5	37.5	25.0
	3	i	r	r	i	s	s	i	s	37.5	37.5	25.0
Balkeshwar Ghat	4	i	r	i	i	s	s	i	s	37.5	50.0	12.5
	5	r	r	r	s	s	i	s	i	37.5	25.0	37.5
Dashara Ghat	6	i	r	i	i	s	s	i	s	37.5	50.0	12.5
	7	i	i	s	s	i	s	i	r	37.5	50.0	12.5
Poiya Ghat	8	i	s	s	i	s	s	r	r	50.0	25.0	25.0
	9	s	s	i	s	i	r	r	i	37.5	37.5	25.0
Moti Mahal Dhobi Ghat	10	s	i	s	i	r	i	r	s	37.5	37.5	25.0
	11	s	s	i	s	i	r	r	r	37.5	25.0	37.5
Mehtab Bagh Station	12	i	s	i	r	i	r	s	s	37.5	37.5	25.0
	13	i	s	s	i	s	i	r	r	37.5	37.5	25.0
Near Aaram Bagh Rambagh	14	r	i	r	s	s	i	s	i	37.5	37.5	25.0
	15	i	r	i	r	s	s	i	s	37.5	37.5	25.0
Cumulative Resistance	Per cent S	26.6	40	26.6	33.3	53.3	46.6	33.3	46.6			
	Per cent i	60	26.6	46.6	40.0	40.0	26.6	33.3	26.6			
	Per cent r	13.3	33.3	26.6	26.6	6.6	26.6	33.3	26.6			

C: Chloramphenicol; A: Ampicillin; T: Tetracycline; G: Gentamicin; Co: Co-trimoxazole; Ci: Cefriaxone; Cu: Cefuroxime; Cf: Ciprofloxacin; s: Susceptible, i: Intermediate; r: Resistant.

High levels of resistance were obtained among the *E. coli* isolates (12.5 – 37.5 per cent) and the cumulative resistance of the *E. coli* isolates to the antibiotics ranged from 6.6 – 33.3 per cent.

Percentile susceptibility to all evaluated antibiotics was shown as Co-trimoxazole (Co), 53.3 per cent Cefriaxone (Ci) 46.6 per cent, Ciprofloxacin (Cf) 46.6 per cent, Ampicillin (A) 40.0 per cent, Gentamicin (G) 33.3 per cent, Cefuroxime (Cu) 33.3 per cent, Chloramphenicol (C) 26.6 per cent, Tetracycline (T) 26.6 per cent respectively. Furthermore, 33.3 per cent *E. coli* isolates was found resistant to Ampicillin (A) and Cefuroxime (Cu) respectively. Later high frequency of isolates resistant to Tetracycline (T), Gentamicin (G) and Cefrioxone (Ci) has been found 26.6 per cent. The cumulative effectiveness of the antibiotics as obtained in this study is Co-trimoxazole > Cefriaxone = Ciprofloxacin > Ampicillin > Gentamicin = Cefuroxime > Chloramphenicol = Tetracycline. Incidence of antibiotic resistance in 5, isolate at Balkeshwar Ghat, Agra and in 11 isolate at Moti Mahal Dhobighat, Agra was comparatively higher than the other isolates (Table 34.3).

Table 34.4: Resistance Pattern among the *E. coli* Isolates

No. of Antibiotics	Resistance Pattern	No. of E.coli	Resistance (per cent)
1. Sensitive	0	0	0
2. Antibiotics	A	2	13.33
	Cf	1	6.66
3. Antibiotics	G, Ci	2	13.33
	T, G	1	6.66
	A, T	1	6.66
	Cu, Cf	2	13.33
	G, Cu	1	6.66
	Co, Cu	1	6.66
	C, T	1	6.66
	A, G	1	6.66
4. Antibiotics	C, A, T	1	6.66
	C, Cu, Cf	1	6.66
Total		15	

C: Chloramphenicol; A: Ampicillin; T: Tetracycline; G: Gentamicin; Co: Co-trimoxazole; Ci: Cefriaxone; Cu: Cefuroxime; Cf: Ciprofloxacin.

Among the bacterial isolates from Yamuna water, three patterns of multiple-drug resistance were encountered and number of antibiotics ranged from one to three (Table 34.4). Of the total 15 *E. coli* isolates, 0 per cent were susceptible to all antibiotics. Of the resistant *E. coli* isolates, 19.99 per cent were resistant against one antibiotics, 66.62 per cent were resistant against two antibiotics and 13.32 per cent were resistant against 3 antibiotics, 0 per cent were resistant against more than three antibiotics. All fifteen isolates showed intermediate resistance to tested antimicrobial agents.

Intermediate resistance to Chloramphenicol, Tetracycline, Gentamicin, Co-trimoxazole, Ampicillin, Cefriaxone, Ciprofloxacin was observed as well. Of the resistant *E. coli* isolates 19.98 per cent were intermediate resistant against two antibiotics, 59.94 per cent were intermediate resistant against three antibiotics and 19.98 per cent were intermediate resistant against four antibiotics (Table 34.5).

Table 33.5: Antibiotic Intermediate Resistance Pattern among the *E. coli* Isolates

No. of Antibiotics	*Resistance Pattern*	*No. of E.coli*	*Resistance (per cent)*
1. Sensitive	0	0	0
2. Antibiotics	Ci, Cf	1	6.66
	C, G	1	6.66
	T, Co	1	6.66
3. Antibiotics	C, T, Co	2	13.33
	C, G, Cu	1	6.66
	A, Co, Cf	1	6.66
	T, Co, Cf	1	6.66
	A, G, Ci	1	6.66
	C, G, Ci	1	6.66
	A, Ci, Cf	1	66
	C, T, Cu	1	6.66
4. Antibiotics	C, T, G, Cu	2	13.33
	C, A, Co, Cu	1	6.66
Total		15	

C: Chloramphenicol; A: Ampicillin; T: Tetracycline; G: Gentamicin; Co: Co-trimoxazole; Ci: Cefriaxone; Cu: Cefuroxime; Cf: Ciprofloxacin.

References

Achudume, A.C. and Olawale, J.T. 2007: Microbial pathogens of public health significance in waste dumps and common sites. *J. Envirn. Biol.*, 28, 151 – 154.

Akman, Y., Ketenoglu, O., Even, H., Krut, L. and Duzenli, S. 2000: Environmental Pollution, Environmental Biology. 1st End. Palme Publication, Ankara.

APHA.1985: In: Standard Methods for Examination of Water and Wastewater, 16th edn. APHA, AWWA, WPCF, Washing.

Bakare, A.A., Lateef, A., Amuda, O.S. and Afolabi, R.O. 2003: The Aquatic toxicity and characterization of chemical and microbiological constituents of water samples from Oba River, Odo-oba, Nigeria, *Asian J. Microbial. Biotechnol. Environ. Sci.*, 5, 11 – 17.

Collins, C.H. and Lyne, M.P. 1976: Microbiological Methods. Butterworth and colpublishers Ltd. London Boston. p. 524.

Coyne, M.S. and Howell, J.M. 1994: The Fecal coliform/fecal streptococcoi ratio (FC/FS) and water quality in the bluegrass region of kentucky. *Soil and Science News and View* 15 (9): pp. 1-3.

El-Fadaly, H., El-Defrawy, M., El-Zawawy, F. and Makia, D. 2001: Chemical and microbiological evaluation of River Nile water in Dakahlia Governorate. *Journal of Environmental Science* 22; 1-18.

Geldreich, E.E. 1982: *J. Water Pollut. Control Fed.* 54, 931.

Jones, J.G., Gardener, S., Simon, B.M. and Pickup, R.W. (1986). Antibiotic resistant bacterial in windermere and remote upllannd trans in the English Lake District. *J. Appl. Environ. Microbiol.* 36: 450-456.

Khan, R.M.K. and Malik, A. 2001: Antibiotic resistance and detection of b-lactamase in bacterial strains of Staphylococci and Escherichia coli isolated from foodstuffs. *World J. Microbiol. Biotechnol.*, 17, 863-868.

Khanna, D.R. 1993: Ecology and pollution of Ganga River.

Kosygin, L., Dhamendra, H. and Gyaneshwari, R. 2007: Pollution status and conservation strategies of Moirang river, Manipur with a note on its aquatic bio-resources. *J. Environ. Biol.*, 28, 669-673.

Lateef, A., Oloke, J.K. and Gueguimkana, E.B. 2005: The prevalence of bacterial resistance in clinical, food, water and some environmental samples in southwest Nigeria. *Environ. Monit. Assess.*, 100, 59-69.

Leeja, L. and Thoppil, J.E. 2007: Antimicrobial activity of methanol extract of *Origanum majorana* L. (Sweet marjoram). *J. Environ. Biol.*, 28, 145-146.

Linton, A.H. 1977: 'Antibiotics, Animal and Man, An appraisal of contentious subject', in M. Woodbine (ed.). Antibiotics and antibiotics in agriculture, Butterworths, London, Boston, pp. 315.

Mueller, J.H. and Hindon 1941: A protein free medium for primary isolation of the gonococcus and meningococcus proceedings of Society of exp. *Biology and Medicine*, 48: 330-333.

Sabae, S.Z. 2004: Monitoring of microbial pollution in the River Nile and the impact of some human activities on its waters. *Proceeding 3rd International Conference on Biological Science.* Tanta University, 28–29, 3: 200 – 214.

Sathware, N.G., Paterl, K.G., Vyas, J.B., Patel, S., Trivedi, M.R., Dave, L.M., Madia, M.M., Kulkarni, P.K., Parikh, D.J. and Saiyed, H.N. 2007: Chromium exposure study in chemical based industry. *J. Environ. Biol.*, 28, 405-408.

Subramanyam, N.S. and Sambamurty, A.V.S.S. 2006: Ecology, 2nd Edn., Narosa Publishing House, New Delhi.

Toroglu, E., Toroglu, S. and Alaeddinoglu, F. 2006: Water pollution in the Aksu river (Kahraman Maras). *Turkish J. Geographical Review*, 4, 93-103.

Toroglu, S., Dincer, S. and Korkmaz, H. 2005: Antibiotic resistance in Gram negative bacteria isolated from Aksu river in (Kahramanmaras) Turkey. *Ann. Microbiol.*, 55, 229-233.

2013, Climate Change Effects on Agriculture and Economy *Pages* ***303–313***
Editors: **D.R. Khanna, A.K. Chopra, Vikas Singh, Rakesh Bhutiani & Gagan Matta**
Published by: **BIOTECH BOOKS, NEW DELHI**

Chapter 35

Mitigation of Water Quantity and Water Quality Challenges in Groundwater of Rajasthan

D.D. Ozha and H.R. Bhatt
Groundwater Department, Jodhpur, Rajasthan

Manifold increase in population, alarmingly increasing industrialization, urbanization and consequently increase in demand of water in our country in general and also in largest State of country *i.e.* Rajasthan in particular has resulted in a rapid depletion in groundwater level and deterioration in water quality due to its over exploitation. In Rajasthan, owing to meager surface water resources, groundwater is only source of dependence and survival, over 70 per cent irrigation and 95 per cent drinking water supply schemes are based on groundwater resources. To meet the demand, the withdrawal of groundwater is more than the recharge to groundwater resources. In fact this has created the problem of depletion and deterioration of groundwater level and its quality. Other geo-chemical factors have resulted in increase of hydrochemical parameters *viz.* salinity, nitrate and fluoride in groundwaters and have adversely affected the lives of inhabitants in general and socio-economic status in particular. The groundwater of western, north-east part of Rajasthan is facing problem of salinity, nitrate and fluoride, whereas central and southern part is generally facing the acute problem of fluoride. Therefore, for mitigating the challenges of sustainable development much efforts must be made for maximum utilization of rain water, revival of traditional water wisdom, aforestation, augmentation of groundwater, dry farming, promotion of sprinkler and drip irrigation, installation of R.O. and defluoridation plants. Above all there is an immense need of inculcate and educate the masses for judicious use of water.

Keywords: *Groundwater, Rainfall, Drought, Irrigation, Rajasthan.*

Introduction

Life on our planet earth is due to water. Demands of water for domestic, irrigation as well as industrial sectors have increased many folds thereby creating water crisis worldwide. It is said that if next world war will be there then it will be for water. Conflicts will be there from street level to International levels.

As arid land of Rajasthan is the most water deficit part on the earth. The largest state of India is also the driest state having only 1.15 per cent of total water resources of entire country. Catering safe drinking water to over 5.65 crores population is a big task before scientists, engineers, administrators and policy planners. Availability of surface water is mainly restricted to canal command areas of Indira Gandhi Nahar Project (IGNP), Chambal and Mahi rivers. Groundwater is only source of dependence in the larger area of the state. The depleting precious groundwater resources and their inferior quality states further inhibit the availability of freshwater assets.

In Rajasthan, drought and famine stays as unwanted guest almost every alternate year. Thus water crisis is of acute nature and needs immediate attention of the water managers. Rajasthan is known for its arid climatic conditions (particularly western parts) and is characterized by low, erratic and unevenly distributed rains causing frequent meteorological, hydrological and agricultural droughts. The overall stage of groundwater development is presently nearly 138 per cent. Quality- wise, more than 25 per cent of the groundwater sources have multiple problems, 16 per cent have excessive fluorides, 15 per cent have excess nitrates and over 9 per cent have excess salinity, thus leaving merely 35 per cent sources as potable. Overall it has 74 per cent country's total habitations affected with two or more quality parameters. There are more than 16,550 fluorosis affected villages in the state out of 32,211 in the whole country which is more than 50 per cent. Similarly, there are more than 14,415 salinity affected villages out of 33,552 in the whole country which is 42 per cent. Likewise, the situation for nitrate toxicity is also very grim. Periodic occurrence of droughts and famines has further aggravated the situation. Therefore, for sustainable socio-economic development, judicious use of this nature's gift is mandatory. In the present communication status of groundwater sources in the state, their quantity and quality problems and mitigative measures will be discussed.

Grim Groundwater Scenario of Rajasthan

The state is occupied by diversified geo-formations and hydro geological conditions. Groundwater occurs in unconsolidated formations (Quaternaries) semi-consolidated formations (Tertiary sandstone, Lathi sand- stone *etc.*) and consolidated rock types. Quaternary alluvium, Lathi sand- stove, Palana sandstone, Borunda limestone and Jodhpur sandstone at places are some of the prominent aquifer systems available in the state and are heavily over-exploited.

Water level is deeper in Western Rajasthan reaching even more than 130 m as in Lathi basin area. Depth to water level is relatively shallower in Eastern Rajasthan. However, rising trends in groundwater levels have been observed in parts of IGNP Canal command areas. Major parts of the state witness rapidly declining water levels.

Severe water level declining districts are – Jaipur, Sikar, Jhunjhunu, Nagaur, Jodhpur, Jalore, Pali, Dausa and Barmer.

Factors Wesponsible for Water Scarcity

The causative factors for water crisis in Rajasthan are as under.

1. Low Rainfall – Arid climate
2. Frequent drought and famine conditions
3. Less availability of freshwater
4. Deep and declining water levels
5. Hard rock area
6. Dispersed population
7. Undependable energy supply
8. Formidable water quality problems
9. Lack of water education

Arid Climatic Conditions

The geographical location of the state has caused arid/semi-arid climatic conditions in Rajasthan. Rainfall is low and erratic. South – westernly monsoon clouds are mostly moistureless especially in Thar Desert. Surface water is under high evapo-transpiration stress. Groundwater recharge is negligible in the western most parts of Rajasthan, where not only rains are scarce but also the groundwater level is generally deep. Soil moisture is usually very low due to aridity. No major drainage system exists in major parts of the Thar Desert. Owing to aridity and geo – formations, groundwater is saline.

Impact of Rainfall Variability on Water Resources

It is quite evident that rainfall is the most vital input in the hydrological cycle and fluctuations in quality and distribution strongly influence surface and subsurface water resources. Groundwater occurs under diverse climatic, physiographic and geological conditions and the subsurface medium through which water filters plays an important role in building up groundwater reserves. In Table 35.1, change in water level over time by district is depicted. Because of the increased overdraft of groundwater from all the potential regions of Western Rajasthan, recharge to the aquifer during normal rainfall periods is inadequate, especially because of the sporadic rainfall distribution patterns and the terrain characteristics, with a major portion of the precipitation being lost as runoff or through evaporation.

Frequent Drought and Famine Conditions

Low and uneven distribution of monsoon rains in Rajasthan causes frequent meteorological and/agriculture droughts in time and space as witnesses especially during 1972-74, 1984-87 and 1998-2002. Western districts are more prone to severe and most severe type of droughts and have been categorized as chronically Drought prone districts.

Table 35.1: Rainfall Variability in different Regions of Rajasthan

Station	*Normal Rainfall (mm)*	*Average No. of Rainy Days*	*Greatest Annual Rainfall (mm)*	*Greatest Annual Rainfall as a % of Normal*	*Year of Greatest Rainfall*	*Greatest 24 hr Rainfall (mm)*	*Date of Greatest 24-hr Rainfall*
			Arid				
Barmer	288.0	14.1	940.0	326	1944	285.7	13.8.1944
Bikaner	289.8	19.0	758.2	262	1917	165.6	25.9.1945
Churu	356.5	20.7	783.8	220	1917	146.1	5.9.1942
Ganganagar	248.4	19.5	674.0	271	1983	251.7	31.8.1928
Jaisalmer	186.2	12.5	583.1	313	1944	129.5	25.6.1961
Jalore	377.2	18.3	1039.4	276	1990	279.4	11.9.1905
Jhunjhunu	399.2	25.5	777.8	195	1956	121.9	14.07.1908
Jodhpur	365.2	20.0	1180.5	323	1917	215.9	12.9.1924
Nagaur	329.9	19.6	1259.0	382	1975	285.0	17.7.1975
Pali	418.3	19.0	1047.0	250	1990	200.0	6.8.1990
Sikar	455.8	29.7	1093.0	240	1977	184.4	25.8.1964
			Semi-Arid				
Ajmer	537.5	31.0	1226.8	228	1917	164.6	31.8.1928
Alwar	667.4	36.0	1260.3	189	1917	289.3	24.9.1904
Bharatpur	651.5	35.8	1382.8	212	1986	228.6	11.8.1916
Bhilwara	682.5	32.0	1304.0	191	1956	216.4	18.9.1950
Bundi	758.6	35.9	1546.6	204	1942	370.3	6.9.1947
Chittorgarh	862.9	33.5	1533.7	178	1944	274.3	20.7.1943
Dungarpur	738.0	36.8	1800.6	244	1937	486.4	30.6.1937
Jaipur	614.4	35.3	1317.0	214	1917	353.6	19.7.1981
Kota	760.9	37.9	1586.5	209	1917	249.2	13.7.1945
S. Madhopur	872.9	37.7	2445.0	280	1942	301.0	16.7.1942
Tonk	669.1	33.0	1513.6	226	1945	246.4	18.8.1945
Udaipur	640.1	34.4	1223.3	191	1917	183.9	18.9.1950
Sirohi	574.2	26.7	1571.6	273	1973	362.7	14.8.1941
			Sub-Humid				
Banswara	952.3	41.9	1977.0	210	1977	558.8	23.7.1957
Jhalawar	975.8	47.8	1708.2	175	1942	252.0	29.6.1945
			Humid				
Mt. Abu	1593.8	52.9	3990.3	250	1944	484.9	14.8.1941

Impact of Drought

Frequent droughts not only have socio-economic impact but also have adverse environmental impacts. Drought situation accelerates sinking of additional groundwater abstraction structures causing diminishing of groundwater resources and eventually decline in groundwater levels. Water crisis is on its peak during spells of drought years affecting millions of human population as well as livestock forcing then to migrate in the adjoining states. During drought and famine which occurs during every alternate or three years, the rate of withdrawal is greater than the recharge rate and leads not only to a decrease in water level but also to a deterioration of the water quality.

Deep and Declining Water Levels

District and block wise long term Water level trend in the state has been computed on the basis of water level fluctuation between the period pre-monsoon, 1984-2009 and short term trend has been computed on the basis of water level fluctuation between the period pre-monsoon 2008-09. The district wise change in water level between pre-monsoon 1984-2009 and pre-monsoon 2008-09 are depicted in Tables 35.2 and 35.3, respectively.

The perusal of Table 35.2 and 35.3 reveals that depletion in groundwater level is very significant in the state. 30 districts in the state showed depleting trend of groundwater level. On the basis of average depletion these districts have been further classified as Most critical (average depletion is > 10 m) Critical (average depletion is between 5 to 10 m) and moderate (average depletion is between 0 to 5 m).

Deeper groundwater levels (reaching upto 130 m) do not allow economic pumping of water for irrigation and other purposes. Increasing draft in major areas is causing lowering of groundwater levels significantly resulting in high cost of pumping. Highest depletion in groundwater table was observed in Jalore district and lowest was in Banswara district.

As per Groundwater Resource Estimation the overall groundwater resource position of the state is as under:-

Net Annual groundwater availability	=	10,563.01 m
Groundwater Draft	=	14,570.40 m
Present groundwater balance	=	4,007.48 m
Stags of groundwater development	=	137.94 per cent

Categorizations

Safe	=	30 Blocks
Semi-critical	=	8 Blocks
Critical	=	34 Blocks
Over- exploited	=	164 Blocks

Table 35.2: District-wise Change in Water Level between Pre-Monsoon (1984-2009)

Sl.No.	*District*	*Average Water Level (m)*		*Average Change in Water Level (m)*
		Pre' 84	*Pre' 09*	*Pre' 84 – Pre' 09*
1.	Ajmer	7.91	16.62	–8.71
2.	Alwar	11.11	24.63	–13.52
3.	Banswara	6.46	7.73	–1.28
4.	Baran	7.63	11.99	–4.36
5.	Barmer	33.74	42.66	–8.92
6.	Bharatpur	6.92	11.48	–4.56
7.	Bhilwara	9.81	14.85	–5.05
8.	Bikaner	71.44	74.82	–3.38
9.	Bundi	9.04	13.28	–4.24
10.	Chittorgarh	11.11	16.09	–4.98
11.	Churu	49.18	54.26	–5.08
12.	Dausa	11.80	26.53	–14.73
13.	Dholpur	8.71	13.66	–4.95
14.	Dungarpur	7.45	11.54	–4.09
15.	Ganganagar	16.51	10.56	5.95
16.	Hanumangarh	20.11	14.94	5.17
17.	Jaipur	11.69	30.66	–18.97
18.	Jaisalmer	60.13	61.97	–1.83
19.	Jalore	13.02	31.02	–18.00
20.	Jhalawar	8.12	10.90	–2.79
21.	Jhunjhunu	30.50	45.86	–15.36
22.	Jodhpur	31.44	42.37	–10.93
23.	Karauli	11.22	18.67	–7.45
24.	Kota	8.49	14.53	–6.04
25.	Nagaur.	27.93	45.92	–17.99
26.	Pali	12.58	23.83	–11.25
27.	Rajsamand	10.29	15.90	–5.62
28.	Sawai Madhopur	9.43	18.77	–9.34
29.	Sikar	27.85	41.23	–13.38
30.	Sirohi	11.91	18.66	–6.75
31.	Tonk	7.12	13.65	–6.53
32.	Udaipur	9.52	12.40	–2.88
	Average	18.13	25.37	–7.24

Table 35.3: District-wise Change in Water Level between Pre-Monsoon (2008-2009)

Sl.No.	District	Average Water Level (m)		Average Change in Water Level (m)
		Pre' 84	Pre' 09	Pre' 84 – Pre' 09
1.	Ajmer	15.68	16.62	–0.94
2.	Alwar	23.82	24.63	–0.81
3.	Banswara	6.37	7.73	–1.36
4.	Baran	11.43	11.99	–0.56
5.	Barmer	41.43	42.66	–1.23
6.	Bharatpur	11.74	11.48	0.25
7.	Bhilwara	13.37	14.85	–1.48
8.	Bikaner	73.52	74.82	–1.30
9.	Bundi	12.93	13.28	–0.34
10.	Chittorgarh	15.66	16.09	–0.42
11.	Churu	54.02	54.26	–0.24
12.	Dausa	25.70	26.53	–0.83
13.	Dholpur	16.48	13.66	2.82
14.	Dungarpur	8.54	11.54	–2.99
15.	Ganganagar	11.01	10.56	0.44
16.	Hanumangarh	15.19	14.94	0.24
17.	Jaipur	29.90	30.66	–0.76
18.	Jaisalmer	61.26	61.97	–0.71
19.	Jalore	28.24	31.02	–2.77
20.	Jhalawar	10.88	10.90	–0.02
21.	Jhunjhunu	43.95	45.86	–1.91
22.	Jodhpur	41.75	42.37	–0.63
23.	Karauli	17.66	18.67	–1.01
24.	Kota	13.08	14.53	–1.45
25.	Nagaur	44.90	45.92	–1.02
26.	Pali	21.03	23.83	–2.80
27.	Rajsamand	12.58	15.90	–3.32
28.	Sawai Madhopur	18.08	18.77	–0.70
29.	Sikar	40.24	41.23	–1.00
30.	Sirohi	14.16	18.66	–4.51
31.	Tonk	13.58	13.65	–0.07
32.	Udaipur	10.68	12.40	–1.72
	Average	24.34	25.37	–1.04

Hard Rock Area

It is obvious that hard rocks have poor storage capacity of groundwater and therefore, witness steep decline in groundwater levels during summers- especially during spells of drought years. These areas are, therefore, under most water crisis and even drinking water is catered by transportation from far off locations by trains and tankers.

Dispersed Population

Population in the desert terrain is dispersed in helmets spreading over several kilometers. Therefore, to cater the water supply of standard quality economically has become a challenging task.

Undependable Energy Supply

Inadequate quality electricity supplies also hinders groundwater pumping/ piped water supply and hence water crisis in the state.

Formidable Water Quality Problems

Quality of water is an important factor in development and use of groundwater resources. In Rajasthan, due to inherited aridity and complexity of sub-surface rocks, the groundwater is characterized by multiple quality problems (Ozha and Golani, 2005). Not only the water is saline but it also contains many dissolved substances that render it unsuitable for the very purpose it is meant for. These substances have either the direct toxic effects on the consumers or have long term indirect adverse effects. The major factors that govern the water quality are salinity, sodicity and alkalinity for waters to be used for agriculture and salinity coupled with nitrate and fluoride for waters to be used as drinking water source.

Chemical quality of aquifer water in the Western Rajasthan, in general and Bharatpur, Ajmer and Jaipur districts in the eastern Rajasthan are saline. High fluoride hazard is found almost all the districts of the state causing disease fluorosis (both dental and skeletal fluorosis). However, situation in Nagaur, Tonk, Sirohi, Jaipur, Jhunjhunu, Jalore, Barmer, Bikaner and Jodhpur is worst where the problem is much severe (Ozha and Khichi, 2007). Increased use of nitrogenous fertilizers and poor sewerage system in the urban agglomerate has caused high nitrates in aquifer water. Industrial and urban pollution has further caused deterioration in the quality of groundwater. Over application of canal water (flood irrigation) and also the water logging with in canal command has also caused soil salinity/alkalinity and contributed to salinity in groundwater. Problem of high fluoride content in groundwaters of Rajasthan has become a serious environmental issue in the field of water quality management and human health. The state ranks second amongst the most endemic fluoride problem areas of the country and shares approximately 10 percent of the world fluorosis problem.

Hydrochemical parameters induced problem of groundwater has adversely affected the socio-economic status of the inhabitants of the state. Area under unsuitable water quality zones of Rajasthan is shown in Table 35.4.

Table 35.4: Area under Unsuitable Water Quality Zones of Rajasthan

Sl.No.	*District*	*Area Sq. km*	*Area Under Unsuitable Water Quality Zone*					
			Salinity		*Nitrate*		*Fluoride*	
			Area	*Per cent*	*Area*	*Per cent*	*Area*	*Per cent*
1.	Ajmer	8481	2663	31.4	1662	19.6	5954	70.2
2.	Alwar	8720	462	5.3	1482	17	671	7.7
3.	Banswara	5037	S.V.	–	31	0.62	373	7.4
4.	Baran	6955	S.V.	–	79	1.14	S.V.	–
5.	Barmer	28387	20172	71.1	12405	43.7	14534	51.2
6.	Bharatpur	5044	2199	43.6	953	18.9	822	16.3
7.	Bhilwara	10455	669	6.4	1035	9.9	4673	44.7
8.	Bikaner	27244	14112	51.8	10189	37.4	19997	73.4
9.	Bundi	5500	99	1.8	512	9.3	726	13.2
10.	Chittorgarh	10856	87	0.8	966	8.9	445	4.1
11.	Churu	16830	10435	62	11781	70	2289	13.6
12.	Dausa	3420	345	10.1	622	18.2	451	13.2
13.	Dholpur	3009	20	0.67	S.V.	–	81	2.7
14.	Dungarpur	3770	5	0.12	162	4.3	1139	30.2
15.	Ganganagar	11604	8134	70.1	3052	26.3	6985	60.2
16.	Hanumangarh	9580	7846	81.9	3113	32.5	3257	34
17.	Jaipur	11061	243	2.2	763	6.9	2931	26.5
18.	Jaisalmer	38401	26573	69.2	7488	19.5	28762	74.9
19.	Jalore	10640	3969	37.3	1596	15	2788	26.2
20.	Jhalawar	6219	S.V.	–	22	0.35	S.V.	–
21.	Jhunjhunu	5928	593	10	791	13.3	1778	30
22.	Jodhpur	22250	7877	35.4	3026	13.6	11882	53.4
23.	Karauli	5039	217	4.3	771	15.3	368	7.3
24.	Kota	5204	S.V.	–	S.V.	–	S.V.	–
25.	Nagaur	17718	6219	35.1	7442	42	7459	42.1
26.	Pali	12357	3596	29.1	259	2.1	3534	28.6
27.	Rajsamand	4635	227	4.9	250	5.4	1242	26.8
28.	S. Madhopur	5021	281	5.6	854	17	346	6.9
29.	Sikar	7881	299	3.8	299	3.8	3223	40.9
30.	Sirohi	5136	128	2.5	241	4.7	2070	40.3
31.	Tonk	7200	216	3	1613	22.4	2578	35.8
32.	Udaipur	12644	63	0.5	556	4.4	1568	12.4
	Total	342226	117755	34.41	74018	21.63	132928	38.84

Mitigation Strategies for Sustainable Future

There is solution for every problem and so is the case with water crisis in the state of Rajasthan. Water management strategies are four fold mainly comprising of following.

Research and Development

- ☆ Water related organizations needs to promote R&D activities jointly with the Universities and other research institutions to access the availability of safe drinking water.
- ☆ Genesis of quality hazards and industrial pollution needs to be studied in order to protect the groundwater resources.

Conservation of Water

- ☆ Policy of reward and punishment needs to be introduced. Financial incentives in terms of electricity bills may be introduced for water savers.
- ☆ Popularization of micro irrigation techniques and made them mandatory in phased manner.
- ☆ Framing of water charges as per quantum of water pumped.
- ☆ Adoption of suitable cropping patterns in the water scarcity area
- ☆ Conjunctive use of groundwater.
- ☆ Community irrigation system.
- ☆ Reuse of domestic wastewater for gardening.
- ☆ For domestic use, judicious use of filtered water, be promoted.
- ☆ Promotion of Rain water harvesting in an integrated manner
- ☆ Recycling of industrial wastewater.

Augmentation of Water Resources

- ☆ Rehabilitation of traditional water harvesting bodies
- ☆ Minimizing evaporation losses of surface water by suitable techniques/ chemicals
- ☆ Arresting the sedimentation and de-silting the tanks and reservoirs
- ☆ Artificial recharge of groundwater by suitable techniques

Regulatory Measures

- ☆ Creation of groundwater sanctuaries.
- ☆ Mass awareness programme and promotion of IEC activities for quality and quantitative aspects of water.
- ☆ Enforcement of effective regulation and management options of groundwater resources.
- ☆ Promotion of participatory management programme.

Referenees

Ozha, D.D and Golani, F.M. 2005: Hydrotoxicants and their effects. *Journal of Env and Health* 22 (3), 76.

Ozha, D.D and Khichi, K.D, 2007: C.M.R.D. Publication, 138.

2013, Climate Change Effects on Agriculture and Economy *Pages* ***315–319***
Editors: **D.R. Khanna, A.K. Chopra, Vikas Singh, Rakesh Bhutiani & Gagan Matta**
Published by: **BIOTECH BOOKS, NEW DELHI**

Chapter 36

Seasonal Trends in Physico-chemical Characteristics of Rural Lake Aheri, Gadchiroli, M.S., India

L.H. Rohankar[1], *N.R. Dahegaonkar*[2], *P.M. Telkhade*[1], *A.N. Lonkar*[3] *and S.B. Zade*[4]

[1]*R.D. Science College, Aheri, Dist. Gadchiroli, Maharashtra*
[2]*Arts, Commerce and Science College, Tukum, Dist. Chandrapur, Maharashtra*
[3]*Nutan A.A.C and Smt. M.H.Wegad Science College, Umrer, Dist. Nagpur, Maharashtra*
[4]*P.G. Department of Zoology, RTM Nagpur University, Nagpur, Maharashtra*

Nature is a great cleansing agent but it has limitations as far as its capacity to assimilate pollutants. The growing load of environmental pollution has resulted into an increase in the degree of civilization damage. Now a day's increased human population and industrial development has put tremendous pressure on the limited freshwater resources. Any damage to this water source by any agency will not only make life miserable but that will also disrupt the aquatic ecosystem. It is therefore necessary to study the quality of water, based on physico-chemical parameters.

The present study deals with the assessment of physico-chemical characteristics of Rural Lake, Aheri. This old lake was constructed during the year 1842-1845 in Gond Raja regime and is now used for domestic, aquaculture as well as irrigation purpose. Many reports were available on the physico-chemical aspects of the freshwater bodies in Maharashtra. But as the Rural lake Aheri is interior, it is intangible to such type of study. Therefore, the lake is undertaken for investigation. The water samples were collected for a year from three sampling stations at regular intervals of one month from February, 2006 to January, 2007. Various physico-chemical parameters including Water Temperature, Conductivity, Transparency, pH, Dissolved Oxygen, Free CO_2, Sulphate and Phosphate were analyzed. The results revealed that there were significant seasonal variations in some physico-chemical parameters and most of the parameters were in normal range and indicated better water quality of lake.

Keywords: *Physico-chemical, Rural Lake, Water quality, Pollution.*

Introduction

Water is the elixir of life that magic substance from which all life springs forth is vital to the existence of all living organisms. Water is one of the most important and precious natural resources. But this valued resource is increasingly being threatened as human population grows and demand more water for domestic purposes and economic activities. The rural lake Aheri is the old lake constructed during the year 1842-1845 in Gond Raja regime and is now used for domestic, aquaculture as well as irrigational purpose. Many reports were available on the physico-chemical aspects of the freshwater bodies in Maharashtra. But as the rural lake Aheri is interior, it is intangible to such type of study. Therefore, the lake is undertaken for investigation.

Materials and Method

Water samples were collected at monthly intervals from the selected sampling sites between 8.00 to 10.00 A.M. from February, 2006 to January, 2007 and brought to the laboratory for various physico-chemical parameters. Water temperature, transparency, pH, dissolved oxygen, free CO_2 were measured on the sampling spot. Standard methods (APHA, 1985) for the examination of water and wastewater were followed for analysis of different physic-chemical parameters.

Results and Discussion

During the study period, water temperature varied from lowest 22.20 °C at site S_3 to highest 34.15 °C at site S_1. The maximum temperature was recorded in the month of May and minimum in December. No marked variation was observed. However, the observed variations in the temperature were closely related to the changes in ambient temperature. pH is a indicative of hydrogen ion concentration and it expresses the intensity of acidity and alkalinity of water. pH value ranged from 7.01 to 8.35, the maximum pH was recorded during monsoon season and minimum in winter. Seasonal fluctuations in pH was also reported by Sreenivasan (1964).

Conductivity has usually been taken up as an indicator for assessing the trophic status of the lake. During the present study, conductivity fluctuated between 0.112 mmhos/cm to 0.415 mmhos/cm. Dakshini and Gupta (1979) in their observations in Bhopal Lake recorded conductivity values between 112 ìmhos/cm to 228 ìmhos/cm.

The lower transparency value showed high pollution load. The higher sediments during rainy season may increase turbidity resulting in decrease in transparency of water. In the present investigation, transparency was recorded in the range of 26.00 to 94.00 cm. The minimum transparency was recorded during monsoon season and maximum during winter. Rama Devi (2007) recorded the minimum transparency in monsoon and maximum in winter season.

Table 36.1: Monthly Values of Physico-chemical Parameters of Rural Lake Aheri during Year 2006-2007

Sl.No.	*Parameters*	*Site S1*		*Site S2*		*Site S3*	
		Minimum	*Maximum*	*Minimum*	*Maximum*	*Minimum*	*Maximum*
1.	Water Temperature	23.20	34.15	23.10	33.85	22.20	33.10
2.	pH	7.10	8.35	7.01	8.14	7.16	8.10
3.	Conductivity	0.20	0.42	0.14	0.36	0.12	0.23
4.	Transparency	28.00	65.00	27.00	59.50	26.00	94.00
5.	Total Alkalinity	199.00	340.00	186.00	288.00	102.00	183.00
6.	T.D. S.	401.00	635.00	316.00	503.00	165.00	392.00
7.	D.O.	4.80	11.17	4.70	8.44	5.00	13.00
8.	Free CO_2	2.02	4.30	2.12	4.30	1.80	5.20
9.	Sulphate	12.05	23.18	24.00	44.00	11.14	36.20
10.	Phosphate	0.230	0.830	0.230	0.603	0.117	0.430
11.	Nitrate	0.320	1.320	0.330	1.100	0.328	0.912

Alkalinity of water is its capacity to neutralize a strong acid and is characterized by the presence of hydroxyl ions. However, a number of bases contribute to the alkalinity but in natural waters carbonate and bicarbonates are considered to be predominant bases. The total alkalinity recorded between 102.6 mg/l to 340 mg/l. Seasonally total alkalinity was recorded maximum in monsoon and minimum in winter; similar observation was reported by Pailwan (2005). He reported maximum alkalinity in summer and minimum in winter and further stated that the fluctuation could be correlated with increase and decreased oscillation of the photosynthetic activities in different seasons.

Water potability depends on the total dissolved solids. TDS denotes various types of minerals present in water. The values of total dissolved solids ranged between 165 mg/l to 635 mg/l. Maximum TDS was recorded in the monsoon and minimum in winter season. Kadam *et al.* (2007) recorded minimum value during winter and maximum during monsoon and further stated that the amount of dissolved solids increases due to the release of decaying matter from aquatic vegetation. In the present investigation, comparatively low winter value of TDS may be attributed to low rate of evaporation and settling of silt. The high TDS values in monsoon are attributed to leaching of soil and silt carried in the lake by the ingressing water lake from the catchment area.

Dissolved oxygen is obviously essential in the metabolism of all aerobics. The main source of D.O. is the direct diffusion from the atmosphere and the photosynthetic process of aquatic autotrophs in any water body. The values of D.O. range recorded between 4.70 to 13.0 mg/l. Low value of dissolved oxygen during summer in the present investigation was probably due to two reasons. In summer at high temperature rate of oxidation of organic matter in water increase and oxygen was consumed in the process, secondly at higher temperature the water had a lesser oxygen holding capacity and surplus oxygen was lost to the atmosphere (Ruttner, 1953; Munawar, 1970; Singh and Sahay, 1979; Saha and Pandit, 1986). During winter at a low temperature high aeration rate and high photosynthetic activity have increases amount of dissolved oxygen.

In the present investigation, the value of free CO_2 ranged between 1.80 mg/l to 5.20 mg/l. Jain and Seethapati (1996) recorded the range of free CO_2 between 0.0 to 5.6 mg/l from Halali reservoir of Vidisha. Kadam *et al.* (2007) during the study of Masoli reservoir, recorded CO_2 in the range of 3.82 mg/l to 6.80 mg/l, 3.81 mg/l to 6.79 mg/l, 3.83 mg/l to 6.79 mg/l and 3.85 mg/l to 6.81 mg/l from site A, B, C and D, respectively. Free CO_2 may be present throughout the year, Ganapathi (1956), Verma (1969) or in few successive months of a year Kaushik and Saksena (1999) and Prasad (1990). In the present investigation, free CO_2 was recorded from all the sites throughout the year.

Values of sulphate recorded between 11.14 mg/l to 44 mg/l. Maximum sulphate concentration was recorded during monsoon and minimum during winter. Angadi *et al.* (2005) reported maximum sulphate value in the month of March and minimum in the month of November.

The phosphate is one of the major elements required by the biota in an ecosystem. It is critical nutrient in the eutrophication process along with nitrates. The values of phosphate recorded between minimum 0.117 mg/l to maximum 0.830 mg/l. The value of phosphate though comparatively less as one would expect in productive water could be attributed to system releasing not only an element considered as a limiting nutrient but also organic phosphorus compound (Rigler, 1956). In the present investigation, the maximum phosphate is recorded during summer and minimum during winter.

Nitrate is indicative of nitrogen richness in an aquatic system primarily attributed to animal origin (Thresh *et al.*, 1944). In the present investigation, nitrate values recorded between 0.320 mg/l to 1.320 mg/l. Minimum value of nitrate was recorded in monsoon and maximum in summer season. Salaskar and Yerragi (1997) also reported high value of nitrate during monsoon and lower in summer season.

References

Angadi, S.B., Shiddamallayya, N. and Patil, P.C. 2005: Limnological studies of Papnash pond, Bidar (Karnataka). *J.Environ.Biol.* 26(2): 213-216.

APHA. 1985: Standard Methods for the Examination of Water and Wastewater, 16th edn. American Public Health Association. Washington D.C.

Dakshini, K.M.M. and Gupta, S.K. 1979: Preliminary observation on the limnology of three freshwater lakes around Delhi State, India, Proc. Indian Natural Science Academy. (345): 564-570.

Ganapati, S.V. 1956: Hydrobiological investigations of Hope reservoir and of Thambarparni River at Papanasum Tirunalveli, Dt. Madras state, *India., Geor. J. Biol.*: 1-2.

Jain, C.K. and Seethapati P.V. 1996: Limnological studies of Kayamkulam Lake. *India.J.Environ.Protection* 16(8): 561-568.

Kadam, M. S., Pampatwar, D. V. and Mali, R. P. (2007): Seasonal variations in different physico-chemical characteristics in Masoli reservoir of Parbhani District. Maharashtra. J.Aqua. Biol., 22 (2): 110-112.

Kaushik, S. and. Saksena, D.N 1999: Physico-chemical limnology of certain water bodies of central India. In: Freshwater ecosystem of India. K. Vijaykumar (Ed). Daya Publishing House, New Delhi: 1-58.

Munnawar, M. 1970: Limnological studies on freshwater ponds of Hyderabad I. The biotope. *Hydrobiologia* 24(1): 127 - 162.

Pailwan, I.F. 2005: Limnology and fisheries potential of perennial tanks of Kolhapur district, Kolhapur. Ph.D. Thesis, Shivaji University, Kolhapur.

Prasad, D.Y. 1990: Primary productivity and energy flow in Upper lake, Bhopal. *Ind. J. Environ. Hlth.* 32(2): 132-142.

Rama Devi, T. 2007: Study of some aspects of Hydrobiology of Alisagar Dam water, Ph.D. Thesis submitted, Swami Ramanand Teerth, Marathwada University, Nanded.

Rigler, F.H. 1956: A trace study of the phosphorous cycle in lake water ecology (37): 550-562

Ruttner, F. 1953: Fundamentals of Limnology. 1st ed. (Transl. by D.G. Frey and F.E. J. Fry). Univ Toronto Press, Toronto. 295.

Saha, L.C. and Pandit, B. 1986: Comparative limnology of Bhagalpur ponds. Comp. Physio. Ecol. 1(14): 213-216.

Salaskar, Pramod and Yergi, S.G. 1997: Studies on water guilty characteristics of Shenala lake, Kalyan, Maharastra, *Ind. J. Aqua. Biol.* 12 (1 and 2): 28 - 31.

Singh, S.B. and Sahay, R. 1979: Study of some limnological features of Jalvanic pond of Ghorakpur. Proc. Nat. Sci. India. 49(B): IV.

Sreenivasan, A. 1964: A Hydrological study of a tropical impoundment, Bhavanisagar Reservoir, Madras, India, for the year 1956-61. *Hydrobiol.* 24 (4): 514-539.

Thresh, J.C., Sucklina, E.V. and Beale, J.P. 1944: The examination of water supplies. (Ed). E. W. Taylor.

Verma, M.N. 1969: Hydrobiological study of a tropical impoundment, Tekanpur reservoir, Gwalior, India, with special reference to the breeding of Indian Carps. *Hydrobiolo.* (34): 358-368.

2013, Climate Change Effects on Agriculture and Economy *Pages* ***321–334***
Editors: **D.R. Khanna, A.K. Chopra, Vikas Singh, Rakesh Bhutiani & Gagan Matta**
Published by: **BIOTECH BOOKS, NEW DELHI**

Chapter 37

Hydrological Profile of Wells at Itwari Sub-Area of Nagpur City

P.R. Bhandari[1], H. Mude[1], A.P. Lambat[1], R.D. Gadewar[1], V.S. Wadhai[2] and R.R. Kamdi[3]

[1]*Sevadal Mahila Mahavidyalaya, Nagpur, Maharashtra*
[2]*Department of Microbiology, Sardar Patel College, Chandrapur, Maharashtra*
[3]*Department of Zoology, Anandniketan College, Warora, Maharashtra*

Well water is a good source of freshwater available on earth. It is important renewable resource having several inherent advantages over surface water. Its wide distribution, negligible evaporation loss and low risk of contamination are some of the advantages. The dependence on well water is increasing in many regions due to limited surface water resources, non perennial rivers and frequent failure of monsoon. In view of this well water resources are often overexploited to meet the increasing demand and thus, the aquifers are heavily stressed. The studies of hydrological profile of samples from different wells have been carried out throughout the year. The physico-chemical analysis *viz*, Temperature, Colour, pH, Conductivity, Alkalinity, Hardness, D.O., TDS, SS, Chloride *etc.* has been done to find out the quality of groundwater.

The pH values were found in the range of 6.8 to 7.8, alkalinity 136 to 342 mg/l, Total Dissolved Solids between 615 to 832 mg/l, Hardness 32 to 216 mg/l. The values of the physico-chemical parameter selected for the present investigation were high during summer than winter and monsoon seasons.

Keywords: *Groundwater, Pollution, Physico-chemical, Seasonal, Industries.*

Introduction

Water is most important of all natural resources known on earth. It is important to all living organisms, most ecological systems, human health, food production and for economic development (Postel *et al.*, 1996). The safety of drinking water is an on going concern within the global village. Water is the prime necessity of life and all organisms on the earth depends on water. Man uses water for many purposes such as drinking, irrigation, industrial purposes, transportation and waste disposal. On an average a human being consumes about 2 liters of water everyday. Water accounts for about 70 per cent of the weight of a human body. About 80 per cent of earth surface is covered by water. Out of estimated 1,011 million cubic km of the total water present on earth, only 33,400 cubic meters of water is available for drinking, agricultural, domestic and industrial consumption. The rest of water is locked up in ocean as salt-water, polar ice caps, glaciers and underground aquifers. The menace of water borne diseases and epidemics still threatens the well being of population, particular in under developed and developing countries. Thus, the quality as well as quantity of water supply is of vital significance for welfare of mankind.

The quality of groundwater is described by its physical, chemical and microbiological properties. These characteristics are inter-linked. Therefore, interpretation of conflation coefficient between water qualities parameters provides good idea about the quality of water and facilitates rapid monitoring method (Stigter *et al.*, 2009). In addition to the direct consumption of water at homes and farms there are many indirect ways in which water affects daily life. The water plays and important role in manufacture of essential commodities, generation of electric power, transportation and recreation industries *etc.* Thus, water can be considered as the most important raw material of civilization because of the fact that without water the man cannot live and industry of human being cannot operate.

With growing population and industrial development, the demand of water is also increasing day by day and hence every country has to take preventive measures to avoid careless pollution and contamination of the available water resources. Temperature is important factor for calculating the solubility of oxygen and carbon dioxide, bicarbonate and carbonate equilibrium. The temperature of drinking water has an influence on its taste. Conductivity is the capacity of water to carry an electrical current and varies both with numbers and types of ions in the solution, which in turn is related to the concentration of ionized substance in water. Chloride ions are generally present in natural and wastewater. The sources of chloride in natural water is due to discharge from chemical industries, dissolution of salt deposits, oil well operation each of these sources may result in contamination of both surface and well water. Sulphate ions usually occur in natural water.

Materials and Method

Sampling Site and Sample Collection

The present investigation was carried out in the Itwari area in Nagpur city (M.S.). This area is one of the biggest whole sale and retail market in Nagpur. The well water from which the samples collected has been used extensively for the drinking purpose and other domestic needs.

Eleven wells have been selected from this region. Water samples from the above said sampling locations were collected in clean containers (Tarson 500 ml capacity) for physico-chemical analysis (APHA, 1985). Temperature of water was recorded in the field itself. Dissolved oxygen of the water samples was fixed in BOD bottles and carried to the laboratory for further analysis. The water samples for the estimation of other parameters were collected in clean bottles, which were previously soaked in 10 per cent nitric acid and thoroughly rinsed with de-ionized water. At each sampling point three replicate samples were collected. A composite sample was prepared by mixing replicate samples together. All the samples were properly sealed and transported to the laboratory for further analysis.

The water samples were collected for post monsoon season was in the month of second week of September and for winter season it was second week of December.

Water Analysis

The following physico-chemical parameters have been selected for the present investigation *viz.*, Temperature, Colour, pH, Conductivity, Alkalinity, Total Hardness, Calcium Hardness, Magnesium Hardness, Dissolved Oxygen (DO), Total Solids, Total Dissolved Solids (TDS), Suspended Solids, Chlorides and Sulphate. All the physico-chemical analysis has been done by standard method (APHA, AWWA and WPCF, 1985; De Lange, 1994; Singh and Chandel, 2006).

Results and Discussion

Samples collected from eleven wells were examined for temperature degree Celsius of water and they have been statistically analyzed at the degree of freedom 10. The values are given in Tables 37.1 to 37.13 and graphically represented in Figures 37.1 to 37.13.

Table 37.1: Temperature of the Water Samples in different Seasons (°C)

Well No.	*Post Monsoon*	*Winter Season*	*Mean*	*ISI Std.*	*T-test*	*P-Value*
1	29.5	27.0	28.25		9.34	0.001
2	29.5	27.0	28.5		9.34	0.001
3	28.0	28.0	28.0		0.0	0.500
4	29.0	28.0	28.5		8.01	0.001
5	28.5	27.5	28.0		8.01	0.001
6	28.0	28.0	28.0	20.0	0.0	5.00
7	28.5	27.0	27.75		14.62	0.001
8	28.8	27.0	27.9		12.53	0.001
9	28.0	28.5	28.25		9.34	0.001
10	28.5	27.5	28.0		8.01	0.001
11	29.0	28.0	28.5		8.01	0.001

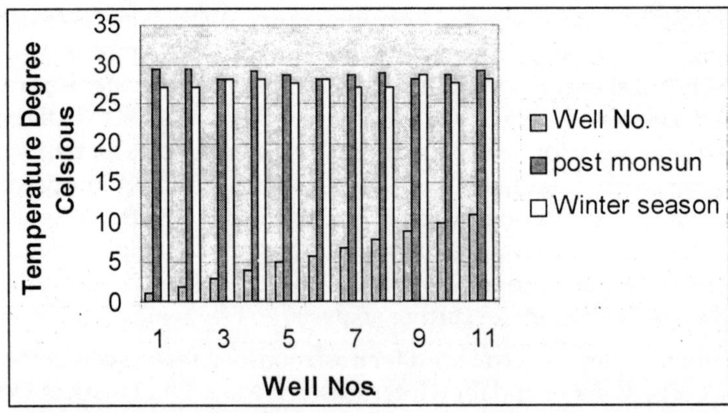

Figure 37.1: Temperature in Post Monsoon and Winter Season

Table 37.2: pH of the Water Samples in different Seasons

Well No.	*Post Monsoon*	*Winter Season*	*Mean*	*ISI Std.*	*T-test*	*P-Value*
1	7.60	6.80	7.20		0.33	0.01
2	7.40	7.86	7.63		0.81	0.50
3	7.20	7.50	7.35		0.14	0.50
4	7.50	7.60	7.55		1.42	0.10
5	7.60	7.70	7.65		0.42	0.50
6	7.70	7.80	7.75	6.5	2.0	0.10
7	7.50	7.60	7.55	To	0.14	0.50
8	7.40	7.50	7.45	8.5	0.14	0.50
9	7.30	7.40	7.50		0.42	0.50
10	7.20	7.70	7.45		2.9	0.02
11	7.10	7.60	7.35		0.88	0.50

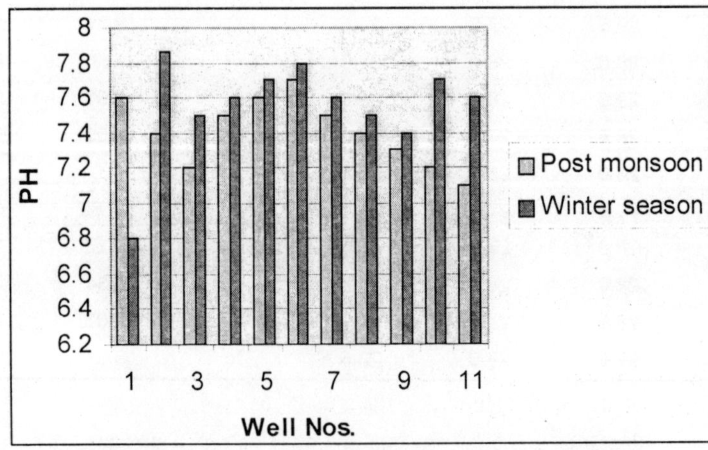

Figure 37.2: pH in Post Monsoon and Winter Season

Table 37.3: Conductivity of the Water Samples in different Seasons (µS/cm)

Well No.	*Post Monsoon*	*Winter Season*	*Mean*
1	0.96	1.02	0.990
2	1.04	1.09	1.065
3	1.00	1.04	1.020
4	1.10	1.19	1.145
5	1.20	1.28	1.240
6	1.03	1.07	1.050
7	1.24	1.30	1.270
8	1.17	1.27	1.220
9	1.10	1.19	1.145
10	1.04	1.09	1.065
11	0.96	1.02	0.990

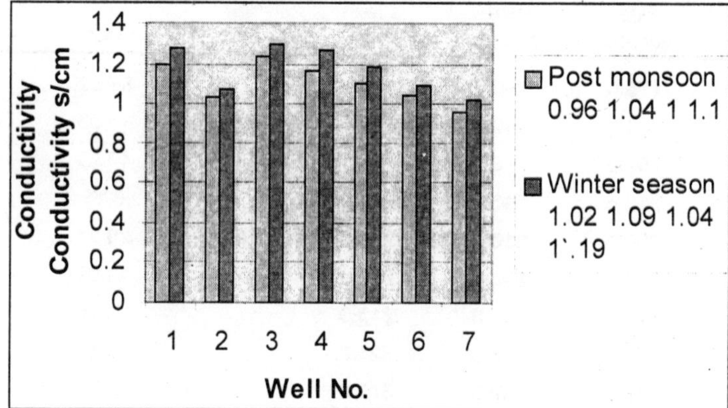

Figure 37.3: Conductivity in Post Monsoon and Winter Season

Table 37.4: Total Alkalinity of the Water Samples in different Seasons (mg/l)

Well No.	*Post Monsoon*	*Winter Season*	*Mean*	*ISI Std.*	*T-test*	*P-Value*
1	136.0	150.0	143.0		11.53	0.001
2	158.0	168.0	163.0		10.57	0.001
3	160.0	166.0	163.0		17.45	0.001
4	172.0	180.0	176.0		8.57	0.001
5	196.0	200.0	198.0		1.42	0.001
6	200.0	210.0	205.0	200.0	1.42	0.001
7	188.0	196.0	192.0		2.63	0.020
8	182.0	190.0	186.0		4.96	0.001
9	180.0	188.0	184.0		5.67	0.001
10	196.0	200.0	198.0		1.41	0.10
11	186.0	192.0	189.0		5.18	0.001

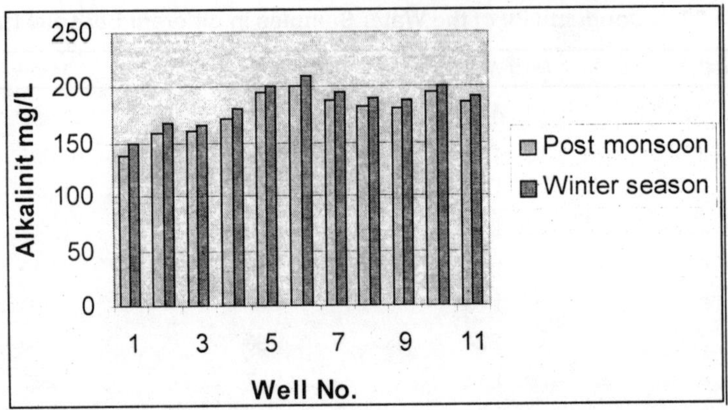

Figure 37.4: Alkalinity in Post Monsoon and Winter Season

Table 37.5: Total Hardness of the Water Samples in different Seasons (mg/l)

Well No.	*Post Monsoon*	*Winter Season*	*Mean*	*ISI Std.*	*T-test*	*P-Value*
1.	176.0	180.0	178.0		86.52	0.001
2	224.0	240.0	232.0		12.03	0.001
3	162.0	170.0	166.0		47.85	0.001
4	218.0	240.0	229.0		9.13	0.001
5	200.0	210.0	205.0		26.91	0.001
6	274.0	290.0	282.0	300.0	12.03	0.001
7	306.0	320.0	313.0		0.53	0.05
8	306.0	336.0	321.0		1.98	0.10
9	312.0	342.0	327.0		2.54	0.02
10	284.0	296.0	290.0		2.38	0.05
11	262.0	272.0	267.0		9.34	0.001

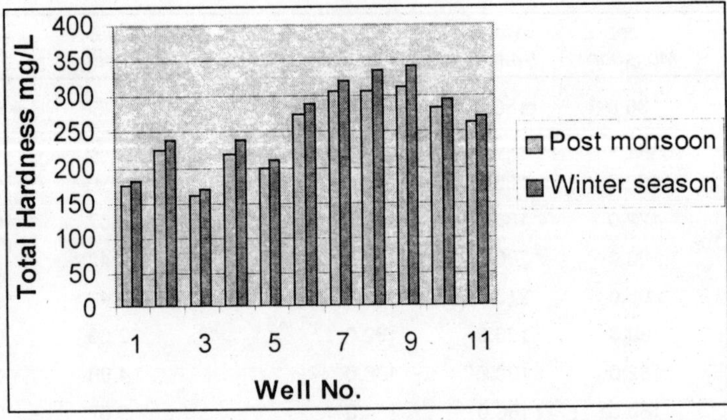

Figure 37.5: Total Hardness in Post Monsoon and Winter Season

Table 37.6: Calcium in the Water Samples in different Seasons (mg/l)

Well No.	*Post Monsoon*	*Winter Season*	*Mean*	*ISI Std.*	*T-test*	*P-Value*
1	130.0	130.0	130.0		0.0	0.50
2	180.0	190.0	185.0		31.19	0.001
3	130.0	130.0	130.0		0.0	0.50
4	132.0	150.0	141.0		10.37	0.001
5	152.0	110.0	131.0		3.77	0.01
6	130.0	120.0	125.0	75.0	14.16	0.001
7	124.0	138.0	131.0		11.33	0.001
8	118.0	140.0	129.0		6.94	0.001
9	112.0	126.0	119.0		8.90	0.001
10	110.0	110.0	110.0		0.0	0.50
11	114.0	120.0	117.0		19.81	0.001

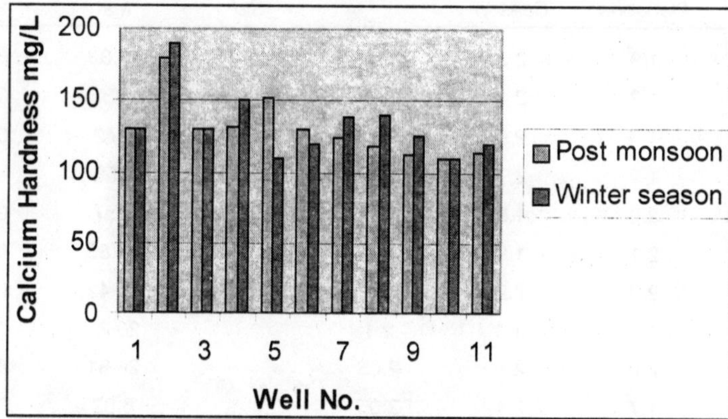

Figure 37.6: Calcium Hardness in Post Monsoon and Winter Season

Table 37.7: Magnesium Hardness of the Water Samples in different Seasons (mg/l)

Well No.	*Post Monsoon*	*Winter Season*	*Mean*	*ISI Std.*	*T-test*	*P-Value*
1	46.0	50.0	48.0		12.76	0.001
2	44.0	50.0	47.0		8.01	0.001
3	32.0	40.0	36.0		12.76	0.001
4	86.0	90.0	88.0		41.13	0.001
5	48.0	60.0	54.0		5.66	0.001
6	144.0	170.0	157.0	30.0	13.81	0.001
7	176.0	182.0	179.0		70.28	0.001
8	188.0	196.0	192.0		57.44	0.001
9	200.0	216.0	208.0		31.50	0.001
10	174.0	186.0	180.0		35.37	0.001
11	148.0	152.0	150.0		85.10	0.001

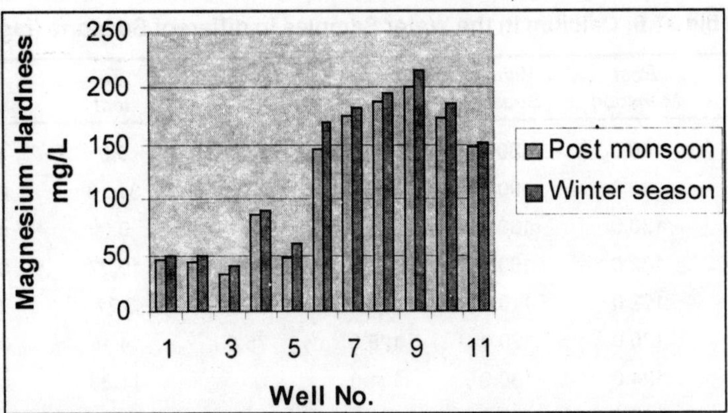

Figure 37.7: Magnesium Hardness in Post Monsoon and Winter Season

Table 37.8: Dissolved Oxygen in the Water Samples in different Seasons (mg/l)

Well No.	*Post Monsoon*	*Winter Season*	*Mean*	*ISI Std.*	*T-test*	*P-Value*
1	1.9	2.5	2.2		13.33	0.001
2	1.8	2.1	1.95		35	0.001
3	1.7	2.9	2.3		6.42	0.001
4	1.9	2.7	2.3		2.58	0.05
5	2.0	1.9	1.95		8.57	0.001
6	2.1	1.9	2.0	4.0 to 6.0	42.85	0.001
7	2.0	2.2	2.1		41.42	0.001
8	1.9	2.3	2.1		20.71	0.001
9	2.0	2.1	2.05		26.81	0.001
10	1.7	2.4	2.05		6.02	0.001
11	1.8	2.3	2.05		95.16	0.001

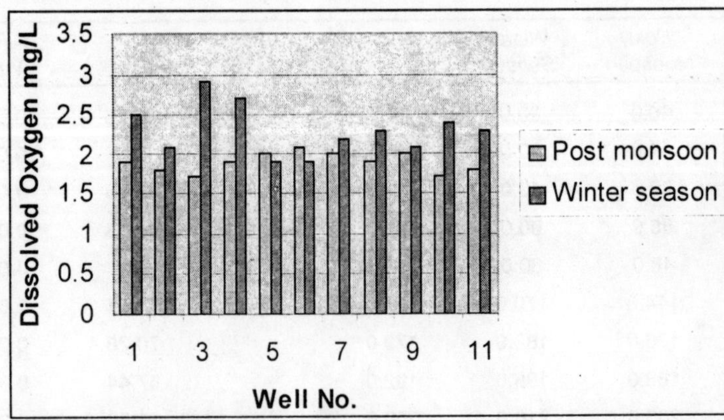

Figure 37.8: Dissolved Oxygen in Post Monsoon and Winter Season

Table 37.9: Total Solids of the Water Samples in different Seasons (mg/l)

Well No.	*Post Monsoon*	*Winter Season*	*Mean*
1	625.0	660.0	642.5
2	678.0	712.0	695.0
3	656.0	682.0	669.0
4	721.0	764.0	742.5
5	774.0	826.0	800.0
6	668.0	698.0	683.0
7	804.0	842.0	823.0
8	758.0	824.0	791.0
9	712.0	770.0	741.0
10	670.0	705.0	687.5
11	617.0	662.0	639.5

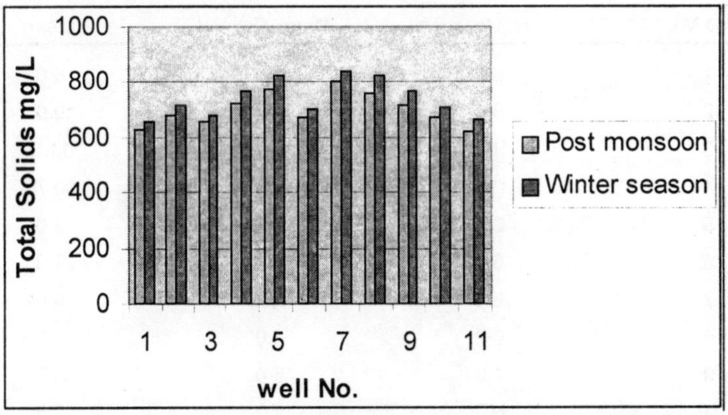

Figure 37.9: Total Solids in Post Monsoon and Winter Season

Table 37.10: Total Dissolved Solids of the Water Samples in different Seasons (mg/l)

Well No.	*Post Monsoon*	*Winter Season*	*Mean*	*ISI Std.*	*T-test*	*P-Value*
1	620.0	655.0	637.5		11.11	0.001
2	670.0	700.0	685.0		17.45	0.001
3	646.0	670.0	658.0		41.98	0.001
4	706.0	756.0	731.0		13.12	0.001
5	772.0	820.0	796.0		17.44	0.001
6	662.0	686.0	674.0	500.0	20.51	0.001
7	796.0	832.0	814.0		24.68	0.001
8	752.0	816.0	784.0		12.55	0.001
9	710.0	765.0	737.5		12.21	0.001
10	670.0	700.0	685.0		17.35	0.001
11	615.0	656.0	635.5		9.35	0.001

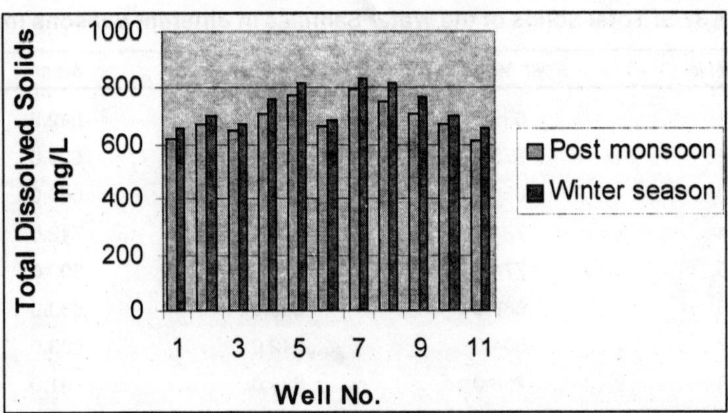

Figure 37.10: Total Dissolved Solids in Post Monsoon and Winter Season

Table 37.11: Suspended Solids in the Water Samples in different Seasons (mg/l)

Well No.	Post Monsoon	Winter Season	Mean
1	5.0	5.0	5.0
2	8.0	10.0	9.0
3	10.0	12.0	11.0
4	15.0	6.0	10.75
5	2.0	4.0	3.0
6	6.0	8.0	7.0
7	8.0	10.0	9.0
8	6.0	8.0	7.0
9	2.0	5.0	3.5
10	ND	5.0	2.5
11	2.0	5.0	3.5

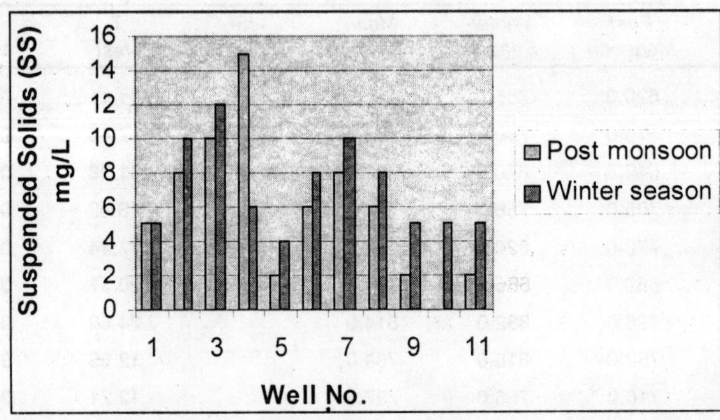

Figure 37.11: Suspended Solids in Post Monsoon and Winter Season

Table 37.12: Chlorides in the Water Samples in different Seasons (mg/l)

Well No.	*Post Monsoon*	*Winter Season*	*Mean*	*ISI Std.*	*T-test*	*P-Value*
1	118.0	126.0	122.0		45.39	0.001
2	120.0	125.0	122.5		72.44	0.001
3	144.0	158.0	151.0		20.04	0.001
4	1260.	135.0	130.5		26.55	0.001
5	174.0	190.0	182.0		12.14	0.001
6	70.0	74.0	72.0	250.0	126.42	0.001
7	152.0	160.0	156.0		33.33	0.001
8	92.0	108.0	100.0		26.54	0.001
9	90.0	100.0	95.0		43.90	0.001
10	86.0	96.0	91.0		45.04	0.001
11	74.0	84.0	76.0		48.44	0.001

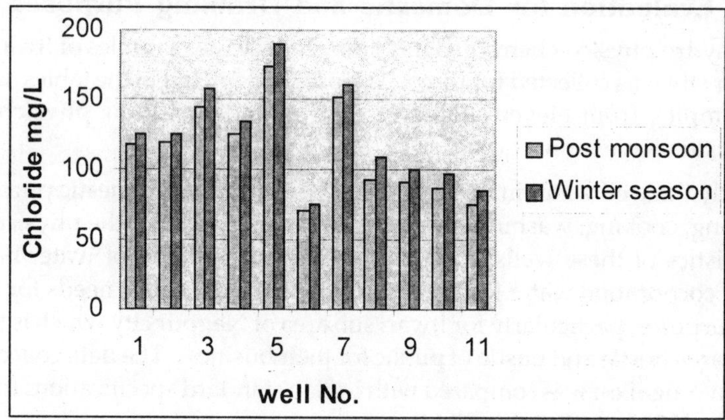

Figure 37.12: Chloride in Post Monsoon and Winter Season

Table 37.13: Sulphate in the Water Samples in different Seasons (mg/l)

Well No.	*Post Monsoon*	*Winter Season*	*Mean*	*ISI Std.*	*T-test*	*P-Value*
1	8.0	10.0	9.0		201.42	0.001
2	18.0	22.0	20.0		92.19	0.001
3	20.0	26.0	23.0		59.90	0.001
4	32.0	36.0	34.0		82.26	0.001
5	36.0	42.0	39.0		52.55	0.001
6	22.0	27.0	24.5	150.0	71.30	0.001
7	26.0	30.0	28.0		86.52	0.001
8	40.0	48.0	44.0		75.17	0.001
9	12.0	16.0	14.0		96.45	0.001
10	10.0	12.0	11.0		198.5	0.001
11	16.0	20.0	18.0		93.61	0.001

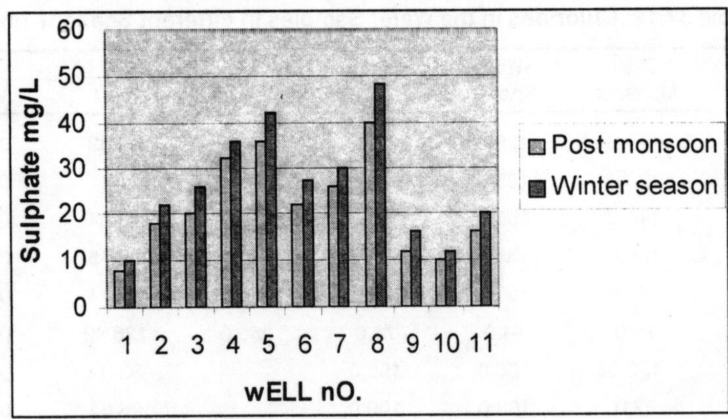

Figure 37.13: Sulphates in Post Monsoon and Winter Season

Results and Discussion

Quality Evaluation for Domestic and Drinking Purposes

The hydro physico-chemical data of open well water samples of Itwari sub area of Nagpur city was collected for three seasons and depicted in the tables and graphs. Water samples from eleven wells were examined and their physico-chemical characteristics have been studied.

Mostly the well water in this region has been used for domestic purpose that is for drinking, cooking, washing *etc*. The reason behind testing the physico chemical characteristics of these wells is mainly to check the quality of water because the supply of corporation water is not enough to satisfy the public needs for their daily routine purposes, particularly for Itwari sub area of Nagpur city which is the highest crowded area hustle and bustle of public for their business. The data collected for the present investigation was compared with Indian standard specifications for drinking water, ISI: - 10500, 1991 (Table 37.14).

The ph value in study area was found to be in the range 7.1 to 7.7 during the post monsoon season and 6.8 to 7.87 during winter season indicating slightly alkaline nature of the well water the other parameters like temperature, conductivity, color and dissolved oxygen were noted to be within the range of standard specification laid down by BIS. The chloride and sulphate content in most samples analysed were within the permissible limit for drinking as well as for domestic use.

The alkalinity value varied from 136 to 200 mg/l during post monsoon and 166 to 210 mg\l during winter season. The slightly high alkalinity values in the winter season may be ascribed to the action of carbonate ions present in the soil (Sinha and Saxena, 2006; Nandintsetseg and Shinoda, 2010). Total hardness in the water during post monsoon season ranges between 260 -306 Mg/l and during winter season it is observed 170 to 342 mg/l. The total hardness is higher than Indian standard acceptable limit. To support the significant high concentration of total hardness there is increase in calcium (range 110-152 mg/l) in the post monsoon season while it was nearly 110-190 mg/l in winter season.

Table 37.14: Indian Standards Specification for Drinking Water (IS: 10500–1991)

Sl.No.	*Substance or Characteristic*	*Method of Test, Cl ref of IS : 3025-1964*	*Other Methods of Test*	*Requirement (Desirable Limit)*
1.	Colour, Hazen units, Max	5	–	10
2.	Odour	7	–	Unobjectionable
3.	Taste	Test at temperature Not lower than 20°C test at several dilution	–	Agreeable
4.	Turbidity	–	Appendix-B	10
5.	Dissolved solids (mg/l) max	12	–	500
6.	pH Value	8	–	6.5 to 8.5
7.	Total hardness (as $CaCO_3$) mg/l, max	16	–	300
8.	Calcium (mg/l) max	33	–	75
9.	Magnesium (mg/l) max	16, 33, 34	–	30
10.	Copper (mg/l) max	36	–	0.05
11.	Iron (mg/l) max	32	–	0.3
12.	Manganese (mg/l) max	35	–	0.1
13.	Chlorides (mg/l) max	24	–	250
14.	Sulphate (mg/l) max	20	–	150
15.	Nitrate (mg/l) max	–	6 of IS : 2488 (part iv) 1974	45
16.	Fluoride (mg/l)	23	–	0.6 to 1.2

It is well known fact that the excess amounts of calcium and magnesium, total dissolved solids *etc.* make the water very hard. The above observed values have been statistically analyzed for the test of significance by calculating the mean, t-test and p-value. It is found that the values which are within the range are significant at 5 per cent level but the p-value of total hardness, calcium hardness and magnesium hardness corresponds to 0.001. It suggests that the value related with hardness of water are most significant therefore, these open wells which are located at Itwari sub area and which are always used for domestic purposes by the people of Nagpur city and other large number of people who are visiting the city particularly to this area for the business purposes exceeds permissible limits of potable water.

The magnesium hardness was analyzed in 11 wells and it was found to be between 32 to 200 mg/l in post monsoon season and 50 to 216 mg/l in winter season. The high levels of calcium and magnesium hardness of water has marked influence on the total hardness of the observed well water. In relation with total hardness, total dissolved solids of water sample varied between post monsoon and winter season. The range of Total Dissolved Solids (TDS) ranges from 615 to 796 mg/l and during winter season it varied from 655 to 832 mg/l.

Conclusion

It is therefore concluded that on the basis of physico-chemical analysis of well waters of Itwari sub area that, well water quality of this area varies from other parts of Nagpur city. The high values of certain chemical species (total hardness) indicate that the water is not suitable for drinking purposes as per Indian Standard Guidelines. It is therefore suggested that the well water in the area chosen for Itwari sub area should be analysed before the use for its suitability for domestic purpose. The data also suggest that treatment is the immediate need for total hardness removal from the existing water sources in the study zone of sub area Itwari of Nagpur city.

References

APHA, AWWA and WPCF. 1985: Standard methods for the examination of water and wastewater, 16th Edition, Washington D.C.

Nandintsetseg, Banzragch and Shinoda, Mansato. 2010: Seasonal change of soil moisture in Magnolia: Its Climatology and Modeling, *Hydrology and Earth System Sciences*, Discussions, 7 (1).

Rani, D. Freeda Gnana. 2006: Hydrochemistry of Groundwater of Thirumanur area,Tamil Nadu (India), *Journal of Environmental Science and Engineering*, 48(3), 199-202.

Sinha, D. K. and Saxena, Ritesh. 2006: Statistical assessment of Underground Drinking Water Contamination and effect of Monsoon at Hasanpur, J. P. Nagar (Uttar Pradesh, India), *Journal of Environmental Science and Engineering*, 48(3), 157-164.

De Lange, E. 1994: Manual for Simple Water Quality Analysis, International Water Tribunal (IWT) Foundation: Amsterdam.

Postel, S.L., Daily, G.C. and Ehrlich, P.R. 1996: Human appropriation of renewable freshwater, Science. 271:785-788.

Stigter, T.Y., Monteiro, J.P., Nunes, L.M., Vieira, J., Cunha, M.C., Ribeiro, L., Nascimento, J. A. and Lucas, H. 2009: Screening of sustainable groundwater sources for integration into a regional drought-prone water supply system, Hydrology and Earth System Sciences Discussions, 6 (1).

Singh, Vijendra and Chandel, C.P. Singh. 2006: Analytical Study Of Heavy Metals Of Industrial Effluents At Jaipur, Rajasthan (India), Journal Of Environmental Science and Engineering, 48(2), 103-108.

2013, Climate Change Effects on Agriculture and Economy *Pages* ***335–344***
Editors: **D.R. Khanna, A.K. Chopra, Vikas Singh, Rakesh Bhutiani & Gagan Matta**
Published by: **BIOTECH BOOKS, NEW DELHI**

Chapter 38

Effect of Heavy Metals on *Hordeum vulgare* and *Linum usitatissimum*

V.S. Dongre

Department of Botany, Sevadal Mahila Mahavidyalaya, Sakkardara square, Nagpur, Maharashtra

There are many complex facets to the trade and environment, when any product is made it has some impact on the environment. In industry, untreated effluent may pollute the water ways and factory chimneys may spew smoke into air. Pollution of environment by heavy metals is a prime importance which harms aquatic as well as terrestrial plants and animal's life systems. The heavy metals cadmium and lead are present in industrial effluent and other polluted water. Once these heavy metals enter in food chain, their biomagnifications takes place from lower to higher organisms. Hence, in the present study the data pulled out from two parameters *i.e.* Germination percentage and seedling growth of *Hordeum vulgare* and *Linum usitatissimum*. Germination percentage in lead and cadmium for both plant species were found to be decreasing in all the treatments as compared to the control. However, the seedling growth for lead in barley was inhibiting and in *Linum* it is stimulating as compared to the control. The seedling growths in cadmium for barley as well as *Linum* were found to be inhibiting to the control. These results indicate the intensity of toxicity of heavy metals in both the plant species.

Keywords: *Pollution, Heavy metals, Hordeum, Linum, Barley, Toxicity.*

Introduction

Under normal conditions, there exists an optimal balance of land, plants, animals and food chain for survival. Such natural balance of ecosystem and environment has been in existence since creation and evolution of different species for their requirements and man emerged as the most important component of almost all the natural ecosystem and environment. Industrial effluent discharged in sewage canals cities causing serious pollution on environment and heavy metals like lead, cadmium, copper, arsenic, mercury, zinc *etc.* are mostly absorbed and get accumulated in various plant parts as free metals which may adversely affects the plant growth, metabolism and reproduction (Barman and Lal,1994).

Heavy metals pollution is widespread due to rapid industrialization and current agricultural practices. These pollution persist in the environment for a longer period of time, as they are not easily degraded by soil micro-organisms and therefore easily absorbed by plants (Raghunath *et al.*, 1999; Gallego *et al.*, 2002).Therefore, heavy metal contamination represents a risk for primary and secondary consumers and ultimately humans (Zeller and Feller, 1999). Heavy metals are mostly absorbed by plants easily and prove toxic to plants that can be observed as growth retardation as a result of alteration in biochemical process like inhibition of enzyme activity, protein penetration and impaired nutrition (Arun *et al.*, 2005; Minaii *et al.*, 2008).

In the recent era, lead (Pb) contamination has gained a considerable attention as a potent environmental pollutant. Significant increase in the Pb content of cultivated soils has been observed near urban and industrial areas where it tends to accumulate in the surface ground layer (Di Toppi and Gabrielli, 1999). Lead (Pb) can also be accumulated in roots, shoots and leaves after being absorbed by plants (Morelli and Scarano, 2001; Joseph *et al.*, 2002). Cadmium mainly is an unnecessary element for both plants and animals and it also has toxic effect for them when its concentration exceeded a limit. Generally, it makes negative effect on their metabolism by doing some reaction with enzymes. Cadmium (Cd) inhibits growth of plant roots (Belimov *et al.*, 2003), uptake of water and nutrients (Boussama *et al.*, 1999) photosynthesis (Krupa and Miniak, 1998) and affecting carbohydrate metabolism and other physiological aspects and thereby reducing yield (Verma and Dubey, 2001; Sandalio *et al.*, 2001).

The present study is designed to investigate the effect of different concentration of $Pb(NO_3)_2$ and $CdSO_4$ (50, 75 and 100 ppm) on seed germination and seedling growth on *Hordeum vulgare* and *Linum usitatissimum*.

Materials and Method

Seeds of *Hordeum* and *Linum* were procured from the grain market. Physically and physiologically similar seed were sorted out and utilized in the present investigations.

Lead and cadmium are known to present in most of the industrial effluent and other polluted water which affects plants adversely. The heavy metals solution of lead and cadmium were made by using $Pb(NO_3)_2$ and $CdSO_4$, respectively. The stock solutions were made after the following calculations.

For Lead

331.23 gm of $Pb(NO_3)_2$ contains 207.21 gm of lead ions.

1 gm Lead ions= X [Amount of $Pb(NO_3)_2$]

$$X = \frac{321.23}{207.21} = 1.5 \text{ gm of } Pb(NO_3)_2$$

Hence, for 1 gm of lead ions takes 1.5 gm of $Pb(NO_3)_2$ to make the stock solution of 1000 ppm in distilled water.

For Cadmium

208.46 gm of $CdSO_4$ contains 112.40 gm of cadmium ions.

1 gm of cadmium ions = X [Amount of $CdSO_4$]

$$X = \frac{208.46}{112.40} = 1.85 \text{ gm of } CdSO_4$$

Hence, for 1 gm of cadmium ions take 1.85 gm of $CdSO_4$ and dissolve in distilled water makes the 1000 ppm stock solution of cadmium ions.

Germination Percentage

Physiologically similar and morphologically uniform seeds of *Hordeum vulgare* and *Linum usitatissimum* were used for the treatment. Seeds were tested for germination prior to the present study which was found to be absolute. Petri dishes and filter paper circles were disinfected with the help of absolute alcohol. The seeds were also rinsed with 1.5 per cent of mercuric chloride solution to avoid the infections. One hundred seeds were used for the treatment with five replicas were maintained. An individual control of tap water was also maintained. Seeds were presoaked in heavy metal solution of different ppm solution as well as tap water for 6 hours in muslin cloth. These presoaked seeds were arranged on the filter paper in petridishes in such a way that they do not stick with each other. The filter paper circles in the experimental petriplates were supplied 3 ml of required concentration of heavy metal solutions per day. The germination count was carried out after 6 days when the length of root tips was about 1.0 to 1.5 cm. The percentages of seed germination were calculated by using following formula for 50, 75 and 100 ppm solution of lead and cadmium ions. This experiment were carried out at room temperature (25 ± 50 °C) and kept in dark.

$$\text{Germination percentage} = \frac{\text{Number of germinating seeds}}{\text{Total number of seeds}} \times 100$$

Seedling Growth

Materials were used are plastic trays, thick blotting papers, seeds of *Hordeum vulgare* (Barley) and *Linum usitatissimum* and plastic slots. One hundred seeds were used with three replicas for both ppm of heavy metal solutions and tap water. The seeds were arranged in the thick blotting paper folders which was disinfected by absolute alcohol and arranged in slots. These slots along with blotting paper containing seeds were kept in the trays. The individual trays were filled with different ppm solution of lead and cadmium and tap water, in such a way that the seeds should not immersed in the solution. The experimental setup provided with the photoperiod of 10 hours at room temperature. Seedling height was measured in centimeters after the emergence of first vegetative leaf. Seedling growths were measured from transition zone of root and shoot on graph paper.

Each experiment was repeated five times and the mean values and standard deviations were then calculated. For all treatments, single factor ANOVA was carried out using Tukey's Test. The analysis of variance indicated a significant ($p<0.05$) difference on the germination of the seeds of varying $Pb(NO_3)_2$ and $CdSO_4$ solution.

Results

The treatment regime of heavy metal solution was observed to consistently decrease the percentage of germination. In 50, 75 and 100 ppm lead nitrate solution the percentage germination was about 60.00, 54.60 and 44.80 per cent, respectively for *Hordeum* seeds. However in 50, 75 and 100 ppm of lead nitrate solution the percentage of germination was reported about 70.00, 60.60 and 48.20 per cent, respectively for *Linum* seeds. However, the tap water (control) the germination was found to be absolute 100 per cent (Figures 38.1 and 38.2).

In 50, 75 and 100 ppm solution of cadmium sulphate, the seed germination percentage was found to be 56.80, 47.80 and 38.60 per cent, respectively for *Hordeum* (Barley). In *Linum* the seed germination was 62.20, 54.00 and 50.40 per cent for 50, 75 and 100 ppm solution of cadmium sulphate, respectively.

The effect of tap water and heavy metal was also studied on seedling height. The seedling height was recorded at the first leaf stage. The appearance of first leaf in *Hordeum* and *Linum* were observed at sixth day. Seedling height is the first step in the establishment of plant in the environment. The effect of heavy metals and seedling growth as found to be inversely proportional to the different ppm solutions in case of both *Hordeum* and *Linum* for lead nitrate solutions. In 50, 75 and 100 ppm were found

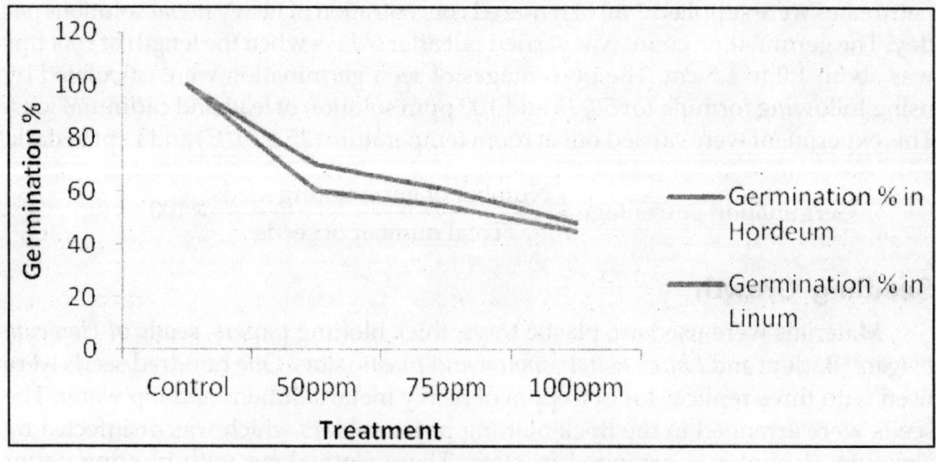

Figure 38.1: Effect of $Pb(NO_3)_2$ on Germination

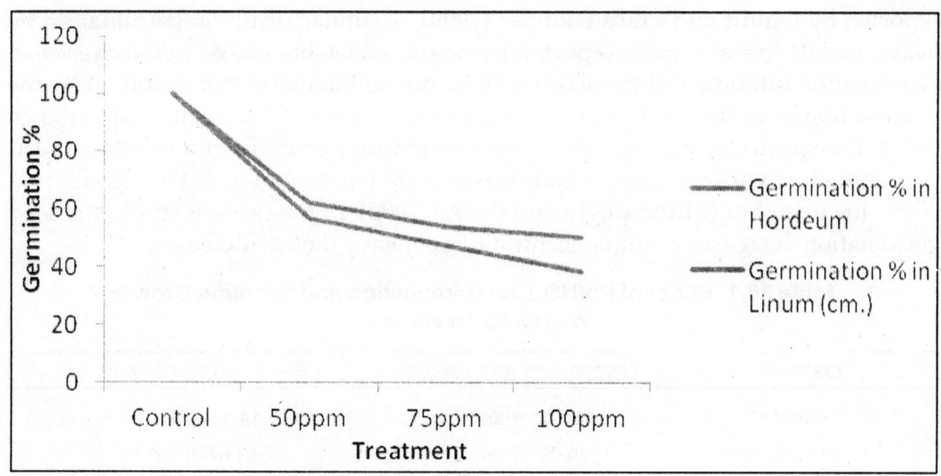

Figure 38.2: Effect of $CdSO_4$ on Germination

to be 13.67, 13.09 and 12.26 cm, respectively and control (tap water) has the 14.24 cm for Barley. Seedling growth of *Linum* in control, 50, 75 and 100ppm of lead nitrate solution were found to be 6.09, 8.34, 7.82 and 7.67 cm, respectively.

The seedling height for different concentrations of cadmium sulphate shows inhibitory as well as stimulatory effect. The seedling growth in Barley seeds was recorded to be 14.08, 13.83, 13.82 and 13.02 cm in control, 50 ppm, 75 ppm and 100 ppm, respectively. However, in *Linum* the seedling height was found to be 6.26, 5.65, 5.38 and 4.92 cm for control, 50 ppm, 75 ppm and 100 ppm, respectively. The correlation coefficient for the seed germination and the seedling height was found to be 0.86 which shows the positive correlation in $Pb(NO_3)_2$ for *Hordeum vulgare*. However in case of *Linum usitatissimum* the correlation coefficient in between the germination and seedling height was reported the negative correlated in $Pb(NO_3)_2$. In cadmium sulphate for different treatment regimes the correlation coefficient for the seed germination and seedling growth was found to be strongly positive.

Discussion

In general the results are consistent with previous reports regarding the effect of respective heavy metals. The results support the idea that tissues covering the embryo play a role in selective penetration of different heavy metals into seeds.

Lead and cadmium toxicity have become an important issue due to their constant increase in the environment. In the present study, seed germination and seedling height decreases in concentrations of lead and cadmium. However, in case of lead treatment of *Linum* for seedling height where the seedling growth increases as compared to the control (Tables 38.1 to 38.4). Lead and cadmium treatment significantly ($p<0.05$) decreased seed germination as compared to the control (Figure 38.1 and 38.2). Seed germination and seedling growth inhibition by heavy metals has also been reported by many other workers (Morzeck and Funicelli,1982).The reduced germination in *Sorghum bicolour* exposed to different concentration of cadmium was reported by Pandit and Prasanakumar (1999). A similar result on germination by heavy metals has also been reported by Shafiq and Iqbal (2005) in *Cassia siamea*. Germination inhibition of the seeds may be due to the metal that absorbed by the embryo tissues of the seeds where it experiences a toxic effect (Akinci and Akinci, 2010). The spectrum and dose-response relationships of inhibitory effects on seed germination of various heavy metals was reported in Arabidopsis (Weiqiang *et al.*, 2005), in cucumber (Munzuroglu and Geckil, 2002). In the present study, the seed germination decreases as the concentrations of heavy metals increases.

Table 38.1: Effect of $Pb(NO_3)_2$ on Germination and Seedling Growth on *Hordeum vulgare*

Treatment	*Germination (per cent)*	*Seedling Growth (cm)*
Control	100.00–+0.00^{a}	14.16–0.15^{e}
50 ppm	60.00 –2.54^{b}	13.67–0.08^{f}
75 ppm	54.60–1.51^{c}	13.09–0.12^{g}
100 ppm	44.80–1.92^{d}	12.26–0.06^{h}

Table 38.2: Effect of $Pb(NO_3)_2$ on Germination and Seedling Growth on *Linum usitatissimum*

Treatment	*Germination (per cent)*	*Seedling Growth (cm)*
Control	100.00–+0.00^{a}	6.09–0.06^{e}
50 ppm	70.00–3.53^{b}	8.34–0.02^{f}
75 ppm	60.60–2.88^{c}	7.82–0.05^{g}
100 ppm	48.20–3.03^{d}	7.67–0.07^{g}

Table 38.3: Effect of $CdSO_4$ on Germination and Seedling Growth on *Hordeum vulgare*

Treatment	*Germination (per cent)*	*Seedling Growth (cm)*
Control	100.00–0.00^{a}	14.08–0.28^{e}
50 ppm	56.80–2.38^{b}	13,83–0.10^{e}
75 ppm	47.80–2.58^{c}	13.82–0.13^{e}
100 ppm	38.60–2.88^{d}	13.02–0.15^{e}

Table 38.4: Effect of $CdSO_4$ on Germination and Seedling Growth on *Linum usitatissimum*

Treatment	*Germination (per cent)*	*Seedling Growth (cm)*
Control	100.00–+0.00^{a}	6.33–0.15^{e}
50 ppm	60.00 – 2.54^{b}	5.69–0.08^{f}
75 ppm	54.60–1.51^{c}	5.34–0,12^{f}
100 ppm	44.80–1.92^{c}	4.93–0.06^{f}

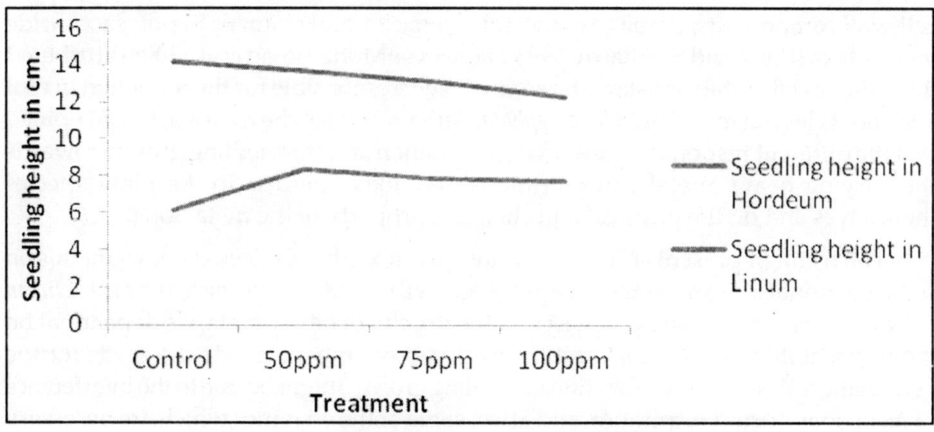

Figure 38.3: Effect of $Pb(NO_3)_2$ on Seedling Height

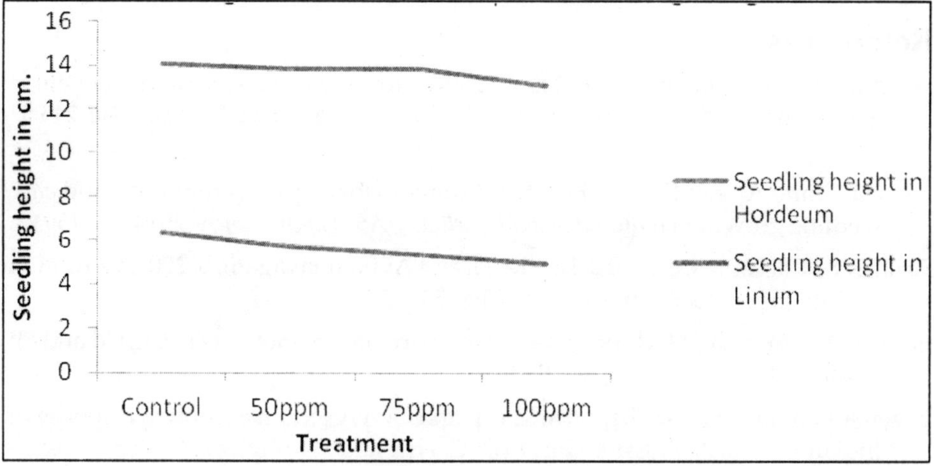

Figure 38.4: Effect of $CdSO_4$ on Seedling Height

The toxicity of heavy metals lead and cadmium on the seedling height was evaluated in different concentrations which was recorded the inhibition whereas in *Linum* when treated with $Pb(NO_3)_2$ was stimulatory (Figures 38.3 and 38.4). Similar growth stimulation was reported by Liu (2002), further he attributed that Pb at low concentrations could promote normal physiological and metabolic activities in plants such as the activities of nitrate reductase, the content of sugar content and chlorophyll of stem and leaves. Inhibition in seedling growth reported by many workers for different heavy metals (Sharma *et al.*, 2011; Milone *et al.*, 2003; Abedin and Meharg, 2002; Begonia *et al.*, 1998). The growth of wheat seedlings requires a great deal of energy, which is supplied primarily by respiration. Respiration not only provide large quantities adenosine triphosphate (ATP) and other metabolites required for the growth of embryos as well as seedling growth (Wang *et al.*, 2001).The inhibition of growth can be attributed in part to the inhibition of mitosis, the reduced synthesis of cell-wall components, damage to the golgi complex and changes in polysaccharide metabolism (Punz and Sieghardt, 1993). However Muhammad *et al.* (2008) attributed that mitotic cell in the meristamatic zone is solely responsible for the reduction in root and shoots length in *Leucaena leucocephala*. Seregin and Kozhevnikova, (2005) opined that the different responses of the seed germination and the seedling growth towards the varying heavy metal concentrations may likely be due to the plant species themselves and on the physical and chemical property of the metal ions.

The result of present study determined the toxicity of selected heavy metals on seed germination and early seedling growth in *Hordeum vulgare* and *Linum usitatissimum*. The findings suggest that the toxicity of heavy metals is dependent on the permeability of seed coat and the response of embryo inside the seeds for the germination. Whereas the inhibition in seedling growth might be due to the interference of metal ions during respiration and affect some of the enzymes which are necessary for the metabolism. However the stimulation in seedling growth of *Linum* for lead nitrate treatment suggest that for growth the concentrations are not detrimental.

References

Abedin, M.J. and Meharg, A.A. 2002: Relative toxicity of arsenite and arsenate on germination and early seedling growth on rice (*Oryza sativa*, L.). *Plant Soil*, 243:57-66.

AkinciI, E. and Akinci, S. 2010: Effect of chromium toxicity on germination and early seedling growth in melon (*Cucumis melo*,L.). *Afr. J of Biotech.*, 9(29):4589-4594.

Arun, K.S., Cervantas, C., Loza-Tavera, H. and Avudainayagam, S. 2005: Chromium toxicity in plants. *Environ. Int.* 31:739-753.

Barman, S.C. and Lal, M.M. 1994: Accumulation of heavy metals (Zn, Cu, Cd and Pb) in soil. *Indian J. Agri. Sci.*, 56:584-91.

Begonia, G.B., Davis, C.D., Begonia, M.F.T. and Gray, C.N. 1998: Growth responses in Indian Mustard (*Brassica juncea* (L) Czern) and its phytoextraction of lead a contaminated soil. Bull. *Environ. Contain. Toxicol.*, 61:38-43.

Belimov, A.A., Safronova, V.I., Tsyganov, V.E., Brisov, A.V., Kozhemyakov, A.P. Stepanok, V.V., Martenson, A.M., Gianinazzi-Pearson, V. and Tikhonovich, I.A. (2003). Genetic variability in tolerance to cadmium and accumulation of heavy metals in pea (*Pisum sativum* L.) Euphytica, 131:25-35.

Boussama, N., Ouariti, O. and Ghorbal, M.H. 1999: Changes in growth and nitrogen assimilation in barley seedlings under cadmium stress. *J. Plant Nutr.*, 22: 731-752.

Di Toppi, S.L. and Gabrielli, R. 1999: Response to cadmium in higher plants. *Environ. Exp. Bot.*, 41: 105-130.

Gallego, J.L.R., Ordonez, A. and Loredo, J. 2002: Investigation of trace element sources from an industrialized area (Aviles, northern Spain) using multivariate statistical methods. *Environ. Int.*, 27:589-596.

Heidari, M. and Sarani, S. 2011: Effect of lead and cadmium on seed germination, seedling growth and antioxidant enzymes activities of Mustard (*Sinapis arvensis* L.). ARPN. *J. of Agri. and Bio. Sci.* 6:44-47.

Joseph, L.U., Andrea, L.C. and Mal, T.K. 2002: Effect of lead contamination on the growth of *Lythrum salicaria*. *Environ. Pollu.*, 120(2): 319-323.

Krupa, Z. and Miniak, M. 1998: The stage of leaf maturity implicates the response of the photosynthetic apparatus to cadmium toxicity. *Plant Sci.*, 138: 149-156.

Liu, D. and Wang, Y. 2002: Effect of Cu and As on germination and seedling growth of crops. *Chin. J Appl. Ecol.*, 13:179-182.

Milone, M.T., Sgherri, C., Clijsters, H. and Navari-izzo, F. (2003) Antioxidative responses of wheat treated with realistic concentration of cadmium. *Environ. Exp Bot.*, 50:265-276.

Minaii, B., Abdollahi and Towfight, Z. 2008: Toxicity of lead acetate on rabbit arteries: a histological evaluation. *Toxicology*, 180:53.

Morelli, E. and Scarano, G. 2001: Synthesis and stability of phytochelatins induced by cadmium and lead in the marine diatoms *Phaeodatylum trieornutum*. *Agro-environ. Prot.*, 52: 83-395.

Morzeck, J.R.E. and Funicelli, N.A. 1982: Effect of znic and lead on germination of *Spartma alterniflora* Loisel seeds at various salinities. *Environ. Exp. Bot.*, 22:23-32.

Muhammad, *I.E.*, Iqbal, M.Z. and Mohammad, A. 2008: Effect of lead and cadmium on germination and seedling growth of *Leucaena leucocephala*. *J. of App. Sci. in Environ. Manage.*, 12(2):61-66.

Munzuroglu, O. and Geckil, H. 2002: Effect of metals on seed germination, root elongation, and coleoptiles and hypocotyls growth in *Triticum aestivum* and *Cucumis sativus*. Arch. *Environ. Cont. Tox.*, 43:203-213.

Pandit, B.R. and Prasannakumar, P.G. (1999). Effect of metals on jowar (*Sorghum bicolour* L.) seedling growth, germination and absorption of elements. *Pollut. Res.*, 18: 307-315.

Punz, W.F. and Sieghardt, H. 1993: The response of roots of herbaceous plant species to heavy metals. *Environ. Exp. Bot.*, 33:85-98.

Raghunath, R., Tripathi, R.M., Kumar, A.V., Sathe, A.P., Khandekar, R.N. and Nambi, K.S.V. 1999: Assessment of Pb, Cd, Cu and Zn exposures of 6 to 10 year old children in Mumbai. *Environ. Res.*, 80: 215-221.

Sandalio, L.M., Dalurzo, H.C., Gomez, M., Romero-Puertas, M.C. and Del Rio, L.A. 2001: Cadmium induced changes in the growth and oxidative metabolism of pea plants. *J Exp. Bot.*, 52: 2115-2126.

Seregin, I.V. and Kozhevnikova, A.D. 2005: Distribution of cadmium, lead, nickel and strontium in imbibing Maize Caryopses. *Russ. J. of Plant Phy.*, 52(4):565-569.

Shafiq, M. and Iqbal, M.Z. 2005: The toxicity effect of heavy metals on germination and seedling growth of *Cassia siamea*, Lamark. *J. of New Seeds*, 7: 95-105.

Sharma, A., Mishra, I.G. and Srivastava, A.K. 2011: Toxicity of heavy metals on Germination and Seedling Growth of *Salicornia brachiata*. *J. Phytol.*, 3(9):33-36.

Verma, S. and Dubey, R.S. 2001: Effect of cadmium on soluble sugar and enzymes of their metabolism in rice. *Biol. Plant.*, 44: 117-123.

Wang, Y.B., Liu, D.Y., Zhang, L. and Guo, H. 2001: Effect of Cu and As and their combination pollution on *Glycine max*. *Chin. J Appl. Ecol.*, 12:117-120.

Zeller, S. and Felller, U. 1999: Long-distance transport of cobalt and nickel in maturing wheat. *European J. Agron.* 10:91-98.

2013, Climate Change Effects on Agriculture and Economy *Pages* ***345–348***
Editors: **D.R. Khanna, A.K. Chopra, Vikas Singh, Rakesh Bhutiani & Gagan Matta**
Published by: **BIOTECH BOOKS, NEW DELHI**

Chapter 39
Effect of Pollution from Garbage of Ballarpur Paper Mill Industry

***Priti S. Suryawanshi*[1], *Aparna S. Dhoble*[2] *and Pravin Charde*[3]**
[1]*Yeshwantrao Chavan Open University, Nashik, Maharashtra*
[2]*Department of Home Science Extension Education, Sevadal Mahila Mahavidyalaya, Sakkardara Square, Nagpur, Maharashtra*
[3]*Sevadal Mahila Mahavidyalaya, Nagpur, Maharashtra*

Industrialization and concentration of population in selected pockets of a country bring in their make, large quantities of industrial and sewage wastes which find their way into either the air or natural water bodies. This has been the case in the industrial waste where the quality of air and water has been continuously deteriorating and all efforts of pollution control have failed to restore the original purity of the air and water. Sewage collection and treatment has always logged behind the increasing population in India, as adequate investment has not been made in their field. Industrial pollution has also made its contribution in a big way with ever increasing industrial production. The main objectives are to study the awareness of pollution and effect of health on residential people. The data were collected from 100 randomly selected people living near Ballarpur Paper Mill with the help of schedule method. Increase in the number of industries and factories due to the industrial revolution, air pollution also have increased significantly. The major problem of respiration was found in respondent's health due to garbage. Tree plantation is the best solution to control the garbage as well as pollution.

Keywords: *Industrial pollution, Awareness, Health, Garbage.*

Introduction

Industrial pollution is pollution which can be directly linked with industry, in contrast to other pollution sources. This form of pollution is one of the leading causes of pollution worldwide. Pollutants given off by various industries and factories are often considered to be one of the prime factors contributing to air, water and soil pollution. According to the Environmental Protection Agency (EPA), it has been estimated that industrial pollution is responsible for almost 50 percent of the pollution present in the United States. There are various wide ranging effects as well as serious consequences, of industrial pollution on the ecological balance of the atmosphere.

Ballarpur Industries Ltd. is an India-based printing paper manufacturing company. It is the largest printing paper manufacturing and exporting company operating in India. It has a strong product portfolio that includes writing and printing paper, industrial paper and specially paper. Apart from enjoying market leadership in India, Ballarpur Industries Ltd. operation are also spread in other prominent international markets. Ballarpur Industries has 6 state of the art manufacturing units spread across India and these are manufacturing facilities strictly conform to environment friendly manufacturing practices.

Pollution emitted from the industries is also one of the major problems. The emissions from various industries contain large amounts of gases such as carbon dioxide, sulphur and nitrogen, among others. These gases when present in elevated levels in the atmosphere often result in various environmental and health hazards such as acid rain, and various skin disorders in individuals. A large quantity of garbage was found which results in health hazards.

The main objective of the present study was to estimate the effect of garbage on people's health and awareness of people about pollution and the type of pollution present due to garbage.

Methodology

The data were collected from 100 randomly selected people residing near Ballarpur Paper Mill with the help of pretested structural interview schedule method. Collected data was tabulated and analyzed. The results have been expressed in percentages.

Results and Discussion

The relationship of man with the environment is necessarily symbiotic; the equilibrium between the two must be maintained at all costs. The environment is mainly a compound form of all the natural elements and living organisms and both are essential for the future existence.

Table 39.1: Awareness of Respondents about Pollution

Sl.No.	*Information about Pollution*	*No. of Respondents*	*Percentage*
1.	Changes in natural environment due to human being	27	27.0
2.	Existence of environmental pollutes	36	36.0
3.	Presence of some aspect in environment results hazard in human life	37	37.0
	Total	100	100.0

Majority of the respondents have knowledge about pollution that there is an existence of environmental pollutants in nature *i.e.* 36.0 percent, whereas presence of some aspect in environment results a hazard in human's life found to be 37.0 percent of respondents aware about it. Some respondents *i.e.* 27.0 percent were expressed that pollution means there is a change in human being.

Table 39.2: Type of Pollution Due to Industries of Respondents

Sl.No.	*Type of Pollution*	*No. of Respondents*	*Percentage*
1.	Water pollution	35	35.0
2.	Air pollution	57	57.0
3.	Soil pollution	8	8.0
	Total	100	100.0

Pollution are of many types, the respondents found 57.0 percent of respondents seem to have an air pollution, 35.0 percent found water pollution and 8.0 percent of the respondents found soil pollution due to Ballarpur Paper Mill.

Table 39.3: Effect of Garbage on Respondent's Health

Sl.No.	*Effect of Health*	*No. of Respondents*	*Percentage*
1.	Problem in Respiration	56	56.0
2.	Skin disease	14	14.0
3.	Food poison	16	16.0
4.	Eye problem	7	7.0
5.	Ear problem	7	7.0
	Total	100	100.0

It is revealed that 56.0 percent of people has problem in respiratory tract, whereas 16.0 percent had food poisoning reported the respondents. Skin diseases were also found *i.e.* 14.0 percent whereas 7.0 percent of the people has eye and ear problem respectively.

Table 39.4: Problem Occur Due to Garbage on Respondents

Sl.No.	*Problems*	*No. of Respondents*	*Percentage*
1.	Problem due to insects	38	38.0
2.	Surrounding environment	11	11.0
3.	More spare occupied	20	20.0
4.	Effect on health	31	31.0
	Total	100	100.0

Many problems occur due to garbage on respondent's health. As the garbage is found more there is a problem due to insects *i.e.* 38.0 percent whereas their health may be affected *i.e.* 31.0 percent suffered from fever, disease like malaria *etc*. More space is occupied due to garbage *i.e.* 20.0 percent reported by respondents. The surrounding environment *i.e.* 11.0 percent people's suffered from odor pollution in environment due to dirty surrounding.

Table 39.5: Solution to Control Garbage by Respondents

Sl.No.	Solutions	No. of Respondents	Percentage
1.	Tree plantation	44	44.0
2.	Recycle of water	20	20.0
3.	Use of modern technique	30	30.0
4.	No information	06	06.0
	Total	100	100.0

The above table reveals that to control the problem of garbage by respondents is to do tree plantation *i.e.* 44.0 percent whereas the use of modern technique should be applied *i.e.* 30.0 percent. Recycle of water *i.e.* 20.0 percent and negligible percentage *i.e.* 6.0 percent people have no idea to control the garbage.

Conclusion

In the present study, 37.0 percent of the respondents feel that presence of some aspect environment results a hazard in human's life. 57.0 percent of respondents reported that there is more air pollution. Due to garbage of Ballarpur Paper Mill, the problem in respiration *i.e.* 56.0 percent found highly affected to people's health whereas 38.0 percent of the people faced the problem due to insects. The solution to overall control of garbage was tree plantation (44.0 percent) reported the respondents.

Acknowledgement

The authors are thankful to the Management of Sevadal Education Society, Nagpur.

References

Mahajan, S.P. 1999: Industrial pollution emissions and Indian standards. McGraw Hill, Tata Publishing Company Ltd., New Delhi, pp.15-17.

Rao, C.S. 2007: Control of specific gaseous pollutants, New Age International Publishers, New Delhi, pp. 227-229.

http://google.com.industrial pollution

http://google.com.ballarpur.pdf

Index

35 A-2 RS-160